The Business of Civil War

JOHNS HOPKINS STUDIES IN THE HISTORY OF TECHNOLOGY
Merritt Roe Smith, *Series Editor*

The Business of Civil War

*Military Mobilization
and the State,
1861–1865*

MARK R. WILSON

The Johns Hopkins University Press
Baltimore

©2005 The Johns Hopkins University Press
All rights reserved. Published 2006
Printed in the United States of America on acid-free paper

Johns Hopkins Paperback edition, 2010
2 4 6 8 9 7 5 3 1

The Johns Hopkins University Press
2715 North Charles Street
Baltimore, Maryland 21218-4363
www.press.jhu.edu

The Library of Congress has catalogued the hardcover edition of this book as follows:

Wilson, Mark (Mark R.), 1970–
The business of civil war : military mobilization and the state, 1861–1865 / Mark Wilson.
p. cm.—(Johns Hopkins studies in the history of technology)
Includes bibliographical references and index.
ISBN 0-8018-8348-2 (hardcover : alk. paper)
1. United States Army—Mobilization—History—Civil War, 1861–1865.
2. United States—History—Civil War, 1861–1865—Economic aspects.
3. Contracting out—United States—History—19th century. I. Title.
II. Series: Johns Hopkins studies in the history of technology (Unnumbered)
E491.W76 2006
973.7'1—dc22   2005029017

A catalog record for this book is available from the British Library.

ISBN 13: 978-0-8018-9820-4
ISBN 10: 0-8018-9820-X

For Diane & Gary

Contents

List of Illustrations ix
Acknowledgments xi

Introduction 1

CHAPTER 1
The Rise and Fall of a Federal Supply System 5

CHAPTER 2
The Formation of a National Bureaucracy 34

CHAPTER 3
The Making of a Mixed Military Economy 72

CHAPTER 4
The Trouble with Contracting 107

CHAPTER 5
The Middleman on Trial 148

CHAPTER 6
The Unacknowledged Militarization of America 191

APPENDIX A
Note on the Value of a Dollar during the Civil War Era 227

APPENDIX B
Leading Northern Military Contractors in Selected Industries 231

APPENDIX C
Note on Data Collection and Record Linkages   237

Notes   241
Essay on Sources   285
Index   295

Illustrations

FIGURES

1.1. The Pennsylvania Volunteers   25
2.1. Map of Treasury Disbursements to Quartermaster's Department Depots, 1856–1860   48
2.2. Group Outside the Office of the Quartermaster General, Washington, D.C.   67
3.1. Map of Treasury Disbursements to Quartermaster's Department Depots, 1861–1865   77
3.2. Filling Cartridges at the United States Arsenal, Watertown, Massachusetts   79
3.3. U.S. Cavalry Stables at Chattanooga, Tennessee   80
3.4. Employees at a Government Horseshoeing Shop   85
4.1. Two Thousand Army Horses Wanted   145
5.1. Running the Machine   177

TABLES

1.1. Procurement Expenditures by Ten Northern States in 1861   13
2.1. U.S. Military Expenditures by Department, 1845–1875   38
2.2. Twenty Leading Procurement Officers during the Civil War   69
4.1. Leading Civil War Contractors   122
B.1. Leading Small Arms Suppliers to the U.S. Army, 1861–1866   231
B.2. Leading Heavy Ordnance Suppliers, 1861–1866   232
B.3. Leading Wagon and Ambulance Suppliers, 1861–1864   232

B.4. Leading Boot and Shoe Suppliers, 1861–1864   233
B.5. Leading Woolen Textile and Blanket Suppliers, 1861–1864   234
B.6. Leading Clothing Suppliers to the U.S. Army, 1861–1864   235
C.1. "Receipts and Expenditures Reports" Used as Data Sources   239

Acknowledgments

It is a pleasure to be able to thank the many people and organizations who helped me to complete this book. The research and writing of my doctoral dissertation, on which this book is based, was supported by grants and fellowships from the University of Chicago, Andrew W. Mellon Foundation, Harvard Business School, State Historical Society of Wisconsin, Illinois Historic Preservation Agency, Indiana Historical Society, and Newcomen Society of the United States. More recently, grants from the Hagley Museum and Library and the Huntington Library allowed me to work in their unique collections.

Many librarians and archivists were generous with their time and advice. At the National Archives, I was assisted by countless staff members, including most of the team in the Old Army room. For helping me navigate their collections and for their advice about the project, I owe special thanks to Michael Stevens at the State Historical Society of Wisconsin, Steven Towne at the Indiana State Archives, Cheryl Schnirring at the Illinois State Historical Library, Frank Conaway at the Regenstein Library of the University of Chicago, and Roger Horowitz at the Hagley.

My teachers, mentors, and colleagues have provided essential encouragement and criticism. At the University of Chicago, I was fortunate to learn from Neil Harris, Julie Saville, Tetsuo Najita, Bill Novak, and Michael Conzen. Amy Dru Stanley and Michael Geyer, who served on my dissertation committee, provided sage advice from the beginning. Kathleen Conzen, my advisor, provided inspiration and support while pushing me to do better work. So did Richard John, whose learning is exceeded only by his generosity. I also owe thanks, for their encouragement along the way, to Jim Grossman, Tom McCraw, Walter Friedman, Don Wright, Gail Radford, Jason Smith, Andrew Godley, Ian McGiver, and Rebekah Mergenthal. For their constructive criticism of parts of early drafts, I thank the members of the Social History Workshop at the University of Chicago; the Newberry Library Seminar on Technology, Politics, and Culture; Harvard Business School; and the Business History Conference.

By plowing through a long early version of this book and offering their considerable insights, Christine Haynes, Gary Wilson, Richard John, and Roe Smith did me a great service. At the Johns Hopkins University Press, Bob Brugger provided much encouragement over the years. More recently, Linda Forlifer patiently helped me to polish the text. I completed the book manuscript during a postdoctoral fellowship at the Olin Institute for Strategic Studies, Harvard University. I thank Steve Rosen for giving me this extraordinary opportunity. For their helpful comments on late versions of the manuscript, I thank Sebastian Rosato, Chris Nowland, Dirk Bönker, Steve Biel, Dan Terris, and Matt Gallman.

I met Christine Haynes just before launching this project, about a decade ago. Thank you, Christine, for all your help and companionship. You have made this book better; more important, you have enriched my life in countless ways. I will always associate this book with our son Oliver, who was born a few months before it was completed and provided plenty of pleasant distractions.

I dedicate this book to my parents, who have encouraged my interest in history and provided me with a lifetime of support.

The Business of Civil War

Introduction

The American Civil War of 1861–65 was a giant economic project. In four years, the national government in the North spent roughly $1.8 billion in 1860 dollars, more than the combined total of all previous U.S. government expenditures.[1] Not only by domestic measures but also in global terms, the war's economic scale was remarkable. The North's war spending amounted to roughly four times the combined French and British outlays for the Crimean War of 1854–56, one of the largest conflicts of the period involving Europe's great powers. Prefiguring developments that would emerge most explosively in the world wars of the twentieth century, the Civil War stood for a half century as the best illustration of the fearsomely destructive potential of military mobilizations conducted by industrialized nation-states.[2] As developments in the Civil War North showed, the tremendous powers of modern armies to kill, coerce, and destroy rested upon prodigious productive efforts on the home front.

The North's unprecedented monetary outlays of 1861–65 translated into a huge amount of matériel. Two-thirds of all U.S. war spending went to pay for goods and services needed to outfit and sustain its forces in the field. In four years, the Union supplied its soldiers with roughly 1 billion rounds of small arms ammunition, 1 million horses and mules, 1.5 million barrels of pork and 100 million pounds of coffee, 6 million woolen blankets, and 10 million pairs of trousers.[3] Given such figures, it is not difficult to conclude that the million-man Northern army was, as one scholar has put it, "the largest, best equipped, best fed, and most powerful war machine ever assembled in the history of the world to that date."[4]

Despite its evident significance, the North's procurement project has long been something of a mystery. This book offers some surprising new answers to old questions about the Civil War and, more generally, about the economic and political development of the United States. Challenging the common assumption that the Union's procurement project was part and parcel of a "spoils system" in which elected officials used government jobs and contracts to reward party faithful, this book shows that the North's supply system was managed by a military bureaucracy that was relatively independent of party politicians and their patronage networks.[5] The organization at the center of this project was the Quartermaster's Department, by far the largest of the army supply bureaus, which relied upon a decentralized national network of depots and officers.[6] One of the most important organizations in nineteenth-century America, the wartime Quartermaster's Department was a major economic and political actor that was not simply a check-writing service for private enterprise or a servant of the White House or Congress. Instead, U.S. quartermasters and their counterparts in other military supply bureaus acted as creative public administrators, driven by ideas and institutional norms and pressures that set them apart from other leading actors in the wartime North.[7] Shaped by a unique education through military institutions and antebellum careers on the western frontier, these officers were at least as important as businessmen or elected officials in the economic and political construction of the Union's procurement system.

The national supply system these army bureaucrats created in the North is best described as a mixed military economy. Far from being a party patronage machine or a triumph of unregulated capitalism, as some historians have suggested, the Union's procurement project relied upon a combination of large-scale public and private operations.[8] Military supply officers were not content to leave the business of war to private enterprise. Certainly, the North bought goods from hundreds of contractors in the private sector. But quartermasters and other supply officers also created substantial wartime public enterprises that went well beyond the system of arsenals and navy yards that had existed on the eve of the war. By the middle of the conflict, the Quartermaster's Department alone employed over 100,000 civilians, far more than any private American business enterprise of the era. Championed by supply officers and war workers alike, the expanded wartime network of public enterprises accounted for a sizable fraction of all goods and services consumed by the Union armies.[9]

When they did purchase from contractors, Northern procurement officers still faced difficult choices that determined the flows of huge sums of money. The main difficulty here was not, as might be expected, that military officers were occasionally pressured by politicians to steer contracts to certain parties. More importantly, supply officers discovered that the economics of Civil War contracting was more complicated than it appeared at first glance. The large orders that they were encouraged to place by the high wartime military demand for goods, many officers concluded, tended to push the bulk of the prime contracting business toward a relatively small group of firms. Although many small enterprises certainly participated directly and indirectly in the supply effort, the technical and financial demands associated with government contracting made many war supply industries more concentrated than many Northerners would have liked.[10] Here again, the dynamics of the Civil War economy were shaped as much or more by bureaucratic institutions as by party patronage or competitive markets in the private sector.

The North's procurement project was also influenced continually by political ideas and pressures, which came from within the military supply bureaus as well as from the outside. During the conflict's first year, several Northern states and localities fought mightily to secure a share of military spending but experienced serious administrative problems that sped the takeover of all significant procurement responsibilities by the national state. Meanwhile, many Northerners cried out against what they saw as mismanagement and illegitimate profit taking in the war economy. Drawing on various strains of a popular political economy that valorized producers and attacked commercial intermediaries, this diverse group of critics—which included military officers as well as workers, politicians, and journalists—used a common vocabulary to condemn so-called middlemen in the war supply arena.

Although this widely held discontent with commercial profit taking in the context of a national war effort never led to a radical overhaul of the procurement system, it had important effects. Many supply officers and war workers used the anti-middleman, pro-producer logic to champion state-run enterprise, promoting a larger public side to the mixed military economy. On the supply system's private side, procurement officers responded to what they saw as a structural bias in the contracting system in favor of large commercial dealers by adopting more informal purchasing schemes that would make it easier for them to buy directly from smaller firms. Finally, during the second half of the war, after new congressional legislation defined contractors as part of the military establishment, military courts began to exact symbolic if not systematic justice by convicting and punishing some suppliers for dereliction of duty. In the end, although the North's economic mobilization certainly relied heavily upon for-profit firms in the private sector, many members of the Northern public joined supply officers and war workers in refusing to embrace the idea of a war economy based on commercial capitalism.

Although it concentrates on the four years of 1861–65, this book suggests that the record of the North's economic mobilization for war represents a challenge to existing narratives of American political development, which have described the American state before the end of the nineteenth century as a creature of the parties.¹¹ Given that the military bureaus that managed the Union's procurement project were largely in place before 1861 and maintained considerable autonomy during the massive wartime public spending project, it seems that the military should figure much more centrally in discussions of the development of national state and bureaucratic institutions of all kinds in America. While late-nineteenth-century innovations in government and business were important, they should be seen in a broader chronological frame that encompasses important developments during the Civil War and before. Much of this earlier story, it now appears, needs to be revised in ways that highlight the activities of military institutions.¹²

For decades, most historians have overlooked the military foundations of modern America. This is not to deny that the demobilization starting in 1865 was remarkably thorough, nor to minimize the novelty and importance of the postwar rise of large industrial corporations in the private sector, which truly transformed the national economy and government-business relations.¹³ Nevertheless, modern American business and government were shaped directly and indirectly by a military model of administration that had been on display in 1861–65.¹⁴ In several arenas in late-nineteenth-century America—including the civil service reform movement, the management of corporate enterprise, and the development of alternative models of political economy by labor and populist organizations, for instance—memories of the Civil War mobilization played an important part.

Even as they have sometimes obsessively remembered and even celebrated the Civil War and other past conflicts, Americans have often downplayed the influence of military institutions upon national political and economic development. Over the long run, one of the effects of this selective memory of war has been to obscure the record of the North's efforts to sustain its armies in 1861–65. Ultimately, that record suggests that democracy and capitalism in America have always been tempered, for better and for worse, by countervailing impulses and institutions. When it mattered most in the nineteenth-century United States, elected officials and business leaders frequently found themselves playing second fiddle to a military bureaucracy largely independent of direct party control and ambivalent about commercial capitalism. Although it may be tempting to bracket this chapter of American history as an irregular interlude, we must also acknowledge that the North's extraordinary wartime project drew upon deep institutional roots and left significant impressions upon the American imagination.

CHAPTER 1

The Rise and Fall of a Federal Supply System

When the North went to war in April 1861, it contested the right of Southern states to secede from the national union. In a fundamental sense, the Civil War represented a struggle over the survival of the American federal system. But secession was not the only test for federalism in 1861. Normally, state governments retained considerable authority independent of Washington. But the national mobilization effort, which demanded unprecedented numbers of soldiers and extraordinary amounts of money, immediately raised difficult questions about the proper division of authority between the national government and the states. As Northerners began in the spring and summer of 1861 to organize a giant new military economy, they confronted the question of how the business of war could be made compatible with the existing federal order.

Both the North and the South, as they mobilized for war, grappled with the question of what roles should be played by the national government and the various states. In the South, as in the North, national supply bureaus performed much of the work. At the same time, however, some Southern states maintained their own independent procurement programs. The largest of these independent state systems was that of North Carolina, where Governor Zebulon Vance insisted on supplying his state's troops directly. By the last year of the war in 1864–65, North Carolina was still running its own independent supply system.[1] Like North Carolina, the various Northern states started the war by mounting their own independent procurement programs. By the summer of 1861, the Northern war economy was remarkably decentralized. Each of the Northern states, in addition to the regular U.S. military bureaus, was buying large quantities of supplies. Orders for thousands of uniforms, for example, were being placed not only by the officers of the Quartermaster's

Department of the U.S Army but also by the quartermaster officers of each of the various states.

This federal military economy, which provided the kind of division of authority that was a hallmark of the American polity, continued to operate in the North into the last months of 1861. But it did not last long. By early 1862, the procurement bureaus of the U.S. Army had taken over nearly all of the important responsibilities for outfitting and sustaining Northern troops. It is tempting to conclude that this change can be explained simply as a shift from a less efficient system to a more efficient one. Certainly, as more troops were dispatched from their home states to various points around the country and began fighting together under national command, it made less sense for individual states to handle procurement. But the development of the Northern military economy during the first months of the Civil War was more than a transition from chaos to coordination; it also involved an important redistribution of political and economic power.

When Northern state governors and their subordinates handed most significant procurement authority over to U.S. officers at the end of 1861, they presided over the demise of a military supply system in which local leadership mattered. Republican governors who had come to power in many Northern states in the 1860 elections lost influence over the biggest government spending project of the century. Some governors resisted this change. But in 1861, when the duration and scale of the war remained uncertain, not even those governors who challenged it could imagine the full significance of the transition to complete U.S. control over war supply. Over the next three and a half years, the North would spend some $2 billion on military goods and services, mostly without the input of governors or other state officials.[2] Across the North, many smaller business firms that filled some of their states' first military orders in 1861 found themselves unable to participate in the national war economy. In short, the transition from what might be called a federal procurement system to a national one took money and power away from local officials and businesses in many parts of the North.[3] In many ways, therefore, it was during the short period from April to December of 1861, when the various states operated their own supply systems, that the Northern war economy corresponded most closely with Republican interests and republican ideals.

Every State a Military Outfitter

On 14 April 1861, after a two-day siege, the small U.S. garrison at Fort Sumter surrendered to the Confederacy. The Civil War had begun. The next day, President Lincoln called for 75,000 state militia to serve three months under U.S. command to fight the South. This was the first of what would become a series of calls upon the Northern states for troops, which became larger and more urgent after the poor showing of Union forces at the First Battle of Bull Run on 21 July. These early calls were met enthusiastically. At the beginning of the year, the regular U.S. Army had consisted of fewer than 16,000 men, most of them posted west of the Mississippi River. By the end of 1861, the North could count roughly 700,000 soldiers, nearly all of them volunteers who enlisted for three years to serve in regiments organized by their states. Although the number of Northern men under arms would grow slightly after 1861 (an estimated total of 2.1 million men—about half the North's men of military age—would serve at some point during the war, but it appears that no more than about a million served at any one time), the mobilization of unprecedented American military manpower had been achieved rather quickly.[4]

Supplying these hundreds of thousands of men was an enormous economic problem. Naturally, soldiers required special equipment, including weapons and knapsacks. Like civilians, they also needed food, shelter, and clothing—but in special forms, including hardtack crackers, tents, and uniforms. Because the Northern armies were so large, procuring these goods would be extraordinarily expensive; it could also be difficult. Under the Constitution, the formal responsibility for this daunting task seemed to fall to the national government. According to section 8 of article I, the U.S. Congress not only has the power "to raise and support armies" and "to provide and maintain a navy" but also "to provide for organizing, arming, and disciplining" any state militia units called up for a national war effort. Nevertheless, during the wars of the first half of the nineteenth century, the various states had done considerable war supply work themselves. So, although the federal supply system that emerged in the North by the summer of 1861 was unplanned, it was not unprecedented.

Although the scale of the war preparations that began suddenly in April 1861 soon dwarfed all other mobilization efforts in American history to that date, earlier mobilizations had established important precedents concerning the wartime roles of the national government and the various states. One of the most important of these precedents, dating from the War of 1812, was the use of "volunteers": a special category of troops (who were in fact paid) raised by the various states but placed under national command. Volunteers, unlike members of the state militias, were not prohibited legally from serving for long periods or leaving their states. During the Mexican War of 1846–48, state-organized regiments of volunteers accounted for roughly two-thirds of the eighty-five thousand soldiers who served. The rest were "regulars"—some veterans, some newly mustered for the war emergency—who were identified with the United States rather than an individual state.[5] During the Civil War, the vast majority of Northern soldiers were volunteers, who served in state-identified regiments, such as the 15th New York Cavalry or the 2nd Minnesota Infantry. State governors not only were responsible for raising volunteer units but also had the constitutional authority to appoint generals and other officers. Thus, the decentralization of power in the American federal system was reflected in the combination of regulars and state volunteer regiments in American armies.

Military procurement, in particular, was also shaped by antebellum precedents. Before 1861, both national authorities and nonnational actors had purchased war supplies. When they raised regiments of volunteers during the earlier part of the century, states paid at least part of the expense of outfitting them and transporting them to rendezvous points before they were mustered into service under national command. Later, states filed claims to be reimbursed for such expenses by the national government. During and after the War of 1812, the United States paid more than $4 million to reimburse the various states for their wartime outlays. During the Mexican War, the national government took responsibility for arming the volunteers, but until the last weeks of the war in 1848, individual volunteers themselves were responsible for buying their own uniforms (using a clothing allowance that supplemented their regular pay). All in all, the history of the relatively small war mobilizations in the United States before the Civil War suggested that the national government had traditionally taken primary responsibility for paying for war, but it was also well established that the various states or even individuals could be given some authority over certain kinds of procurement. In the spring and summer of 1861, exactly how much and what kind of procurement authority might be wielded by various parties remained an open question.[6]

When the Civil War broke out, state officials were confused about who was supposed to provide what for the masses of new recruits. Immediately after Lincoln's call for seventy-five thousand troops on 15 April 1861, governors across the North cabled Washington with queries about supply. On 17 April, for instance, New York Governor Edwin D. Morgan telegraphed Secretary of War Simon Cameron to ask, "Will General Government uniform and equip the volunteer militia raised by this state?" On 19 April, Wisconsin Governor

Alexander Randall wrote Cameron, wondering "whether the clothing or uniforms of men and officers are to be furnished by the State or by the Govt."[7] At this time, neither the national government nor the states had on hand enough uniforms or other military supplies to outfit the new regiments. The key question was who was responsible for acquiring the new equipment.

During the first weeks of the war, the North stumbled into a decentralized procurement system in which a variety of organizations and individuals purchased military supplies. In a few cases, Cameron and other top officials in Washington enlisted their personal friends to be special procurement agents for the national government. At the same time, some of the first new regiments were supplied in part by the colonels who led them. And in some Northern cities, voluntary associations of local elites took it upon themselves to buy equipment.[8] But the two sets of procurement organizations that emerged as the most important in the North were the supply bureaus of the U.S. military (which had been equipping the small force of regulars in peacetime) and their counterparts in the various states.

By the end of April 1861, it was clear that both the states and the U.S. military would be responsible for outfitting the new troops. On 23 April, Cameron ordered the Quartermaster's Department of the U.S. Army to procure uniforms and blankets for the volunteer regiments of the various states. At the same time, however, Cameron answered the queries of the governors by telling them that they would indeed be asked to provide their own troops with uniforms and other goods. Cameron informed Pennsylvania Governor Andrew Curtin that the United States could not outfit many of the new volunteer regiments because "just at present we have not the supplies." The states, therefore, would have to act. Quickly, state quartermasters and commissaries began to order thousands of uniforms, boots, tents, and other goods. Through the summer, Washington continued to advise state officials to outfit their own troops. In a meeting on 10 July, for instance, Cameron told New York officials that they should keep clothing and equipping their regiments but that the United States would provide wagons and horses.[9]

As they began to order tens of thousands of dollars worth of military goods, state officials hoped that the state monies they were spending would eventually be reimbursed by the United States. During the first days of the war, most state legislatures had authorized war bond issues to pay for mobilization. (The larger states, such as New York, Pennsylvania, and Ohio, each issued $3 million in bonds.)[10] Cameron's War Department soon informed the states that Washington would accept ultimate financial responsibility, as it had in earlier wars. As early as 27 April, one Wisconsin purchasing agent in the East wrote back to Governor Randall to assure him that "[t]he State will be reimbursed by the Government." This policy was soon confirmed in new congressional legislation. Signed into law by Lincoln on 27 July, the "indemnification act" promised that the national government would reimburse the states for "expenses properly incurred . . . for enrolling, subsisting, clothing, supplying, arming, equipping, paying, and transporting its troops . . . to be settled upon proper vouchers, to be filed and passed upon by the proper accounting officers of the Treasury." Until such settlements, which in some cases took many years to finalize, the states had to rely on their own bond issues or other resources to pay for the supplies.[11]

Even before they were assured of having their outlays refunded, officials in each of the Northern states plunged into the work of outfitting their regiments of volunteers. In most states, procurement authority was based in commissary and quartermaster departments, each led by a general officer and his assistants. In theory, these departments had been operating before 1861 as part of the state militia system. But by the time the Civil War began, state militias and their supply departments were in serious disrepair. Indiana's quartermaster department in 1859, for example, had spent a total of four hundred dollars, split between rent of an old storehouse and a part-time salary for the quartermaster general.[12] When the war emergency arose in April 1861, the states' formerly anemic supply bureaus grew suddenly into robust economic actors.

It was not unprecedented for states to oversee complex projects that required large expenditures. Between the 1810s and the 1840s, several American states had built major canals and railroads. To manage these projects, which involved hundreds of separate contracts, states had set up powerful coordinating boards or commissions. Although the heyday of state construction projects was over by the 1850s, some states were still spending large sums on the upkeep of older systems. The main artery of the Erie Canal system in New York was completed in 1825, for instance, but by the 1850s the state was still awarding dozens of canal maintenance contracts worth hundreds of thousands of dollars a year.[13]

But even states that could look to the canal boards or other significant administrative models from the antebellum era were not likely in 1861 to create large war supply bureaucracies that operated by strict rules and formal procedures. State canal boards may have developed standard administrative methods over the years, but they were still influenced heavily by the partisan politics of the day, in which patronage networks and personal authority remained central.[14] Furthermore, because the war emergency created demands that were especially urgent, state officials had little time to establish full-blown bureaus that would apply a comprehensive set of rules to military contracting. Instead, in most states, Republican governors tapped local elites to lead the state supply bureaus or serve as purchasing agents outside the state.

Across the Northern states, most leading supply officers were well-established businessmen and lawyers, some of whom enjoyed close personal connections to the Republican governors. One of Ohio's chief quartermasters, George B. Wright, was a forty-six-year-old lawyer and railroad director when the war began. A Republican, Wright was well acquainted with Ohio Governor William Dennison before the governor called him to Columbus in the summer of 1861. Ohio's commissary general, the fifty-two-year-old Columbus Delano, was a prominent lawyer who had attended the 1860 Republican convention in Chicago that had nominated Lincoln. In Illinois, the top procurement posts were filled by John Wood, a sixty-two-year-old former governor, and John Williams, a fifty-three-year-old merchant and banker, who ranked as one of the wealthiest residents of the state capital city of Springfield.[15] Such men, well connected in the worlds of business and party politics, were positioned to use personal contacts and patronage networks in the service of a speedy economic mobilization for war.

As the various Northern states began to outfit their new regiments, they each prepared to spend hundreds of thousands of dollars on military goods. Republican governors and their subordinates, men who were familiar with local and regional businesses, managed these large purchases. As these state procurement officials went to work in the spring and summer of 1861, the North saw the emergence of a federal war economy. Although U.S. military officials were working simultaneously to procure supplies for volunteers as well as regulars, state authorities had considerable power over the spending to equip their first regiments. In this early military economy, officials in each state filled their orders by looking first to local sources.

Making and Buying Close to Home

Outfitting the new volunteer regiments was expensive. In 1861, each of the Northern states spent between several hundred thousand and several million dollars on military supplies. Although spending patterns varied from state to state, the array of goods being purchased was similar everywhere. Each state spent large sums on uniforms. Most states also placed large orders for equipment—or "equipage," as it was sometimes called—including knapsacks, haversacks, and canteens. State officials also purchased food to subsist recruits until they were mustered into the U.S. service. Some states also bought weapons and horses. The bills for all of these items mounted quickly. Across the North in 1861, the several states together spent a total of about $25 million on military goods.

Spending these large sums of money entailed difficult choices. Governors and other state officials confronted a series of critical decisions about the kind of military economy they would create. One basic choice concerned what exactly should be purchased. Another was what would later be called the "make or buy" decision: should the state produce what it required in public establishments, or should it purchase from the private sector? If officials decided to buy, they were faced with more decisions. From whom should they attempt to buy? And how should they structure their purchases? Should they try to acquire all the knapsacks they wanted for the moment in a single transaction with one source, for example, or should they divide the order into several parts? In 1861, officials in the several Northern states used a wide variety of procurement methods. In general, however, they relied most heavily on local sources, some public-run but most private. Not surprisingly, state officials tended to fill military orders by dealing with businesses and workers within their states.

Given that no state was in the business of producing military supplies or much of anything else in 1861 and given that even the big state canal and railroad projects of the early nineteenth century had relied heavily on contracting with private firms, it might seem highly unlikely that state procurement officials would choose to make war supplies instead of buying them. In fact, however, several Northern states did manufacture some war goods in public establishments. By the summer of 1861, a variety of state-run military factories had sprung up across the North to complement the larger, long-established national facilities such as the Springfield Armory and the Philadelphia Navy Yard.

At the state level in 1861, public military manufacturing establishments included garment-making operations that engaged tailors and seamstresses; ammunition manufactories that employed dozens of men, women, and children; and prison workshops. Of the early state-run clothing projects, the largest was Pennsylvania's, based at the Girard House Hotel in Philadelphia. The state's quartermaster general, R. C. Hale, started this operation only days after war was declared. Quickly, the Girard House became a hive of activity, where some one hundred cutters rushed to fashion textiles into the proper shapes. Hundreds of tailors and seamstresses took up the cut cloth and sewed it into garments. Some of them worked with the fifty sewing machines set up at the hotel, and others worked in other commercial buildings or in private homes. In less than a month, before state authorities decided to close it and contract

Table 1.1 Military Procurement Expenditures by Ten Northern States in 1861

State	Population in 1860 (millions)	Spending by category, in thousands of nominal dollars				
		Clothing, blankets	Equipment, arms	Food	Other	Total
New York	3.9	876	883	558	178	2,495
Pennsylvania	2.9	n.d.	n.d.	n.d.	n.d.	2,517
Ohio	2.3	1,025	737	274	307	2,343
Illinois	1.7	2,728	855	84	363	4,030
Indiana	1.4	541	299	94	416	1,350
Massachusetts	1.2	1,014	1,254	209	850	3,327
Wisconsin	0.8	708	230	167	265	1,370
Michigan	0.7	226	51	66	79	422
New Jersey	0.7	448	195	50	209	902
Connecticut	0.5	n.d.	n.d.	n.d.	n.d.	1,230

Sources: Calculated from "Report of the Majority of the Select Committee," *Documents of the Assembly of the State of New York,* 85th sess. (Albany, 1862), 7:567–83, 7:604–9; "Report of the Auditor General on the Finances of the Commonwealth of Pennsylvania for the Year Ending November 30, 1861," in Pennsylvania, *Reports of the Heads of Departments* (Harrisburg, 1861); "Report of the Auditor General . . . for the Year Ending November 30, 1862," Pennsylvania, *Reports of the Heads of Departments* (Harrisburg, 1863); "Report of Commissary General" and "Report of the Quartermaster General," *Ohio Executive Documents,* pt. 1 (Columbus, 1862), 540, 563–601; *Journal of the Constitutional Convention of Illinois, Convened at Springfield, January 7, 1862* (Springfield, 1862), 363, 403; "Report of the Board of Army Auditors," *Illinois Reports to the General Assembly* (Springfield, 1865), 2:593–808; "Report of John H. Vajen, Quarter-Master General," and "Report of Ashael Stone, Commissary General," in *Documents of the General Assembly of Indiana,* 42nd sess., pt. 2 (Indianapolis, 1863), 1:656–57, 1:803–6; *Report of the Adjutant General of the State of Indiana* (Indianapolis, 1869), 1:431–36; "Report of the Auditor of Accounts of the Commonwealth of Massachusetts for the Year Ending December 31, 1861," and "Report of the Master of Ordnance," *Public Docs.* 6–7, *Public Documents of Massachusetts . . . for the Year 1861,* vols. 1–2 (Boston, 1862); *Report of the Quartermaster General of the State of Wisconsin* (Madison, 1862), 7–11; E. R. Wadsworth to Gov. Alexander Randall, 4 Jan. 1862, box 12, ser. 49, State Historical Society of Wisconsin; "Annual Report of the Quarter-Master General," *Joint Documents of the State of Michigan for the Year 1861* (Lansing, 1862), 2; *Annual Report of the Quartermaster-General of the State of New Jersey for the Year 1861* (Jersey City, 1862), 5–8; John Niven, *Connecticut for the Union: The Role of the State in the Civil War* (New Haven: Yale University Press, 1965), 407.
Note: n.d., no data.

instead with private firms for any further requirements, the Girard House operation created some ten thousand garments. Elsewhere in the North, at least a few other states created smaller uniform-making operations. Wisconsin, for example, was paying fifty seamstresses to make shirts and drawers by June 1861 and continued to employ a few shirtmakers through the end of the year.[16]

In Illinois, Ohio, and Indiana—all western states distant from the U.S. military arsenals in the East—state officials created public facilities to make bullets and cannonballs. These ordnance operations proved to be the most durable of all the state-run establishments opened in 1861. Illinois officials set up a state arsenal in Springfield only days after the war began. Employment and production at the Illinois facility peaked in August 1861, when some 140 workers were making between 25,000 and 50,000 musket and rifle cartridges a day. By the time the Illinois state arsenal shut its doors in the winter of 1861–62, its workers had made more than 4.6 million rounds of small arms ammunition and over 32,000 projectiles for heavy guns. At a similar facility in Ohio, a work force that peaked at 260 people in 1861 turned out more than 2.4 million rounds of musket ammunition by the end of the year. In Indiana, where Governor Oliver Morton supported the quick expansion of a small trial facility into a large public arsenal, the several hundred workers employed by October 1861 were making nearly 100,000 rounds of small arms ammunition per day.[17]

Most of the workers in these state-run ordnance plants were women and children. As hundreds of men in each state enlisted in the new volunteer regiments, some of their wives, sisters, or younger brothers labored to supply them with ammunition. According to Illinois's quartermaster general, by hiring "a large number of children, both boys and girls, from eight to sixteen years of age," the state allowed them to contribute to the war effort while supporting their families. Through the ammunition-making work, he continued, children also "acquired habits of industry and became accustomed to a discipline that will have its salutary effect upon the formation of their characters." But there were also more practical reasons for the states to employ such workers. Hermann Sturm, the former German army officer who supervised the Indiana arsenal, explained that he hired mainly women and girls to make cartridges because "female labor was cheapest, and in my opinion, best adapted to the lighter work." Certainly, women were cheaper to hire than men. In the Illinois arsenal, at the high point of production in the summer of 1861, average wages were about forty-five cents a day for female employees and sixty-four cents a day for males. At a time when men generally earned about a dollar a day for common labor, hiring children and women for light manufacturing work allowed the states, like employers in the private sector, to economize on labor.[18]

Northern states also engaged prisoners to make military supplies. On the eve of the war, many state prisons contained manufactories, which were run either by private contractors or by the states themselves.[19] At the Massachusetts state prison in Charlestown, contractors had been employing inmates before the war to make simple shoes, some of which had been shipped to plantations in the South. When the beginning of the war shut down the Southern trade, many of these inmates stood idle. Massachusetts prison officials responded by putting a force of some 140 prisoners to work on military goods. By the end of 1861, the Charlestown prison had provided Massachusetts with roughly $26,000 worth of canteens, shoes, and other items. In Pennsylvania, the state penitentiary at Allegheny City (outside Pittsburgh) made $28,000 worth of boots and shoes for the troops in 1861. In Wisconsin, state prison commissioner Hans Christian Heg championed a small military manufacturing program in which inmates made dozens of garments and shoes.[20]

Although many state officials proved willing to establish public facilities to make military supplies, they also purchased from businesses in the private sector. In fact, during 1861 the bulk of military spending by most states for all kinds of goods (except ammunition) was devoted to purchases rather than public manufacture. This meant that state officials working to supply their regiments became concerned mainly by the questions of from whom and how to buy.

In some cases, officials found that the equipment required by the new volunteer regiments could not be acquired easily from sources within the boundaries of their own state. This problem led several Northern governors to appoint special traveling agents. In April 1861, no state held large stocks of rifles or other suitable weapons. Some state governors were unwilling to wait for the Ordnance Department of the U.S. Army to arm their regiments of volunteers. So, during the first weeks of the war, several of the traveling state agents took on the job of buying small arms in Europe or from importers in New York City. Governor John Andrew of Massachusetts sent Francis B. Crowninshield to England, where the venerable merchant quickly spent more than $350,000 on over 19,000 Enfield rifles and 10,000 sets of equipment. The state of New York spent a similar sum on European weapons acquired by the New York City arms-importing firm of Schuyler, Hartley & Graham. In Indiana, governor Morton himself ordered one thousand Enfield rifles (at thirty dollars apiece) from Schuyler's firm. But Morton also dispatched Robert Dale Owen, the prominent reformer, as a purchasing agent. Owen supplemented Indiana's existing stocks of small arms—which amounted to only 6,000 usable muskets and rifles when the war started in April 1861—by purchasing from Schuyler, Hartley, & Graham in New York and Samuel Buckley & Company in Liverpool, England. Owen, who like Schuyler, Hartley & Graham purchased in Europe on behalf of the U.S. military as well as for his own state, bought close to $900,000 worth of arms and other supplies on behalf of Indiana and the national government in 1861.[21]

Like Indiana, several other western states sent agents east—toward the nation's industrial core—during the first weeks of the war to hunt for supplies. Ohio and Wisconsin, like other states, looked into buying weapons directly rather than waiting for Ordnance Department officials to send them. But such efforts did not always lead to purchases. A team of Wisconsin commissioners visited several eastern cities and then concluded that since, "among all the establishments visited, not a single manufacturer could be found who would contract to furnish the arms in any reasonable time," it was better to wait for deliveries from the national government.²²

Even when traveling state agents located supplies or willing manufacturers, they sometimes had trouble coming up with the cash they needed to secure the goods. In June 1861 William Levering, an Indiana state agent, succeeded in engaging Philadelphia awning maker John Welsh to make hundreds of tents for Indiana volunteers. But Welsh soon ordered his employees to stop work on the Indiana tents and turn to more recent orders from Pennsylvania until he saw some money. "[O]n account of our failure to pay as agreed upon," Levering wrote back to Indianapolis on 4 July, Welsh "is very much annoyed and embarrassed, which renders our hitherto social intercourse, somewhat unpleasant." (After these difficulties were overcome, Indiana eventually received more than five thousand tents worth over fifty thousand dollars from Welsh, who ranked as the state's second-leading supplier in dollar terms.) Other state agents had similar troubles. Wisconsin Assistant Quartermaster William Mears, who hurried to New York and Boston in the summer of 1861 to look for enough textiles, finished garments, and equipment to outfit five new infantry regiments and five batteries of artillery, soon wondered whether he was on a fool's errand. Suitable goods were scarce, and those firms that held them expected to be paid—something that Wisconsin would be unable to do in the foreseeable future, even if the United States quickly reimbursed even half or three-quarters of its expenses (a best-case scenario). "How can they expect us," Mears asked in a letter sent back to state quartermaster general William W. Tredway in Madison, "to clothe [and] equip five regiments when our treasury is empty[?]"²³

Stymied by shaky finances and picked-over markets for military goods, many traveling state agents in 1861 returned home empty-handed. In the meantime, however, most Northern states had begun to fill the bulk of their requirements for military supplies (other than arms) by contracting with firms close to home. In effect, state officials transferred much of the work of securing scarce materials and finished goods to private businesses located within their borders. One potential advantage of such a procurement policy was that it could tap existing commercial networks, expertise, and competition in a way that dispatching official purchasing agents did not. Above all, however, it was good politics. Buying close to home meant that state officials sent procurement dollars, fronted by the states themselves but ultimately reimbursable by the national government, to their constituents. By buying locally, governors and other state officials could help to keep more profits and jobs within state borders; they might also bolster their own political power with military money.

Many Northern states bought military supplies from local firms as a matter of state policy. As Pennsylvania Quartermaster General R. C. Hale put it in his departmental report for 1861, "It has been my policy, for the encouragement of our own manufacturers, and that the money raised on the credit of the State might be spent among our own people, to procure everything in Pennsylvania, where it was possible to do so." Even in western states, which had fewer factories, state officials did their best to order from local businesses. In the summer of 1861, Wisconsin Quartermaster General Tredway recalled a few months later, he had been "determined to supply the troops with Wisconsin shoes," as opposed to footwear purchased elsewhere. In Indiana, according to the state quartermaster general, Governor Morton had made it clear "that he preferred that the contracts, as far as was consistent with the interest of the State, should remain in the State."[24]

From the point of view of potential contractors, it seemed easiest to sell directly to one's own state. Even for a large and successful business, filling orders for other states might be possible only if it served as a subcontractor. Such was the assumption made during the early weeks of the war by the partners of L. J. and I. Phillips, leading New York City dealers in men's headwear, when they offered to supply Wisconsin troops with caps as a subcontractor for an in-state firm (with which they promised to split any profits). After Wisconsin officials protested that such an arrangement would be improper, the Phillips company wrote Governor Randall in late May 1861 to explain, "We had been informed that the authorities of your state were not in favor of giving the orders for uniforms, etc., to any but citizens of the state."[25] This, notwithstanding the response of state officials to the company's first inquiry, was an accurate assessment of the situation: Wisconsin and other Northern states did generally steer purchases to in-state firms.

To be sure, many Northern states bought some goods from firms outside their borders. Some states' orders crossed political boundaries in ways that reflected existing regional patterns of trade. Wisconsin bought many of its tents from ship chandlers in nearby Chicago, as well as Milwaukee. Indiana purchased uniforms from several leading Cincinnati firms, including Stadler Bros. and Heidelbach, Seasongood, and Company, as well as from Indianapolis clothiers such as Glaser Bros., which itself was a branch of a Cincinnati house.[26] State officials also purchased from sources outside the state when the goods in question were specialty items that simply were not available locally. Weapons were not the only example of such goods. Many western states looked to out-of-state companies for rubber blankets, a particularly exotic item. One of the few sources for rubber blankets in 1861 was the Rubber Clothing Company, which had offices in New York and Boston. During the months that the various states outfitted their own troops, this firm did thousands of dollars in business with Massachusetts, Indiana, Michigan, and Wisconsin.

Even if there were important exceptions, however, state officials as a rule filled most of their procurement needs in 1861 by buying close to home. Massachusetts, which spent more relative to its size than did any other state, relied almost exclusively on Boston firms. One of the only states to spend large sums on cavalry horses (which elsewhere were usually procured by regimental colonels, individuals, or the U.S. Army), Massachusetts paid $435,000 for horses in 1861 to Benjamin Cheney and Moses Colman, both well-established Boston businessmen. Cheney was a successful New England stagecoach operator (whose business eventually became part of the American Express Company); Colman was superintendent of the Metropolitan Railroad, a horse-powered urban transport line in Boston. Massachusetts bought most of its clothing and blankets in 1861 from Pierce Bros. & Company and Whitten, Hopkins & Company, two Boston dry goods wholesalers that each supplied over $300,000 worth of goods. Smaller orders were also filled by local enterprises. Of Massachusetts's top ten suppliers of goods and services other than arms and transportation, nine were based in Boston. The tenth, the state's leading supplier of infantry equipment (including knapsacks and cartridge boxes), was Samuel Walker & Company, of Milford, Massachusetts.[27]

The state of New York, of course, purchased largely from firms based in New York City, the nation's leading commercial center. One of these was the giant dry goods wholesaling and retailing company of A. T. Stewart, already one of the nation's richest men, who sold some $60,000 worth of clothing and blankets directly to his state in 1861. In Pennsylvania, nine of the state's ten leading suppliers—of uniforms, knapsacks, food, and other goods—were from Philadelphia or Harrisburg. In Michigan, the state's leading suppliers were Detroit firms. Two Detroit clothing houses, Samuel Sykes & Company and E. S. Heineman & Company, filled orders for uniforms and blankets worth a total of $180,000. Trowbridge, Wilcox, & Company, a Detroit company that outfitted ships plying the Great Lakes, sold Michigan $24,000 worth of tents (made from the same sort of canvas used for ships' sails). In Wisconsin, the supplier who did the largest dollar volume of business with the state was Marcus Kohner, a clothing merchant from Madison. During the second half of 1861, Kohner sold Wisconsin nearly $100,000 worth of uniforms—some of which were produced by seamstresses in Madison and Milwaukee with material furnished by the state, but most of which were made in New York City under the supervision of Marcus's brother Joseph.[28]

For state officials, buying locally was often simply easier than the alternative. In the spring and summer of 1861, state supply department offices were bursting with chaotic activity. In Springfield, Illinois, one assistant quartermaster recalled later that most business "was done in a crowded office, in which loud talking and confusion frequently prevailed, and was often, especially during the first few months, continued till late in the night." In such an environment, local businessmen who already knew state officials and could easily visit their offices had a distinct advantage. For Marcus Kohner, Wisconsin's leading supplier of uniforms, it was a short walk from his store in Madison to the state supply departments. "I went to the Quartermaster General's office two or three times every day" during the summer of 1861, he recalled a few months later. It was far easier for state officials to rely upon familiar faces, such as Kohner in Wisconsin, than it was for them to buy from distant and unfamiliar out-of-state businesses.[29]

In many cases, state officials implicitly favored local firms by choosing to buy without advertising in advance for proposals or bids. According to Illinois Quartermaster General John Wood, advertising was "too inconvenient and impracticable for adoption" during the first months of the war and "personal intercourse with the contractors" was the best way of securing good supplies. Wood's counterpart in Wisconsin, William Tredway, said that he normally solicited proposals by "inviting parties who were dealers in the articles required, to furnish samples" and propose prices. Both of these officers insisted that, although they did not advertise, the targeted invitations to bid—especially when combined with the many unsolicited offers that rained down upon their offices in 1861—created more than enough competition to keep prices down. Tredway claimed later that all military goods were "supplied under active competition"; Wood recalled that his office was "frequently over crowded with competitors when contracts were to be made, and multitudes of letters have been received making offers to supply the several articles."[30] There can be little doubt that there was indeed competition for many state orders in 1861. But by failing to advertise and by relying largely upon "personal intercourse with the contractors," state officials favored the local firms that enjoyed the easiest access to state supply departments and the closest relationships with state officials.

In some eastern states, where public administration had a slightly longer history, state officials adopted procurement practices that were somewhat more formal. In New York, the state's Military Board—headed by Governor Morgan—began to advertise for bids only days after the beginning of the war. When it made its first set of large purchases in late April 1861, the state gave the entire order for twelve thousand uniforms to Brooks Brothers, the New York

City ready-made clothing manufacturer and wholesaler that was one of the largest in the nation. Other prospective suppliers complained about this winner-take-all award, especially after it became clear that Brooks Brothers had misled state officials when it promised it could make all of the garments with heavy wool broadcloth. So when New York made a second large purchase of uniforms in May, state officials divided the order. After examining seventy-five bids naming prices between $15.88 and $19.33, the officials decided to set a fixed price of $18.00 per uniform—because, as Governor Morgan explained soon afterward, "they believed that was a fair price." They then divided the $270,000 order among six firms, which together could presumably fill the order more quickly than could a single company. According to Military Board members, the six contractors they selected were "all large houses," which were "known to be in the clothing business and to be responsible houses"; in making their choices, the Military Board had "consulted the public interest and nothing else."[31]

Even in New York, where the procurement process was more formal than in most other Northern states, personal connections between state officials and prospective military suppliers proved to be important. To be sure, some of the six firms that shared New York's large May order for uniforms—including Devlin, Hudson & Company and A. & G. A. Arnoux, both of New York City—were large, well-established enterprises with the capacity to deliver quickly. Devlin, Hudson & Company, like Brooks Brothers, was one of the nation's largest manufacturers of ready-made clothing by the 1850s, when it had more than $1 million in annual sales.[32] But New York was full of big, successful businesses. When Military Board members decided to fix a common price and divide the May clothing order among several firms instead of giving it to a low bidder or bidders, they gave themselves the power to choose among a large number of qualified firms. Such a situation invited favoritism, if not outright corruption. Although Military Board members claimed they thought only of "the public interest" when they chose suppliers, there is evidence that private interests also prevailed. Military Board member D. Floyd Jones admitted later, for instance, that F. B. Baldwin & Company of New York City was chosen as one of the six winners of the May clothing order because "I knew the family of Mr. Baldwin."[33]

Personal connections and private interests seem to have determined the identity of New York's top contractor in dollar terms, Thomas C. Smith. On 20 April, only days after the beginning of the war, New York Commissary General Benjamin Welch Jr. agreed to pay Smith more than $250,000 for twenty-eight thousand sets of infantry equipment (including knapsacks, haversacks, cartridge boxes, and other items). Smith, a master mason and construction contractor from New York City, apparently had no special capacity to manufacture military goods. After signing the contract, Smith proceeded to subcontract much of the work. He had many of the haversacks and knapsacks made by Peddie and Morrison, a large trunk-making factory across the river in Newark, New Jersey. For leather cartridge boxes, for which New York paid him $1.50 each, Smith subcontracted with Betts and Nichols, a New York City saddlery firm that agreed to make them for $1.12 each. As these arrangements suggest, it was not clear why the state should have contracted with Smith rather than with firms accustomed to making goods out of leather and other heavy materials. But Smith's success in winning this large order became easier for New Yorkers to understand when it was revealed that his silent partner, carriage and coach manufacturer Joseph Godwin, was a close friend of Commissary General Welch.[34]

Private and public interests became even more confused in Indiana, where top state supply officials chose to buy from companies in which they held personal stakes. The state's commissary general, Isaiah Mansur, was the owner of a large pork-packing business in Indianapolis. During the first weeks of the war, when he was charged with feeding the new recruits, Mansur ordered meat from his own company. The Indiana state assembly soon passed a resolution calling on Mansur to resign; he did so on 29 May.[35] Meanwhile, the state's other top supply officer, Quartermaster General John Vajen, also mixed private and public business. Vajen bought some five thousand dollars worth of goods from his own hardware company, a decision he later justified by claiming that he had always charged Indiana the "lowest price."[36]

Besides using Indiana state funds to order from his own company, Vajen also mixed state accounts and personal accounts and ran his office in a way that encouraged kickbacks and bribes. After Governor Morton invited a special U.S. congressional committee on war contracts to visit Indiana in March 1862, Vajen and other Indiana officials and contractors were called to testify. Vajen told the committee that he had dipped into his personal accounts to pay military suppliers and later reimbursed himself from public funds, so that the state's accounts were constantly mixed with his own. He did not record all of these transactions, said Vajen: "All I know is that I expended all I received, and perhaps a little more." Meanwhile, some of the state's leading contractors, including woolens manufacturer J. C. Geisendorff, told the congressional investigators that Indiana Assistant Quartermaster Frank Murphy had told suppliers that they needed to hand over a cash payment amounting to 5 or 10 percent of the value of their orders. Vajen claimed that he had always considered these payments to be donations "for the benefit of the soldiers" rather than bribes, but he acknowledged he had no record of how much money had come in or where it had gone. Presented with this record of financial irresponsibility and aware that Vajen had also bought from his own company, congressional investigators concluded in 1862 that Indiana's quartermaster general had displayed during the previous year "[a]n unpardonable eagerness to make the misfortunes of the nation the source of personal aggrandizement."[37]

In Illinois, the state that spent more on procurement in 1861 than any other, Governor Richard Yates steered at least one large contract to a personal friend. During the summer and fall, Illinois bought most of its uniforms and blankets—nearly $1 million worth in the end—from Cole and Hopkins, a Cincinnati dry goods wholesaler that was working in conjunction with A. T. Stewart & Company, the immense dry goods concern in New York City. But in late November, the Cincinnati firm was surprised to learn from Illinois state agent Martin Cassell that the state might send future clothing orders to other firms. Although Cassell told Cole and Hopkins that Illinois was simply opening up the process to "honorable competition," the truth was that Governor Yates had privately asked state supply officials to buy from a friend who ran a small tailoring business in Philadelphia. In early November, Yates had informed Illinois Quartermaster General John Wood that "Mr. Sarmiento of Philadelphia, who is still here [in Springfield] assisting me in my sickness, is an old friend of mine." Referring to military spending as "patronage," the governor continued: "I feel very desirous that you would give him at least a portion of what may be needed provided he can furnish you as cheaply and upon as good terms as others." A few days later, Cassell wrote from Washington to assure Yates that he would go to Philadelphia to contract with Sarmiento's firm. In December, Illinois placed a large order for clothing (worth about $125,000 in the end) with the governor's friend.[38]

Across the Northern states in 1861, many state officials used their procurement powers to steer military purchases to men they knew. In some cases, state officials chose to buy from their friends even if it meant bypassing more qualified suppliers. In other cases, they actually bought from their own companies. This does not mean that state supply officials paid no attention to quality and price or that the part of the Northern war economy created by the various states was purely a patronage machine used to benefit the Republican Party and its friends. In some cases, if not much of the time, state officials put aside personal and parochial interests. During the summer and fall of 1861, for example, Wisconsin officials wrote to their counterparts in Illinois and at the main U.S. Army supply depot in Philadelphia to ask for lists of reliable contractors and recommendations on prices and quality standards.[39] Nevertheless, in the federal war economy that state officials helped to build in 1861, much depended on party connections and personal relationships.

The Contested Transition to a National War Economy

As the various states proceeded with their own supply efforts during the first part of the war, it was not difficult for many Northerners to imagine the potential benefits of centralizing procurement authority. Officials in every state knew that, during the spring and summer of 1861, state agents and national officials had competed with one another over scarce supplies. When Indiana agent Robert Dale Owen visited A. T. Stewart & Company in New York City in August 1861, he found that the company had already "ransacked the whole city for blankets for the general [U.S.] Government." Around the same time, Wisconsin agents had gone east to buy weapons, only to find that "many of the Agents from other states had already canvassed the market thoroughly and obtained all that could be had on reasonable terms." This state of affairs, in which a dozen teams of state agents repeated the same pilgrimages to the same merchants and manufacturers one after the other, evidently made little sense. Even if this direct competition for the same lots of goods did not end up inflating the prices paid by the public, the agents were repeating one another's work needlessly. Such redundancy of function suggested that a more centralized, national procurement system would be more efficient.[40]

While each state outfitted its own troops, moreover, it was difficult for the North to enforce the kind of design and quality standards that were desirable in large national armies. When a dozen different states bought clothing from dozens of different suppliers without enforcing the same specifications, uniforms across the Northern armies became something less than uniform. At the First Battle of Bull Run in July 1861, the multiple colors worn by troops on both sides caused tragic mistakes.[41] Days later, Quartermaster General Montgomery C. Meigs told New York Governor Morgan that any uniforms his state ordered in the future should follow U.S. Army design standards. Uniformity would not only prevent friendly fire accidents, Meigs suggested, but would make it difficult for Confederate forces to single out Northern units believed to be especially strong or weak. "It is better that an enemy should have nothing to indicate to him the character of the Regiment he is to direct his attack against. Let all be to him regulars, all volunteers."[42] Over the following weeks, as Meigs's bureau and its sister supply departments assumed responsibility for procurement, they created a more centralized supply system that achieved the kind of uniformity desired by the quartermaster general.

The uniforms and other military goods purchased by state officials not only failed to meet a single standard for color and design but sometimes turned out to be of poor quality. Both New York and Pennsylvania, the most populous states in the North, were rocked in 1861 by scandals concerning poor uniforms. In New York, state officials had given their first large order for twelve thousand uniforms to Brooks Brothers because that firm was the only one to promise to keep to its original bid while making the garments from broadcloth, a heavy wool fabric that was expensive and difficult to secure during the first weeks of the war. Soon after, however, Brooks Brothers had informed the state that it could not find enough broadcloth to fill the order. After New York officials agreed to allow the firm to substitute other materials so long as they were "of equal value," Brooks Brothers went ahead with the order, and state inspectors accepted the garments it supplied. But it soon became apparent that many of the uniforms were of poor quality; as a legislative report concluded later, they had been "badly cut, badly sewed and made up." Although Brooks Brothers agreed in a settlement with the state to replace the defective items, the history of this transaction suggested that New York officials had trouble enforcing quality standards.[43]

In Pennsylvania, there were even more problems with uniforms. First, Pennsylvanians heard reports that some of the garments produced by the state-run operation at the Girard House in Philadelphia were no good. Although these early complaints may have exaggerated the extent of the problem, as many as 3,000 of an early batch of 20,000 uniforms were found to be unusable. Pennsylvania soldiers found that garments purchased from private firms were also wanting. The 2,000 uniforms that state agent Charles Neal purchased (at the bargain price of ten dollars apiece) from the Pittsburgh firm of Frowenfield and Bros. turned out to be so bad that they were virtually useless. Like residents of other Northern states, Pennsylvanians were exposed to a steady stream of reports of grand jury hearings and legislative investigations into alleged fraud and corruption. In the end, the Pennsylvania investigations, like similar ones conducted in Michigan, ended up absolving most contractors and public officials.[44] But the cases of evident wrongdoing combined with a general atmosphere of scandal were enough to convince many Northerners that mismanagement of the states' supply efforts harmed their soldiers in the field.

Pennsylvania and New York were hardly the only states awash with military supply scandals in 1861. Virtually every Northern state saw its top supply officials and contractors—after they had been working only a few weeks—targeted by the press or legislators for alleged misdeeds. In Indiana, as noted above, both the commissary general and the quartermaster were forced out.

Fig. 1.1. "The Pennsylvania Volunteers." In this July 1861 drawing, Pennsylvania soldiers are compelled to stand in tight formations to prevent female camp visitors from seeing their backsides, which are exposed because their days-old uniforms are already in rags. "Dedicated to the Girard House Contractors, and Governor Curtin," the cartoon blames state officials for outfitting Pennsylvania's first regiments with substandard goods. As reports of such problems spread across the Union, the word *shoddy* entered the popular lexicon. *Source:* Department of Special Collections, University of Chicago Library.

In Wisconsin, a legislative committee with a Democratic majority deemed the state's Republican-directed military expenditures "careless and extravagant." Wisconsin officials, the committee claimed, had received too little for the state bonds they floated to raise funds and then paid too much for supplies; the office of Assistant Quartermaster General James Holton in Milwaukee, according to the legislators, was "a crude and chaotic mass of bungled papers, that would not pass muster among business men." In Ohio, the Republican governor and his top supply officials were denounced by one leading Cincinnati newspaper for their "reckless extravagance" and "disgraceful" management of the military economy. "[C]orrupt and swindling contracts," declared the *Daily Commercial* in June 1861, "were given out repeatedly" in Ohio.[45]

By the autumn of 1861, it was difficult for officials in the various Northern states to claim that the business of war supply was being conducted with success and efficiency.[46] State agents were stumbling over one another in the big eastern cities, competing for the same goods and driving up prices. Nearly every state administration was contending with accusations of serious mismanagement or fraud. Beyond this, it was clear that the longer the war lasted and the bigger it became, the more difficult it would be for states to supply their own troops. For a state to outfit a few of its first regiments before they departed for the front was one thing, but for a state to sustain troops that had already begun to move from point to point across the eastern half of the continent was another thing entirely. After Northern forces fell back in disarray at Bull Run, it seemed likely that the conflict would stretch into the next year. As the Northern armies grew larger and the volunteer regiments marched farther away from their home states, the federal war economy that had grown up in the spring and summer of 1861 looked more and more unworkable.

As late as September 1861, when Meigs informed Governor William Sprague of Rhode Island that the state would be reimbursed for future purchases of clothing and equipment so long as they were "at prices not exceeding United States rates," the federal supply system was still authorized by officials in Washington.[47] Over the next few weeks, however, the War Department began to order states to stop buying and to transfer procurement authority to U.S. military officers. In some states, including Ohio, officials offered little resistance to this change. In September, Meigs had ordered veteran U.S. Army quartermaster Frederick Myers to go to Columbus, Ohio, where Myers was to "act in conjunction with Governor Dennison." On 7 November, Ohio state quartermaster George Wright wrote Meigs to say that the state was ready to give Myers full control over supply. "You will readily see," Wright suggested to Meigs, "the advantage of such an arrangement in preventing complication and confusion of accounts, and competition between the United States and State authorities. It seems to me the whole business should be under one general management." Meigs was pleased that Ohio officials saw the benefits of a more centralized system. As he told Myers, "The sooner the whole returns to the control of the United States the better."[48]

Not every Northern state, however, proved so willing to hand over purchasing authority. Because the U.S. Army by this time was operating large procurement depots in Cincinnati, Philadelphia, and New York City, the change to national control was easier to swallow for state officials in Ohio, Pennsylvania, and New York. In these states, even if governors and their subordinates lost direct authority over spending, the large U.S. depots were likely to sign hundreds of contracts with local firms. In other states, however, officials anticipated that the transition might diminish military spending within their states. During the last months and weeks of 1861, officials in several Northern states told Cameron and Meigs that they wanted to keep outfitting their own volunteers.

In the autumn of 1861, after War Department officials in Washington told state officials to stop buying military supplies, the national officers scolded any states that were slow to follow the order. By early October, Maine's Governor Israel Washburn was still spending procurement money that would ultimately be reimbursed by Washington, telling military officials that it was his right to do so. Even after the War Department asked him to stop, Washburn continued to authorize purchases of blankets and other supplies. This annoyed Meigs, who believed that state and national supply officers were driving up prices by competing to buy the same scarce goods. On 12 October, Meigs wrote to Washburn to insist that the time for state purchasing—once "gladly availed of" by the War Department—had now passed. By insisting on their so-called "right" to keep buying, continued Meigs, governors would effectively make war supplies more expensive and undermine a national war effort that sought to preserve the Union. "With great respect to the Governor," Meigs wrote Washburn, "I must beg leave to suggest that to persist in purchasing as a state right, for the United States, and to be paid for by the United States, what the Government Agents inform him the United States does not want, may savor a little of the very source of all our troubles."[49] For a governor to say that retaining procurement authority was a "state right," in other words, was to invoke the very doctrine that had led to the war in the first place.

Such nationalist arguments may have convinced some state officials, but others continued to resist giving up all of their control over procurement. In Indiana, Governor Morton asked the War Department as early as August 1861 to send U.S. commissary and quartermaster officers to help outfit and sustain the new volunteer regiments. But once the U.S. officers arrived and began to take away his power to direct spending, Morton realized that he had made a mistake. By 1 September, Morton was already complaining to U.S. Secretary of State William H. Seward that mobilization in Indiana was slow because "[t]he U.S. Quartermaster and Commissary here deny my authority to order expenditures of any kind in the organization of troops."[50] Morton was particularly displeased with Alexander Montgomery, the veteran U.S. quartermaster who had been sent to Indianapolis. "The public service has suffered and will suffer much more," Morton told Seward by telegraph,

> from the conflict between the State and Federal authorities here. Major Montgomery will not tolerate the least interference on my part, and on all occasions, repudiates my authority. . . . As the matter stands now there is no unity of action, it is a double headed concern, each head being independent of the other. While those [procurement] departments were under the control of the state, every thing was ready when it was demanded, and our regiments all went out in good condition.
>
> The public does not know but what I control these things as I did before, and I am held responsible for failures over which I have no control.
>
> It is but justice and I believe the public interest would be greatly subserved by placing those departments under the control of State authorities here.[51]

Morton wanted his procurement powers back. But he would be unable to regain them. Across the North during the autumn of 1861, U.S. supply officers were relieving state officials of their authority over military spending.

Even in Indiana, where the energetic Governor Morton lobbied hard to retain his powers over spending, most purchasing responsibilities came to rest with national officers. In October, after Morton and Montgomery refused to heed requests from Washington that they try to work with one another, Morton did succeed in getting Cameron to remove the U.S. quartermaster. But this did not mean that Morton would take back spending authority—only that a new U.S. officer would have to be sent to Indianapolis. Montgomery's replacement was James Ekin, a former Pittsburgh businessman who had solid Republican credentials. Morton, pleased with the change, soon told Meigs that Ekin was "a very efficient officer and I earnestly hope he will be allowed to remain here."[52] But even having a friendly U.S. quartermaster in Indianapolis did not allow Morton free rein. When Morton asked Washington for permission to order overcoats, Meigs refused. Just as he had done with Maine's Governor Washburn, Meigs told the Indiana governor that he and other state officials were paying too much for supplies and driving up prices. "It is time this is ended," Meigs told Morton. "The people cannot afford to pay at this rate." For the sake of economy, Morton would have to cede all contracting authority to Ekin, who was the local representative of a centralized, national supply network.[53]

Of all the Northern states, Illinois proved to be the most resistant to giving up its procurement powers. In late September 1861, Meigs sent a veteran

U.S. quartermaster named Asher Eddy to Springfield, the state capital. Meigs ordered Eddy to install himself in the state quartermaster department, "taking charge of all its expenditures in supplying and equipping volunteers. You will report yourself to the governor of Illinois, and assist the state authorities." Even before he arrived in Springfield, Eddy may have realized that there might be a great deal of difference between "taking charge" and acting to "assist the state authorities." He soon found that Illinois officials expected him to do little more than sit back and watch them work. By 16 October, only days after he went to Springfield, Eddy wrote back to Meigs to report that Governor Yates "wishes for the state to go on making contracts as she pleases, and for our department to stand by, money in hand, and pay the debts. In other words the U.S. quartermaster is to be an automaton in the hands of the state."[54]

Because neither Eddy nor his superiors in Washington were ready to allow Illinois officials to spend reimbursable funds as they pleased after October 1861, the state became a site of conflict between state and national authorities. After arguing with Illinois officials for days, Eddy wrote Governor Yates on 9 November to insist formally that the state give up all responsibilities for supply. Although state authorities might believe that Illinois "can and will take better care of her troops than the United States," wrote Eddy, they were wrong. "I recommend the transfer of the quartermaster's department to my hands wholly," Eddy continued, "from a sense of duty to the United States and a desire to do all in my power to 'assist the state authorities.'" Yates ignored this letter. Over the next month, Illinois officials continued to buy large quantities of uniforms and cavalry equipment. On 9 December, Yates finally took the time to inform Eddy that the state did not intend to stop buying these goods. To justify this decision, Yates cited the indemnification act that had been passed in July. "I avail myself," declared Yates, "of the right conferred by act of Congress upon each state to furnish such supplies, because I think this state can better provide for her troops." Thus, as late as December 1861, Illinois—alone among the Northern states—was still refusing to transfer procurement authority to U.S. officers.[55]

As soon as he received Yates's letter of 9 December, Eddy told the governor that it simply was not acceptable for the state to keep buying military supplies. "It is contrary to the wishes of the authorities at Washington that this state should continue to act for the United States," Eddy informed Yates, "when the latter has her own appointed agents." He would therefore refuse, Eddy announced, to recognize any subsequent purchases by Illinois as legitimate expenditures reimbursable by the U.S. Treasury. After weeks of bickering, Eddy had finally thrown down the gauntlet: the national government, which by this time had already advanced more than $4 million in mobilization reimbursements to the various Northern states, would not pay for any more Illinois purchases.⁵⁶ Next, Eddy went public, placing advertisements in Illinois newspapers declaring that his office would now take over procurement authority. Still, Yates resisted. On 12 December, he wrote to ask the members of the Illinois congressional delegation in Washington to "[s]ee immediately that the policy of the state is sustained by the General Government." But even with an Illinois man in the White House, Yates was not able to keep his state from joining the rest of the Northern states. After Meigs prodded Cameron to uphold Eddy's authority, the secretary of war on 16 December finally ordered Yates directly to stop purchasing. Only then, at the very end of the year, did Illinois finally give up control over procurement.⁵⁷

The dispute between Eddy and Yates, which attracted national interest during the winter of 1861–62, showed that the locus of procurement authority was not a matter of consensus and that it mattered a great deal. According to a report drawn up quickly by Meigs in Washington, Illinois had paid nearly $734,000 for clothing in December alone; this was about 20 percent more than the U.S. Army would have paid for the same goods. When Meigs's report became public, Yates responded by saying that the state had paid high prices to secure goods of superior quality and that the War Department had not ordered him expressly to stop purchasing until December. In the end, Washington agreed to pay for the December clothing orders, but the sum to be refunded was $130,000 less than Illinois had actually paid for the goods, to reflect the difference in prices calculated in the Meigs report.⁵⁸

But the criticism of Yates and his state's procurement expenditures kept coming. At an Illinois constitutional convention in January 1862, in which Democrats stood in the majority, they attacked the military expenditures of the Republican governor, Yates, and his subordinates. The convention's committee on finance noted that, although the state legislature had authorized the issue of only $2 million in state bonds, procurement officials had spent nearly twice this amount. "[T]he expenditures made by the state of Illinois, for war purposes," the finance committee concluded, "have, for the most part, been made, not only without authority of law, but in direct opposition to the wish of the general government." Meanwhile, in Washington, on the floor of the U.S. House, prominent Illinois Democrat William A. Richardson attacked his own state's Republican-led executive branch. "I fear that in the State of Illinois," Richardson told his fellow congressmen in late January, "the entire tribe of State officials have engaged in plundering the public Treasury."⁵⁹ Although Governor Yates and other Illinois officials had managed to retain control over spending longer than had their counterparts in other Northern states, they now found themselves being denounced at home and abroad for having done so.

When Cameron demanded on 16 December that Illinois stop buying military goods, the period of major procurement spending by the Northern states came to a close. This is not to say that no state spent money for military purposes after January 1862. As the war went on, state and local governments raised huge sums for recruitment bounties paid to soldiers upon enlistment. For local and state governments, which generated a large fraction of the nearly $500 million worth of bounties paid in the North, the raising of these bonuses—along with direct payments to soldiers' families—became the central problem in the wartime military economy.[60] Furthermore, even state-level supply work did not disappear completely. During the summer of 1862, Lincoln's sudden call for 600,000 new troops led several states to re-enter the supply business for a brief time. Among the most active of all states after the first year of the war was New Jersey, which (for reasons that remain unclear) continued to spend large sums on horses and arms, in some cases with special permission from Meigs and other U.S. military authorities.[61] Border states, including Missouri and Kentucky, were also especially active because they outfitted many state militia units (not mustered into U.S. service) raised after 1861 to respond to Confederate raids. In other cases, as in Indiana, state governments supported the work of state "sanitary" boards, which—like the national, nongovernmental U.S. Sanitary Commission and U.S. Christian Commission—was devoted in part to securing hospital supplies.[62]

A few state-run manufactories created in 1861 continued producing for the better part of the war. Ohio's state arsenal continued to make over a million rounds of ammunition a month until it closed in August 1863; Indiana's arsenal ran even longer, until the spring of 1864. (Congress had authorized the construction of new U.S. arsenals in Indianapolis and Columbus, as well as Rock Island, Illinois, in July 1862, but these did not begin to operate until after the end of the Civil War.) Indiana also ran a state bakery outside Indianapolis that made bread for soldiers and their families throughout the war.[63] Across the North as a whole, however, procurement expenditures by state governments were very small after December 1861. During the last three years of the war, state quartermasters and commissaries spent much of their time preparing official claims for reimbursement of their 1861 outlays by the U.S. Treasury.[64] By the beginning of the new year in January 1862, when Edwin Stanton replaced Cameron as secretary of war, the various Northern states were largely out of the military procurement business. Stanton told Massachusetts Governor Andrew in August 1862, as the new manpower mobilization drive proceeded, that the "experience of last year produced too many frightful evils to renew the experiment" of having each state equip its own troops.⁶⁵

THE STORY OF THE RISE AND FALL of the federal military economy that equipped the North's first regiments in 1861 suggests some of the challenges that a major war mobilization posed to the traditional American political order. During the first months of the war, when both the national government and the various state governments handled procurement, the economic mobilization was characterized by the kind of decentralization and division of authority that were at the heart of American federalism. While U.S. officers in Washington and at supply depots around the country bought some goods for Northern troops, state governors and their subordinates also enjoyed power over military spending, which they sent mainly to businesses and workers within their states. This was a style of war mobilization that was compatible with the traditions of American republicanism. But it survived less than a year. The fall of this federal war economy, which ended with the transfer of all significant procurement authority from the states to U.S. officers, was both predictable and extraordinary.

It may be easy to understand how the end of state procurement activity in 1861 can be explained as a rational step toward a more efficient war economy. But it is equally important to see how seriously this development challenged existing dynamics of political and economic power in the United States. Clearly, there were many good practical reasons for national authorities to coordinate all war supply work. But, however rational the centralization of procurement authority might have been, it also represented a potentially radical redistribution of political and economic authority. It was potentially radical not because the Constitution did not technically permit national authorities to direct war efforts (in fact it did so), but because in practice the war economy was a high-stakes project that was initially co-managed by the various states. Some Northern governors, including Morton in Indiana and Yates in Illinois, resisted the U.S. takeover of procurement in late 1861 because they knew that it would strip power and money from themselves, their subordinates, the Republican Party, and businesses and workers within their states. If the Civil War had ended in early 1862, these losses would have been small. But because the war actually expanded in 1862 and lasted into 1865, the monopolization of war supply control by U.S. officials turned out to be an enormously important transfer of money and power away from localities and states. In mid-nineteenth-century America, this takeover by national authorities represented a departure from the standing federal order and the republican traditions that had helped to create and sustain it.

At the end of the war's first year, one possible solution to the political and economic problem of Northern war procurement had already emerged and disappeared. By the beginning of 1862, a federal military economy had been replaced with a national one. Governors and other state officials had been displaced by national officers. Exactly how big a difference this would make in the ongoing Northern mobilization effort would depend in large part on who these national officers were and what kinds of choices they made about how to manage the business of war. In fact, these managers of the Northern war economy represented one of the most unusual groups of powerful political and economic actors in American history.

CHAPTER 2

The Formation of a National Bureaucracy

When state military supply departments and governors gave up their procurement authority in late 1861, the Northern war economy became national. At state capitals such as Albany and Columbus, U.S. military officers arrived to take charge of the work that state authorities had handled previously. At the same time, national officers at the War Department in Washington and army supply depots in major Northern cities—including Philadelphia, New York City, Cincinnati, and Saint Louis—consolidated their control over procurement. By the beginning of 1862, most state capitals no longer served as important war supply centers and state governments no longer purchased significant quantities of military goods. This transition, accepted quickly in some states but resisted for weeks in others, gave control of the Northern procurement project to officers of the U.S. government.

The shape of the North's national war economy would now depend on the identity of the national officers who ran it and the sort of procurement policies they chose to pursue. The Republican Party, a national organization with members at all levels of government, was apparently in a good position to manage the transition to national control in a way that might have limited the significance of the end of procurement by the various states. The secretary of war in 1861, Simon Cameron, was himself a patronage-minded Republican politician from Pennsylvania. Although he insisted by the end of the year that all states cede control to U.S. officers, Cameron knew it was not good party politics to alienate Republican governors. Meanwhile, President Lincoln and other top Republicans, having taken control of the White House in early 1861, were using their patronage powers to give hundreds of U.S. government jobs across the North to loyal Republicans.[1] If the Republican Party determined the identity of the U.S. officers who took full control of procurement by the beginning of 1862, the difference between a federal war economy and a national one might be minimal.

In fact, however, elected Republican Party officials ended up having surprisingly little direct control over the North's military economy during the Civil War. Instead, veteran military officers managed the national war economy that emerged by early 1862. These officers staffed U.S. military supply departments, which were longstanding bureaucratic institutions that traced their roots back to the beginning of the century. To be sure, as these supply departments ballooned in size during the war years, they came to employ dozens of Republican-appointed officers. But the most powerful Northern procurement directors were men like Asher Eddy, the regular army officer who had struggled with Illinois supply officials in late 1861. Reared in Rhode Island by a widowed mother, Eddy had become facile enough with mathematics as a boy to allow him to enter the U.S. Military Academy at West Point, from which he graduated in 1844. After teaching math at West Point as a young lieutenant, Eddy had spent the 1850s at posts in Florida and California. When he was dispatched to Illinois in 1861 by Quartermaster General Meigs, Eddy had already spent two decades in the army officer corps.[2]

To understand the workings of the North's war economy during the Civil War, one must look back to the antebellum era, when Eddy and his peers were educated. Well before 1861, the U.S. military supply bureaus were among the most stable, most bureaucratic, and most important governmental institutions in America. After the end of the Mexican War of 1846–48, the army's Quartermaster's Department had become the primary manager of the United States's new continental empire. During the years before the Civil War, most army quartermasters were stationed at far-flung posts in the great West, where they handled small-scale procurement and long-distance logistics. In 1861, after years in the West, these supply officers headed east to take charge of the Northern military economy. Members of an unusual American elite, formed largely at West Point and in the antebellum West, these officers ran the Civil War economy according to principles they had learned on the job of supervising continental logistics. Although they worked side by side with new, Republican-appointed volunteer officers in 1861–65, the career officers carried into the war a distinctive understanding of the American state. Thanks in part to their antebellum experiences, the managers of the North's war supply system were unusually sympathetic to bureaucratic standards and public entrepreneurship.

Institutional Origins and the Development of the Quartermaster's Department

The project of Civil War mobilization that started in April 1861 quickly overwhelmed the existing capacities of American government. This was one reason that the various Northern states began to outfit their own regiments of volunteers during the first months of the conflict: neither U.S. military departments in Washington nor anyone else in the country was prepared at that point to handle the job. The new war, which would soon become far larger than any previous conflict in America, required new capacities, new people, and new solutions. It stands to reason that this unprecedented crisis would also require new institutions—new rules and new organizations that could manage the challenge of mobilization. But to a surprising degree, the North's massive, four-year procurement project was managed by existing institutions that grew in size but retained the basic operating procedures and even many of the key personnel that had been in place before the war. This was possible only because during the antebellum decades the United States had created a small but robust military bureaucracy.

At first glance, the idea that there was a significant military bureaucracy in the early United States seems strange. The United States, in comparison to European nations, had a small standing army and navy—it included only about twenty-five thousand soldiers and sailors by 1860, when the peacetime British military had nearly ten times as many men and France and other continental powers had even more.[3] Partly because the U.S. military was relatively tiny, the entire national governmental apparatus in America was smaller and cheaper than its European counterparts. At the same time, public opinion in antebellum America tended to be hostile to both standing armies and bureaucratic organizations and regulations.[4] Not without some good reasons, therefore, many historians have located the origins of the modern American military and modern bureaucratic administration in the Progressive Era, around the beginning of the twentieth century.

Despite its small size, however, the military was a critically important part of the national state in early America. Most of the money taken in by the early U.S. government (during this era, mainly at customs houses and from land sales) was channeled to the army and navy. During the decades before the Civil War, the military regularly accounted for well over half of the total national government expenditures.[5] If the early American military accounted for a large fraction of the national budget, it was also given large responsibilities. Between the War of 1812 and the beginning of the Mexican War in 1846, for instance, the army played a leading role in the national government's efforts to push tens of thousands of indigenous people off lands wanted by European-American settlers or speculators. If certain groups of Indians resisted, as did the Seminoles in Florida in the late 1830s, the army was expected to settle the issue by force. But the military also handled less violent transactions. Until 1849, the army oversaw the Indian Department, which was responsible for handling U.S. relations with Native Americans all over the country. Meanwhile, early American military engineers and ordnance officers helped to develop important new technologies, including interchangeable machine parts and railroads. So even if it was relatively small, the early U.S. military served as an important political and economic actor during the decades before the Civil War.[6]

Although field commanders and their forces normally serve as the main characters in military histories, these soldiers on the line depend upon the staff departments responsible for keeping them supplied. Of all of the staff bureaus in the nineteenth-century American military, none was more important than the army's Quartermaster's Department. By the 1850s, the Quartermaster's Department was spending over a third of the total army budget—far more than any other military department. The Quartermaster's Department was the biggest and most expensive bureau because its duties were so comprehensive. While the Ordnance Department procured arms and the Subsistence Department supplied food, the Quartermaster's Department handled almost every other aspect of procurement and logistics. Quartermasters were responsible for acquiring, storing, and issuing a wide range of goods, including uniforms, tents, wagons, horses and mules, hardware, lumber, and fuel. They handled the transport of all of these goods, as well as those procured by the other bureaus. In short, quartermasters were the chief managers of the business of war.

Quartermasters, who had been part of European armies for decades, supplied American troops during the Revolutionary War. But the bureau that managed the Northern supply system during the Civil War traced its origins to the reorganization of the American military in 1818. In that year, Congress and Secretary of War John Calhoun created a new system of staff bureaus—including the Quartermaster's Department, the Ordnance Department, and other supply organizations—that would last for decades. From the beginning, the quartermaster general in Washington and his officers across the country were charged with handling a wide range of supply jobs. In 1842, the Quartermaster's Department became even more important when it absorbed the bureau of the commissary general of purchases. Based in Philadelphia, this civilian-led department had been responsible for procuring military uniforms, boots, knapsacks, and other items of equipage; now, it came under direct military control.

Table 2.1 U.S. Military Expenditures by Department, 1845–1875

	Expenditures (in millions of nominal dollars)							
	Army Departments							
Fiscal Year	Quartermaster	Subsistence	Ordnance	Pay	Other	Total army	Total navy	Total, all U.S. govt.
1845	1.0	0.9	1.0	2.2	0.7	5.8	6.3	22.9
1846	2.3	1.0	1.4	4.4	1.7	10.8	6.5	27.8
1847	20.7	2.3	2.0	7.8	5.5	38.3	7.9	57.3
1848	14.3	4.2	1.4	8.5	−2.9	25.5	9.4	45.4
1849	5.8	n.d.	1.2	7.6	0.3	14.9	9.8	45.1
1850	4.1	n.d.	1.2	2.6	1.5	9.4	7.9	39.5
1851	5.7	n.d.	1.1	2.7	2.3	11.8	9.0	47.7
1852	2.6	n.d.	1.2	2.7	1.7	8.2	9.0	44.2
1853	2.8	1.3	1.0	2.7	2.1	9.9	10.9	48.2
1854	4.0	1.8	1.0	2.7	2.2	11.7	10.8	58.0
1855	5.2	1.3	1.2	2.9	4.2	14.8	13.3	59.7
1856	7.0	1.6	1.4	4.0	2.9	16.9	14.1	69.6
1857	6.8	1.9	1.5	4.6	4.5	19.3	12.7	67.8
1858	9.3	2.5	1.9	4.6	7.2	25.5	14.0	74.2
1859	10.6	1.5	1.6	5.0	4.5	23.2	14.6	69.1
1860	6.5	1.6	1.5	5.4	1.4	16.4	11.5	63.1
1861	9.3	4.7	—[a]	4.1	4.9	23.0	12.4	66.5
1862	174.2	48.8	38.8[a]	106.7	25.9	394.4	42.7	474.8
1863	234.2	69.5	42.3	216.4	36.9	599.3	63.2	714.7
1864	305.4	98.7	38.5	218.1	30.1	690.8	85.7	865.3
1865	440.2	144.8	43.1	344.0	59.2	1,031.3	122.6	1,297.6
1866	53.2	7.7	16.6	192.5	14.5	284.5	43.3	520.8
1867	39.0	10.9	5.5	42.8	−3.0	95.2	31.0	357.5
1868	36.5	8.0	3.1	60.7	14.9	123.2	25.8	377.3
1869	22.0	8.9	2.8	38.8	6.0	78.5	20.0	322.9
1870	23.0	4.9	2.4	23.8	3.6	57.7	21.8	309.7
1871	12.5	3.9	1.6	18.5	−0.7	35.8	19.4	292.2
1872	12.5	2.9	1.9	12.7	5.4	35.4	21.3	277.5
1873	13.4	n.d.	2.2	13.2	17.5	46.3	23.5	290.3
1874	13.2	n.d.	2.9	13.3	12.9	42.3	30.9	302.6
1875	12.4	n.d.	2.0	13.5	13.2	41.1	21.5	274.6

Sources: Total U.S., War Dept., and Navy Dept. figures from U.S. Department of Commerce, *Historical Statistics of the United States: Colonial Times to 1970*, pt. 2 (Washington, D.C.: GPO, 1975), 1114–15. Army department figures from U.S. War Dept., *Annual Report[s] of the Secretary of War* (Washington, D.C.: 1845–76). Unfortunately, the Subsistence Dept. rarely provided budget data in its annual reports. For 1845–65, therefore, I compiled subsistence expenditures listed in annual U.S. Treasury "Receipts and Expenditures" reports. (These are available in the U.S. Serial Set; for a list of the documents and a discussion, see app. C.) The "Receipts and Expenditures" reports were also used to compile Quartermaster's Department figures for fiscal 1862–66 because the processing of accounts within the Quartermaster's Department itself was too slow to allow for accurate annual data during those years. The 1866–72 Subsistence Dept. data is taken from "drawn from Treasury" amounts listed in "Public Property Sold by the War Department," House Exec. Doc. 200, 42nd Congr., 2nd Sess. (1872), ser. 1513. Because the "Other" data were created by subtracting the major departmental numbers from the total War Dept. figure and because the various numbers came from different sources, a few "Other" figures are negative.

[a] Ordnance Dept. expenditures for fiscal 1861 and 1862 were reported together as a combined figure in the 1862 *Annual Report of the Secretary of War*.

When the Mexican War started in 1846, the Quartermaster's Department was clearly the main overseer of American military procurement and logistics.[7]

The early Quartermaster's Department enjoyed a continuity of organizational leadership that was remarkable for any time and place. Thomas S. Jesup, a veteran of the War of 1812, took charge of the new Quartermaster's Department in 1818 and remained its leader until his death in 1860, less than one year before the start of the Civil War. Few national government officials in history have served so long as head of a single important executive department. Born in 1788, Jesup had joined the U.S. Army in 1808, before his twentieth birthday. The War of 1812 represented an important opportunity for the young soldier: before the end of the conflict, Jesup had been promoted to major and led the 25th U.S. Infantry. When he became quartermaster general in 1818, during Monroe's first term as president, Jesup was only twenty-nine.[8]

An ambitious leader, the new quartermaster general proved to be remarkably energetic, both as a commander of troops in the field and as an administrator behind his desk in Washington. Just after his appointment in 1818, Jesup wrote Secretary of War Calhoun to say that he knew his office was "one of high responsibility" and that he would have to build the new bureau from the bottom up. In doing so, Jesup wrote, he would need "to introduce system into a Department, hitherto without arrangement, without organization." But with Calhoun's support, he continued, "I have no doubt of making the Quartermaster's Department in our service, what it is in all European services, the first department in the Army."[9] As he built and supervised the Quartermaster's Department over the next four decades, Jesup succeeded in making his bureau the prime mover of the antebellum army.

Although he left Washington for Florida to lead troops during the Second Seminole War in the late 1830s, Jesup was even more important as an organization builder. As his reference to his desire to "introduce system" in his 1818 letter to Calhoun would suggest, Jesup promoted regular procedures. Soon after taking office, Jesup required each quartermaster officer at each supply depot around the country to submit monthly and quarterly reports to the Washington office. By 1825, quartermasters communicated with Washington and one another using thirty-seven standard paper forms. They were forbidden to accept gifts from or have financial interests in firms that served as military contractors. In Washington, Jesup and his small clerical staff subjected officers' reports to close scrutiny, routinely ordering quartermasters to resubmit reports and other forms according to exact standards.[10] During his forty-two-year tenure as quartermaster general, Jesup built and ran a bureaucratic national organization during a time when such a thing was virtually unheard of in American government and business. Although he is seldom recognized as such, Jesup ranks as a pioneer in the field of bureaucratic management in the United States.[11]

As Jesup's career suggests, not every aspect of early American government was controlled by the principles of rotation in office and the distribution of spoils to supporters of the winning political party, even after the Jacksonian Democrats took the spoils system to new heights starting in the 1830s. Instead, the sphere of early national administration featured what the historian Leonard D. White once called a "dual system," in which one part of the national government was affected greatly by shifts in the political winds, while another part remained much more stable. Although party officials controlled many jobs at U.S. customs houses and post offices, they had less direct influence over the federal courts, the Treasury Department in Washington, and the military bureaus. At the Treasury Department, comptrollers and auditors enjoyed tenure, as did some departmental clerks, who began to be subjected to examinations under an 1853 law. In the War Department and Navy Department, military surgeons faced rigorous exams before receiving their commissions, and staff bureau chiefs and their clerks could serve for decades at a time. Commissary General George Gibson—the army's top officer in charge of the procurement of food—headed his bureau from 1818 to 1860. Jesup's chief clerk, William A. Gordon, started in 1824 and served for thirty-seven years.[12]

Although Jesup's bureau and other governmental organizations on the more stable side of the "dual system" were never completely insulated from party politics, the long tenures of their leaders advanced their autonomy. Certainly, even the War Department—which, after all, was led by a civilian appointed by the president—was subject to partisan controls. For instance, the secretary of war had the privilege of naming the newspapers his department would pay to advertise its procurement needs. Because most newspapers in this era were connected to a particular party, this amounted to a modest patronage power. Still, Jesup's long tenure allowed him to accumulate power and knowledge that gave him and his bureau considerable independence, even from the army's top line officers, the secretary of war, and the president. This independence bothered line officers and civilian officials alike. During the Mexican War, after he clashed with Jesup, President Polk complained of the autonomy of the army supply departments and their practice of promoting officers from within the bureaus on the basis of seniority. A decade later, Secretary of War John B. Floyd used his first annual report to argue that the Quartermaster's Department and other staff bureaus were too independent.[13]

Built up over several decades by Jesup in Washington, the antebellum Quartermaster's Department was a national organization staffed by officers posted at supply depots across the country. In his 1818 letter to Calhoun, Jesup remarked that he intended to recruit "young, active, and intelligent subaltern officers" for his bureau.[14] Although these officers' responsibilities were considerable, their numbers were small. In July 1838, Congress passed new legislation that reorganized and expanded Jesup's bureau by authorizing the appointment of thirty commissioned quartermaster officers at the rank of captain or above. Although more than half of these 1838 appointees were dead by 1861, among the original thirty were nine officers who would serve as top quartermasters for the North during the Civil War: Daniel D. Tompkins, Charles Thomas, Thomas Swords, George H. Crosman, David H. Vinton, Osborn Cross, Robert E. Clary, Edwin B. Babbitt, and Ebenezer S. Sibley. By 1855, there were thirty-seven commissioned quartermaster officers, compared with thirty-six permanent officers in the Ordnance Department and twelve in the Subsistence Department.[15]

By the 1840s and 1850s, most quartermasters—and, increasingly, most army officers overall—were graduates of the national military academy at West Point, New York. Created in 1802, the academy became especially distinctive after 1817, when Superintendent Sylvanus Thayer remodeled it along the lines of L'École Polytechnique, the national engineering school in France. Run by the Army Corps of Engineers, West Point offered a four-year curriculum (expanded to five years in 1854) that stressed French, mathematics, science, and engineering. In an era when most other American colleges and universities concentrated on teaching ancient languages, history, and theology, this course of study was unusual. West Point cadets, who were appointed by the president and members of Congress from each state, came from all over the country, but most were the sons of relatively successful and well-connected professionals, merchants, or civil servants. Nearly half of the cadets who entered West Point failed out, often because of poor grades in math; those who survived were regularly ranked according to strict quantitative measures. In any given year, West Point produced only about forty or fifty graduates. But because the antebellum army was small, this number was more than enough to fill the peacetime officer corps: by 1860, over three-quarters of army officers were West Point graduates.[16]

Naturally, many of the North's chief supply officers during the Civil War had attended West Point several decades before. Like other West Point cadets, many of them owed their appointments to their families' better-than-average political connections, which provided the necessary recommendations from congressmen. Future Quartermaster General Montgomery Meigs grew up in the 1820s as a child of unusual privilege. The son of a prominent Philadelphia physician with solid ties to the Democratic Party, Meigs was able to study at the University of Pennsylvania as a teenager and secure recommendations from powerful men. Another future top Northern quartermaster, Thomas Swords, was ten years older than Meigs but shared a similar background. The son of a successful New York City book publisher (whose own father had been a British army officer who served in the Seven Years' War), Swords was able to spend two years at Columbia College before entering West Point.[17]

In comparison to many of their peers, Meigs and Swords were unusually fortunate and urbane. Many men who would become top Union supply officers came from families in more modest circumstances, for whom West Point, with its free tuition, represented one of the only ways for their talented sons to continue their educations and attain economic security and social respectability. Justus McKinstry of Michigan, whose first efforts to enter West Point began when he was just fourteen, was recommended by one teacher as "a gentleman and a scholar." But another supporter of McKinstry's application made it clear that the young man's father was then living in "narrow circumstances." Samuel Holabird of Connecticut, who would eventually become a top supply officer at New Orleans during the Civil War, in 1843 was a sixteen-year-old who was described by a distant relative as "the son of a farmer of this place who is too poor to aid his son."[18] James Donaldson, a West Point classmate of Meigs, never knew his father, a Baltimore lawyer and politician who had been killed at age thirty-two in the War of 1812. As the seventeen-year-old Donaldson attempted to secure a West Point appointment in 1831, his mother wrote the War Department to explain that her "son is soon to begin life, and such is my situation that I cannot materially assist him." Another member of the West Point class of 1836 and future top Union quartermaster, Robert Allen, was also raised by a widow. In 1830, his teachers described Allen, who walked more than four miles from his rural Maryland home to school, as a young man "exceedingly amiable, and ambitious to excel," whose "progress in learning was truly astonishing." But as a fatherless boy from a family without great wealth, they explained, "he certainly has not the means to pursue the study of any one of the learned professions." Allen was "just the sort of youth," his teachers suggested, "the institution of West Point was intended to benefit."[19]

For the young West Point graduate in early America, the army offered a steady job but not one that provided quick rewards or comforts. A newly minted officer who left the academy as a second lieutenant of infantry when he was twenty-one, for instance, might serve a dozen years before he was promoted to captain. After this, it became even more difficult to advance in rank: colonels were rarely younger than fifty. Young officers were compelled to move every few years from post to post all over the country, often in places that offered few comforts or charms. Although military officers might enjoy relatively high social status in the eyes of at least some of their compatriots, the soldiers they commanded were widely regarded as the dregs of antebellum American society. At many army posts, desertion, heavy drinking, and disease were common problems. By the 1830s, it was common for lieutenants and captains—many of whom were already investing in real estate or engaging in other side businesses while they were full-time officers—to resign their commissions to pursue work they thought might be more rewarding for themselves and their families. (Among the many West Point graduates who ended up resigning their commissions during the antebellum years were several men who would become top generals during the Civil War, including George McClellan, Ulysses S. Grant, and William T. Sherman.)[20]

For a lieutenant in an infantry or artillery regiment, the prospect of becoming a member of the Quartermaster's Department or another staff bureau offered several attractions. Most important, the staff bureaus represented another avenue to advancement in rank, something most young officers wanted desperately. It was also possible that a job in one of the staff bureaus could offer more comfortable living conditions for a young officer—and perhaps for his family as well, since officers (unlike enlisted men) were allowed to bring wives and children with them to their posts. Although many staff officers and line officers shared the same frontier forts, quartermasters and commissaries were more likely to be stationed at supply depots in larger towns. As many young officers knew from serving as acting assistant supply officers when they were lieutenants, quartermasters enjoyed unique powers that came with overseeing the disbursement of public funds, buying and shipping supplies, and employing civilian clerks and laborers. Some young officers may have been reluctant to take a position that would lessen their chances of leading large forces in battle, but the slow rate of promotion in the antebellum army and the tangible advantages of working as a supply officer ensured that Jesup had plenty of applicants for any opening in his organization.

By the mid-1840s, Jesup had been leading the Quartermaster's Department for nearly three decades. The bureau's staff of some three dozen permanent officers, many of whom had graduated from West Point and served several years as lieutenants on the line before they joined the bureau, coordinated procurement and logistics from a network of depots and forts across the country. It was this organization that produced many of the men who would manage the North's military economy during the Civil War. Those men were schooled not only by their years at West Point and their contact with the system created by Jesup during the first decades of the century but also by their experiences in the new American West. Perhaps even more than their first years in the army or their work during the Mexican War of 1846–48, it was the challenge of handling procurement and logistics across the vast new territories seized in that conflict that provided the North's future military supply chiefs with their formative education in the business of war.

The Logisticians of Continental Empire, 1848–1861

The Civil War was fought mainly across territory east of the Mississippi River. But for many of the army officers who eventually led or supplied Civil War armies, the trans-Mississippi West was an important schooling ground. For the Quartermaster's Department, the nation's territorial gains in the aftermath of the Mexican War of 1846–48 meant extraordinary new challenges. Just at the moment when railroads had begun to reduce the difficulty of army logistics and the hardships of army life, the American military was thrown far beyond the reach of the new railroads and other existing transport networks. As much as it had ever been, the army was separated from much of the rest of the country in time and space. During the fifteen years before the Civil War, the Quartermaster's Department worked to manage a new continental empire. At supply depots from Kansas to Utah to New Mexico to California, quartermasters grappling with this new challenge became even more important as political and economic actors than they had been before 1846.

During the two-year Mexican War itself, the Quartermaster's Department and its officers handled procurement and logistics for a U.S. invading force as large as forty-five thousand men at any given time. This was a continental operation. Quartermasters at depots from Philadelphia to New Orleans to Mexico City to San Diego struggled to acquire and transport enough supplies to sustain the American forces. They were concerned largely with mobilizing muscle power: to support the conquest of Northern Mexico and the invasion of Mexico City, U.S. quartermasters purchased no fewer than 11,549 horses, 22,907 mules, 16,288 oxen, and 6,886 wagons. At the army's Schuylkill Arsenal in Philadelphia—the public uniform and equipage manufactory that the Quartermaster's Department had taken over from the defunct Commissary General of Purchases only four years before—the civilian labor force of seamstresses and tailors grew to four thousand, ten times its normal peacetime level. During the war, the Schuylkill Arsenal began to make footwear and tents, along with clothing.[21]

Much of the work of war supply in Mexico was handled at the regimental level by lieutenants—including a young Ulysses S. Grant—who served as acting assistant quartermasters or commissaries. But the Quartermaster's Department also increased its staff of full-time officers by fourteen during the war years. Several of the new officers who joined the bureau in 1846 and 1847—including Alexander Montgomery, Robert Allen, James Belger, James L. Donaldson, Langdon C. Easton, Justus McKinstry, and Stewart Van Vliet—would end up working as top quartermasters for the North during the Civil War.[22]

The Mexican War experiences of Thomas Swords, the New York book publisher's son who became a top army quartermaster, suggest both the extraordinary logistical challenges posed by the war itself and the future problems that would face the Quartermaster's Department after 1848. Swords graduated from West Point in 1829 at the age of twenty-two. As a lieutenant and then captain and assistant quartermaster during the 1830s, he was stationed at frontier posts in Missouri and Kansas, where he supervised the construction of barracks and storehouses. Accompanied at these posts by his wife Charlotte, Swords passed his leisure hours by reading newspapers, Charles Dickens novels, and histories of Mexico.[23] Finding himself at the key supply depot of Fort Leavenworth at the beginning of the Mexican War in 1846, Swords became chief quartermaster to the Army of the West under General Stephen W. Kearney. Before Kearney's force of three thousand men started to march southwest from Kansas toward the Pacific, Swords provided it with ample transportation in the form of 459 horses, 3,658 mules, 14,904 oxen, and 1,556 wagons. Over the course of seven weeks during the summer of 1846, Swords rode with Kearney and his army to Santa Fe to claim New Mexico as a new territory of the United States.[24]

Soon after arriving in Santa Fe in mid-1846, Swords and Kearney, along with five companies of dragoons (cavalry), left for the Pacific Coast. After reaching San Diego in December, they realized that there were few local sources for clothing or other goods the men wanted. Swords responded to this problem by chartering a commercial ship to take him to the Sandwich Islands (Hawaii), located in the middle of the Pacific Ocean, about twenty-two hundred miles to the southwest. After eighteen days at sea, Swords reached Honolulu, where he bought garments and construction materials on U.S. government credit. He then sailed back with his goods to San Diego, landing safely in February 1847 and distributing his supplies.

At the end of May 1847, Swords departed the coast with Kearney, who had decided to head back to Fort Leavenworth by going first northward through California and then east via the Oregon Trail. In the Sierra Nevada of eastern California in June, Swords found himself overseeing the burial of the remains

of the Donner Party. A group of emigrants from Illinois who had become trapped there over the winter, the Donner Party had included eighty-seven members, nearly half of whom perished of cold and hunger before rescuers arrived in early 1847. As he disposed of the dead, Swords was forced to confront the horrors of bodies mutilated by cannibalism. Swords and his companions finally arrived back in Fort Leavenworth on 22 August 1847, over a year after they had departed. On the return trip from San Diego, they had traveled overland 2,152 miles in eighty-three days.[25] Although this hardly matched the pace of the eighty-day trip around the world that the French novelist Jules Verne would imagine a generation later, Swords had accomplished some impressive feats of travel and logistical support.

As Swords's Mexican War experiences suggest, the territorial gains secured by the United States at the end of the conflict in 1848 transformed the role of the American military. The size of the nation suddenly expanded by 1.2 million square miles (an area over two-thirds as large as the entire country had been in 1844), much of it in rough terrain as yet unreachable by any form of freight transport other than wagon train. Congress, always stingy with appropriations, acknowledged the new circumstances by lifting the authorized peacetime strength of the army from eight thousand in 1848 to sixteen thousand by 1860. By the 1850s, most of the army was stationed west of the Mississippi, where it often stood as the most important institution in the region. In addition to fighting small conflicts around the West that killed several hundred Indians and a handful of U.S. soldiers during this decade, the army also served as the region's primary explorer, surveyor, and road builder.[26]

In the West of the 1850s, the Quartermaster's Department proved to be especially important. In many localities, quartermasters and other staff officers became leading economic actors. These officers were channels for much of the national government's spending in the West, which was about ten times higher per capita than in the rest of the country. The army injected considerable sums of much-wanted cash into western communities. In 1855, for example, Quartermaster General Jesup ordered David Vinton, then the supply chief in Saint Louis, to procure "a good safe" before sending $25,000 in silver and $175,000 in gold by wagon to the quartermaster's depot at Santa Fe. The presence of an army supply depot in any western community transformed the local economy. In antebellum San Antonio, for instance, as many as 150 civilians worked for the local quartermaster. The quartermaster and commissaries bought substantial quantities of flour, corn, and animal feed from local farmers and merchants. By the 1850s, the army was spending about $1.75 million per year in the New

Mexico territory and $1 million a year in the state of Texas. In these huge but sparsely populated regions, this was big money.[27]

But much of what the army needed could not be found locally in the West. Uniforms, for example, were made by army tailors and seamstresses in Philadelphia; rifles came from the national armories at Harpers Ferry, Virginia, and Springfield, Massachusetts.[28] To deliver these and other goods to soldiers concentrated in the West, the antebellum Quartermaster's Department devoted much of its energies to the problem of transport. As it had before the Mexican War, Jesup's office in Washington coordinated the operations of a network of depots across the country. But after 1848, the geography of supply changed considerably. By the 1850s, the department's main depots included not only older centers such as those at Philadelphia and Saint Louis but also far-flung western sites such as Santa Fe and San Francisco. Fort Leavenworth, which was fed by Saint Louis as well as western Missouri and eastern Kansas, was the army's single most important forwarding depot during the 1850s. It supplied main regional depots such as Santa Fe and smaller forts around the great western plains by shipping goods via wagon trains. On the Pacific Coast, San Francisco (or, technically, the army depot in nearby Benecia) served as the main supply center.

After 1848, the cost of military transport in the West—even without any war—was so great that it threatened to bankrupt the U.S. government. The annual expenditures of the Quartermaster's Department jumped from $871,000 in fiscal 1844 to more than $4,295,000 in 1850. This meant that Jesup's bureau alone now accounted for over 13 percent of total national outlays, other than treaty payments and debt service. The Quartermaster's Department spent most of this cash on transportation in the West. By 1856, when the annual budget of Jesup's bureau stood at $7 million, two-thirds of the money was being spent on transport contracts, horses, and animal feed.[29] Jesup and other military officials in Washington, as they submitted larger and larger budget estimates to Congress during the 1850s, heard plenty of complaints about excessive spending. But such were the costs of the new continental empire.

Quartermaster General Jesup and other military leaders, who were keenly aware of the unprecedented new costs of military logistics, justified them to Congress by suggesting that the result of the Mexican War had made the United States into a European-style imperial power. Such a change demanded bigger military budgets; in particular, it required high expenditures on logistics. Just after the end of the Mexican War in 1848, Jesup reminded Congress of the implications of "[t]he vast increase in our territories, with the long sea voyages necessary to reach a portion of them. . . . We are now in precisely the condition

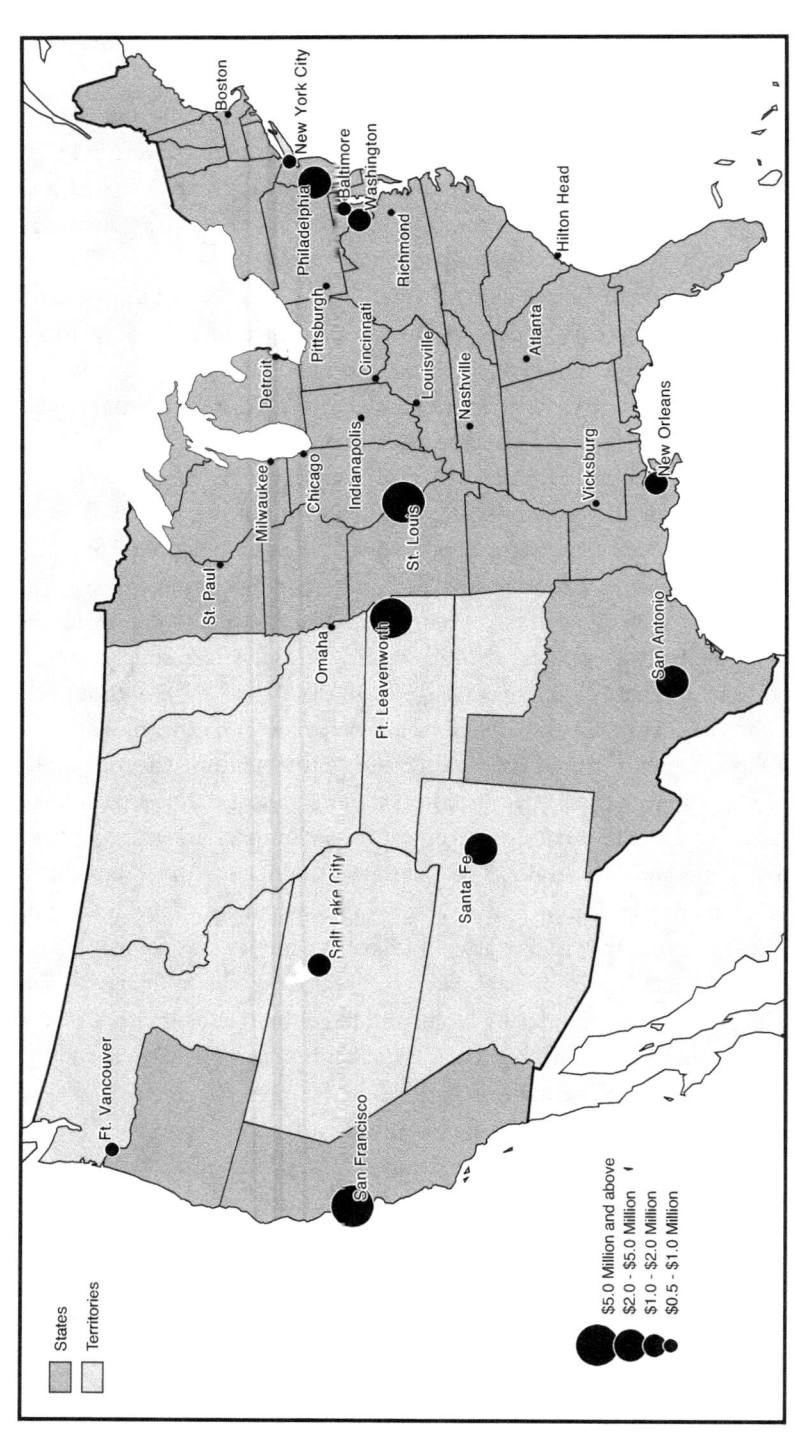

Fig. 2.1. Treasury Disbursements to Quartermaster's Department Depots, 1856–1860. *Sources*: U.S. Treasury "Receipts and Expenditures" reports, linked to information about supply officers' posts. See appendix C. *Credit*: Map by Patrick Jones, Cartographic Laboratory, Department of Geography and Earth

of the great maritime nations of Europe, who have distant colonies." But it was not only maritime transport that was so costly. In 1855, Jesup justified his bureau's large budgets by claiming that "our army, small as it is, covers more ground, and carries on more extended operations, than the armies of all the nations of continental Europe, west of Russia, including all their colonies, in addition to their European territories. Now, operations upon a scale of such magnitude cannot be carried on without a heavy expenditure of money." A year later, in his final report as secretary of war, Jefferson Davis supported Jesup by arguing that it was not unreasonable for the United States to spend about $3 million a year (or nearly two hundred dollars per soldier) on military transport. "When we take into view the fact that our small army covers a territory equal to the whole of Europe, from Russia to France, including those two powerful empires," wrote Davis in 1856, "and that its operations are carried on in sections of the country without resources, it seems not so strange that the expenditure should be so much, but that it is not greater."[30]

During the 1850s, the need for communications and transport in the new West promoted the rise of new kinds of businesses in America, some of which became leading military contractors. Although the army could have chosen to maintain and operate its own large shipping fleets and wagon trains, it was cheaper to contract out for at least some of its needs with private firms. During the 1850s, the largest provider of ocean-going shipping services to the Quartermaster's Department was the Pacific Mail Steamship Company (PMSC). As its name suggested, one of the original missions of this corporation, established in 1848 with the support of Congress, was to deliver mail to and from San Francisco or other points on the Pacific Coast. But thanks to the California gold rush of 1849, the PMSC quickly built up a large commercial business that amounted to $2.5 million in annual revenues by the mid-1850s. For the PMSC, even relatively large postal and military contracts were only a small fraction of the commercial side of its business. But the emergence of the PMSC and similar companies allowed the military to bypass old barriers of time and space. Troops and supplies, as well as miners or settlers, could now move from New York to San Francisco—by making connections with ships serving either side of the Panama isthmus—in a little over a month. Using the PMSC and its sister companies, the U.S. Army now enjoyed a relatively fast and reliable means of projecting military power to the Pacific Coast.[31]

Many parts of the new American West of the 1850s, however, lay hundreds of miles from ports. Unfortunately for the army, they were also far beyond the reach of the railroad system, which had not yet advanced much beyond the Mississippi River. At this time, the hundreds of miles separating Fort Leavenworth or San Francisco from points on the interior such as Santa Fe or Salt Lake City could be traversed only by human or animal muscles. Because muscle-powered overland transport is so slow and expensive, military logistics in the West became the army's primary challenge and the national government's heaviest new fiscal burden during the 1850s.

Most of the wagon trains that supplied military posts across the West were run by private companies instead of by the Quartermaster's Department itself. This need not have been the case, since the end of the Mexican War in 1848 had left the army with considerable transport capacity in its own hands. At Fort Leavenworth over the winter of 1848–49, for instance, Quartermaster Langdon C. Easton found himself caring for 3,862 oxen and 2,762 mules, which would pull hundreds of supply wagons across the plains in 1849. As late as 1851, Edwin B. Babbitt, then chief quartermaster at San Antonio, was supplying army posts around Texas with a force of 752 government-owned wagons, each of which was pulled by six mules. Employing teamsters paid directly by the Quartermaster's Department, Babbitt sent nearly 90 percent of all the army freight moved in Texas in 1851 with these public wagon trains.

By the early 1850s, however, Jesup and several of his subordinates had decided that it would be best to rely on private companies for most of the army's overland transport requirements. Facing ballooning transport costs, including the huge expense of feeding thousands of draft animals, Jesup expected that contracting out would be cheaper. In theory, wagon freighting firms should have been able to cut overall costs by keeping their animals and workers employed more steadily as they served a growing base of commercial customers in addition to more erratic (but potentially large) military demands. In practice, this would prove to be a difficult business. But there is no doubt that, by deciding to contract out, Jesup stimulated the already-growing overland freighting industry, based along the Missouri River near Fort Leavenworth and Kansas City.[32]

Of all the military industries in the United States during the 1850s, overland freighting was the biggest. And after 1854, when Jesup decided to consolidate many smaller freighting contracts into large, comprehensive agreements covering transport across a few huge regions for two years at a time instead of one, the top transport contractors became even bigger. In Texas, leading San Antonio businessman George T. Howard moved into large-scale wagon freighting and became the army's leading prime contractor in the region, handling the lion's share of the some sixty thousand dollars a year Texas quartermasters paid for overland transport during the 1850s.[33] Far larger, however, was the freighting company that came to dominate army transport across most of the great western plains: Russell, Majors, and Waddell (RM&W).

Based in Lexington, Missouri, RM&W filled the army's first giant consolidated freighting contract in 1855 by employing a force of 600 teamsters and wagon-masters, 500 wagons, and 7,500 oxen. As an employer of labor and capital, RM&W already stood among the largest enterprises in the United States. After taking in as much as $300,000 in profits in 1855 and 1856, the company was ready to expand even more. When the army sent a major military expedition to Utah in 1857, RM&W—easily the largest military contractor in America during the decade before the Civil War—became the leading partner of the Quartermaster's Department.[34]

As the largest U.S. army operations during the decade before the Civil War, the Utah Expeditions of 1857–58 highlighted the difficulties quartermasters faced in coordinating logistics in the new West. The dispute that inspired these expeditions was already raging by the time President James Buchanan moved into the White House in March 1857, when newspaper reports and official communications with Washington suggested that the authority of U.S. officials and judges in Utah Territory was being disregarded by Brigham Young, who was both territorial governor and the religious leader of the Mormon settlers who had recently come to dominate the region. President Buchanan soon began to make plans to replace Young with a new territorial governor; at the end of May, he ordered the War Department to prepare a military expedition that could enforce this decision.[35]

Although the Utah expedition of 1857 was not as large an effort as the Mexican War had been a decade before, the problem of moving hundreds of men and their baggage overland in short order from eastern Kansas to Utah tested the capacities of the small peacetime army. Quartermasters had become accustomed to having the winter months to prepare their shipments before they were loaded onto RM&W wagons for delivery across the plains during the spring and summer. By autumn, cold weather made overland travel dangerous. When Jesup and his subordinates were told in late May to prepare for the army's biggest operation in a decade, the quartermasters knew that they were fighting the clock. But the quartermaster general, in typically grand language, refused to accept failure. "Impress upon all the Officers of the Department the absolute necessity of overcoming all obstacles," Jesup urged Quartermaster Charles Thomas, who was then working in Saint Louis and Fort Leavenworth to secure supplies. No doubt thinking that his bureau must avoid the logistical failures that had killed thousands of British and French soldiers during the Crimean War of 1854–56, Jesup called upon his quartermasters to make a heroic effort. "[W]hat ordinary men think impossible," the quartermaster general declared, "men of genius and true intrepidity of intellect find only difficult. All difficulties must be conquered, even those presented by time itself."³⁶

As Jesup and his quartermasters prepared for the Utah expedition of 1857, they came face to face with the central problem of the American military economy during the years before the Civil War: overland transport across the great western plains. In the end, the 1857 expedition consisted of only thirteen hundred men, or about a tenth of the peacetime army. But sending even a small force to Salt Lake City was a difficult project that threw Jesup's bureau into a sudden burst of activity. In Philadelphia, depot chief Edwin Babbitt worked furiously to procure and ship dozens of tents and wagons, as well as uniforms and equipment for the soldiers.³⁷ Meanwhile, officers at the central Quartermaster's Department depots in Saint Louis and Fort Leavenworth worked quickly to procure transportation for the expedition. Although the army would rely on RM&W to handle some of the transport (it paid the company at least $540,000 for its efforts connected with the 1857 expedition), quartermasters also paid for large numbers of mule-powered wagon teams owned and operated by the army itself. Eventually, the senior quartermasters working in Saint Louis and Fort Leavenworth—including Charles Thomas, George Crosman, Stewart Van Vliet, and Thomas Brent—outfitted the 1857 expedition with 370 wagons and more than 2,400 mules to pull them. By the end of June, the 1,300 soldiers, along with 120 laundresses and several hundred teamsters and other civilian employees who had been assembling at Fort Leavenworth, had begun to ride and walk with their mules and wagons toward the Rockies. In early September, the force reached Fort Laramie (in present-day Wyoming). Riding ahead to Salt Lake City, Van Vliet informed Young of the impending arrival of the expedition but failed to convince the territorial governor to accept it with open arms. On 15 September, Young declared martial law in Utah territory and promised that any invasion would be met with stiff resistance.³⁸

Although Van Vliet and his colleagues at the Quartermaster's Department depots in Fort Leavenworth and Saint Louis had completed the task of outfitting the expedition, the supply officers assigned to travel with the force now found their work increasingly difficult. One of those officers was Assistant Quartermaster John H. Dickerson, an 1847 West Point graduate who had been overseeing military road improvements in Nebraska territory when he landed a spot in Jesup's bureau in 1856. At the end of September 1857, Dickerson and the rest of the expedition were located about 150 miles northeast of the Great Salt Lake and about 200 miles west of the nearest major army depot at Fort

Laramie. With winter approaching fast, Dickerson knew that the expedition was at risk. He soon saw things go from bad to worse. On 4 October, a small group of Mormons managed to torch three of the expedition's RM&W wagon trains, destroying 72 wagons and their contents—which included 92,000 pounds of bacon, 167,000 pounds of flour, 9,000 pounds of coffee, 13,000 envelopes, and 9,000 steel pens. In mid-October, part of the expedition, still hoping to reach Salt Lake City, became stuck in a snowstorm that killed dozens of animals. By the beginning of November, it was evident that the force could not reasonably hope to reach its goal and had to move quickly to find shelter for the winter. The chosen spot, called Camp Scott, was located about 100 miles northeast of Salt Lake City. Along the way, Dickerson and other members of the force contended with temperatures as low as negative forty-four degrees Fahrenheit. When Dickerson took stock of the losses upon reaching Camp Scott, he calculated that at least 134 horses, 588 mules, and 3,000 head of oxen or cattle had died along the way.[39]

With Dickerson and the rest of Johnston's force stuck for the winter at Camp Scott, President Buchanan and the War Department decided to assemble a new, reinforcing expedition of some four thousand men that could depart Fort Leavenworth in the spring of 1858. The Quartermaster's Department now worked to outfit this larger force. RM&W, the giant freighting contractor that had lost nearly $140,000 worth of wagons and oxen in the 1857 expedition, was again called upon to haul supplies. In January 1858, Jesup signed a new comprehensive two-year freighting contract with RM&W that provided for the delivery of as much as 15 million pounds of supplies per year, a maximum three times higher than in the previous contract. RM&W responded by expanding its operations, building a new depot in Nebraska City, and augmenting its force to include some 3,500 wagons, 40,000 oxen, 1,000 mules, and 4,000 men.[40] At the same time, the Quartermaster's Department hurried to procure more goods than it had during the previous year. Army manufactures and purchases for the new expedition included 24,000 pairs of boots and shoes, 17,000 pairs of trousers, 14,000 shirts, 300 wagons, 4,000 mules, 1,500 cavalry horses, and 200,000 iron horseshoes.[41]

The Utah expeditions turned out to be a major military and financial embarrassment for the U.S. government and its military contractors. In the spring of 1858, when the second expedition had left Fort Leavenworth, the new Utah territorial governor, Alfred Cumming, was negotiating with Young. In June 1858, after a commission sent by President Buchanan confirmed that the Utah settlers would not resist U.S. forces, about twenty-four hundred troops (including some who had wintered at Camp Scott and others who had come that year) marched through a mostly abandoned Salt Lake City. The military crisis ended without bloodshed, and Secretary of War Floyd and Jesup now asked Congress for $6.7 million—on top of the large sums already required to support the army in the West during peacetime—to pay for the 1857 and 1858 expeditions. Because the U.S. economy had taken a sharp downturn in 1857 while the first expedition was on its way to Camp Scott, the huge bills for a war that never happened seemed especially outrageous. Even before the news of a peaceful settlement in Utah reached Washington, congressmen from both parties were criticizing the recent military expenditures. In the House, Representative Lawrence Branch, a North Carolina Democrat, declared during the April debate over the deficiency appropriations request that army spending over the last two years had been "monstrous, and entirely unjustified." Across the aisle, Republican Owen Lovejoy of Illinois accused the Buchanan administration of squandering public funds on "its favorite pet contractors." The operations of 1857 and 1858 would come to be known in history, declared Lovejoy, "as the war of the plunderers and contractors, instead of the Utah War."[42]

In the end, the Utah expeditions did more than create consternation in Congress over the $6.7 million Quartermaster's Department deficiency bill for a war that never happened: they also bankrupted the army's leading contractor and took down the secretary of war. After the destruction of its wagon trains and the onset of the Panic of 1857, RM&W was counting on huge military orders for Utah in 1858 and 1859 to sustain its large operations. Although the U.S. Treasury did pay the company $3.1 million in 1858 and another $1.5 million by 1860, RM&W was so far in debt that it resorted to desperate measures. In 1858 and 1859, Secretary of War Floyd—acting against the orders of President Buchanan—provided RM&W with so-called "acceptances," paper promises of future payment from the government with which the company could raise money. Then, in 1860, an Interior Department clerk (who was a distant relation of Floyd's wife) handed RM&W some $870,000 worth of state bonds that were being held in trust in a government safe for various Indian tribes in accordance with treaty settlements. By the time the secretary of the interior discovered the theft, in December 1860, Abraham Lincoln had been elected president and South Carolina was urging other Southern states to join it in seceding from the Union. Secretary of War Floyd, a Virginian, had just signed a controversial order sending cannon from Pittsburgh to Texas. In January 1861, Buchanan was forced to remove the secretary of war. RM&W, now disgraced as well as bankrupt, ceased to exist; in much of the Northern press, Floyd was vilified.[43]

Although the Utah crisis had been resolved without a direct military confrontation, by the end of 1860 it had depleted the U.S. Treasury, bankrupted the nation's leading military contractor, and caused the resignation of the secretary of war. Coming at a time when the nation threatened to come apart, these events undermined the prestige of the U.S. government and its War Department at a critical moment, doing nothing to discourage the secession of Southern states in early 1861. Whether or not the Utah expeditions helped to cause the Civil War, their history illustrates the state of the American military economy during the 1850s. For Quartermaster General Jesup and his staff, the operations in 1857 and 1858 confirmed that the army's main challenge after the Mexican War was projecting power across the huge territories the nation had gained in 1848. Before the arrival of railroads to the West, which would not come until well after the Civil War, even relatively small operations such as the Utah adventures or expeditions against groups of Indians could prove to be extraordinarily expensive and difficult. John Dickerson and other supply officers with inside knowledge of the 1857 march knew that they had been lucky to avoid having dozens of men die from cold or starvation. Even the greatest private contractor in the field of overland transport, RM&W, had collapsed under the strains generated by the army's logistical needs in the West. In the years leading to the Civil War, the West's unforgiving terrain and vast places stood as the fundamental challenge in the imaginations of Quartermaster's Department officers and other military planners.

As it turned out, of course, these officers faced the military challenge of a lifetime in 1861–65 in the East instead of the West. But the West was where they came of age. Between 1848 and 1861, the army's chief quartermasters on the Pacific coast, working out of the main depot at Benecia outside San Francisco, included David Vinton, Robert Allen, and Thomas Swords. All three would be chief quartermasters at large Northern depots during the Civil War. Several of the chief supply officers assigned to the Utah expeditions—including Charles Thomas, George Crosman, Stewart Van Vliet, and John Dickerson—also served as leading procurement officers for the Union in 1861–65. Most of the other veteran army quartermasters who supervised large procurement and logistical operations for the North during the Civil War—including Morris Miller, William Myers, and Asher Eddy—had spent a good part of the 1850s at posts in California, Oregon, New Mexico, Texas, and the Great Plains. Although many new officers would join the Quartermaster's Department and other staff bureaus during the Civil War, it was the corps of veteran officers who had risen through the ranks in the West who became the leading managers of the Northern military economy.

Their years in the West during the antebellum era pushed these veteran quartermasters to become even more independent of party politicians and eastern businessmen than they otherwise might have been. Already members of a small professional military elite who shared the common bond of a West Point education and training in the ways of Jesup's continental supply bureaucracy, these officers learned in the antebellum West to wield economic and political power at a long distance from Philadelphia, New York, and Washington. Although they took orders from Jesup and local commanding officers, quartermasters in the West had few other superiors or even peers. At most western depots other than the one outside San Francisco, the army stood as the richest and most powerful institution around; as the army's main business agents, quartermasters became accustomed to having their way. In many western locales, they themselves employed more civilians—clerks, teamsters, craftsmen, and so forth—than any private firm. And although quartermasters contracted out with a variety of private businesses in the antebellum West, the remarkable rise and fall during the late 1850s of RM&W—the nation's biggest military contractor—did not make army officers inclined to depend upon capitalist enterprise to solve all of their problems.

Veteran army supply officers in mid-nineteenth-century America were no more intelligent or ambitious than their counterparts in law, business, and politics. But during the fifteen years before the Civil War, they did grapple with unique challenges in the West that set them apart from other Americans. Swords's journey of several thousand miles during the Mexican War was hardly a common experience. The trials of Van Vliet and Dickerson during their work on the Utah expeditions were only slightly less remarkable. For these men (and in some cases, for their wives as well), the West meant discomfort and danger, but it also offered them an opportunity to test some of the material and social limits under which most Americans lived. In 1861, many of these officers rushed back to the East to take leading roles in the Northern armies. For many veteran quartermasters, who took charge of central supply depots across the North, the Civil War presented the opportunity to oversee procurement on an entirely new scale.

Managers of the Civil War Economy

From 1861 to 1865, it would be the regular senior officers from the antebellum Quartermaster's Department built by Jesup who were the chief managers of the North's procurement project. As they struggled with this giant problem, these career officers were joined by new colleagues, including young West Point graduates and many volunteers who left the private sector in 1861 to join the Northern armies. Over the next four years, the army supply bureaus that managed the Northern military economy would be led by a combination of veterans and volunteers. The two groups differed sharply because of the veterans' unique experiences at West Point and in the antebellum West and in their political persuasions. Several of these volunteers became powerful supply administrators who spent huge sums of money. In the end, however, it was the regulars who had the greatest influence over the flow of military dollars and the structure of the North's supply system.

During the Civil War, just as it had during the antebellum era, the Quartermaster's Department continued to have larger budgets than any other military bureau or indeed any other part of the American national government. But because the scale of military procurement increased so sharply in 1861–65, only now did the Quartermaster's Department become a true economic giant. During the war years, military expenditures were twenty to thirty times higher than they had been before 1861 and the military accounted for over 90 percent of all U.S. government spending. In four years of war, the Quartermaster's Department spent more than $1 billion. The other major army supply departments, the Subsistence and Ordnance bureaus, together spent only half this amount; all Navy Department spending was not much more than $300 million (see table 2.1).

As usual, quartermasters were responsible for all transport and for procuring most of what the army needed other than weapons, food, and medicine. Before the Civil War, when there were no more than about fifteen thousand men in the army, Jesup's department had struggled to handle its work with a staff of fewer than forty commissioned officers along with perhaps as many as two hundred part-time, acting assistant officers and agents who were not assigned permanently to the bureau. During the Civil War, when the size of the Northern armies approached 1 million men at any given time, the Quartermaster's Department had to expand considerably. It did so mainly by taking on dozens of volunteers, recently occupied in civilian life, to join the small core of veteran supply officers who had spent much of the antebellum era in the West.

During the first weeks of the war in 1861, when many of the new regiments of volunteers were being outfitted by their own states, Congress was already working to expand the Quartermaster's Department and other military staff bureaus. In August 1861, Congress increased the authorized number of permanent regular officers in the Quartermaster's Department to sixty-four, nearly double the antebellum figure. But the most significant increase in the size of the bureau came from the appointment of dozens of volunteer officers. By May 1862, the Quartermaster's Department staff included at least

150 volunteer assistant quartermasters (with the rank of captain), as well as 49 regular quartermasters (with the rank of captain or above).

Some of these high-ranking supply officers oversaw logistics for whole armies or other large units of soldiers in the field, while others were assigned to central supply depots well behind the lines, where they worked on procurement. There were also lower-ranking supply officers in the field: each regiment had a quartermaster-sergeant and a regimental quartermaster with the rank of lieutenant, appointed by the regiment's colonel. During the Civil War, when most regiments were volunteer regiments identified with particular states, the overwhelming majority of these lower-ranking supply officers were volunteers. But among the higher-ranking quartermasters who handled large-scale logistics in the field or procurement at the major depots, the ratio of regulars to volunteers was more even.[44]

As Northern officials worked to fill dozens of new positions in the Quartermaster's Department, they faced the question of what kind of qualifications might make a good military supply officer. Most people who discussed the question agreed that, because military supply work involved lots of financial transactions, the best volunteer quartermasters would be men with good business sense. Charles W. Moulton, a lawyer from Toledo, Ohio, was recommended to Secretary of War Cameron in May 1861 as someone with commercial smarts. According to his congressman, Moulton was "a good lawyer—and among . . . the best business men of our city with executive abilities of a high order and would be an invaluable man in the Quartermasters Department or in any position requiring business talent of the first order." A few months later, a Bostonian wrote to the Quartermaster General's office to suggest that, if the army could not spare a veteran supply officer to head up the depot in his city, the best sort of civilian would be an officer from a railroad company. Railroad officials were well suited to handle military procurement and logistics, this correspondent argued, because they were accustomed to the bureaucratic management of complex enterprises. "Their habituation to system, accountability and economy," he argued, "renders them more nearly suited to take the place of the regular Army quartermaster, than men of any other class."[45]

In practice, railroad managers, lawyers, and others with considerable business experience did indeed become leading officers in the Quartermaster's Department during the Civil War. William G. Le Duc, a forty-year-old lawyer from Minneapolis, managed to get a commission as assistant quartermaster of volunteers after he traveled to Washington in early 1862 bearing letters of recommendation from the Minnesota delegation to Congress. For much of the rest of the war, Le Duc helped to oversee the business of mule-drawn wagon transport for Northern armies in the East. One of the most important of all the Northern managers of logistics was Lewis B. Parsons, a son of a wealthy Buffalo businessman and a graduate of Harvard Law School. By the 1850s, Parsons was working in Cincinnati and Saint Louis as a chief officer of the Ohio & Mississippi Railroad. (One of Parsons's fellow railroad managers at the Ohio & Mississippi was George McClellan, the West Point graduate who would become the Union's top general during the first part of the Civil War.) In their wartime work as quartermasters, these middle-aged lawyers had the opportunity to apply their business talents to the largest economic enterprise of the century.[46]

But if several of the new volunteer officers who joined the Quartermaster's Department in 1861 and 1862 were seasoned lawyers and businessmen who could lay claim to special skills, many owed their wartime positions more directly to political connections. Charles Moulton, the lawyer from Toledo who became one of the most important volunteer quartermasters in the wartime North through his procurement work in Cincinnati, was a brother-in-law of Senator John Sherman.[47] James Ekin, the forty-two-year-old volunteer who replaced Quartermaster Alexander Montgomery in Indianapolis in 1861 after Governor Morton complained he could not get along with the old veteran, was most remarkable not as a business leader but rather as someone with impeccable Republican credentials. A resident of Pittsburgh, Ekin had attended the 1860 Republican convention in Chicago that had nominated Lincoln and the very first national convention of the Republican Party in 1856.

Many of these men were more than party hacks: Ekin, for example, performed well enough as Indianapolis depot chief during the first part of the war to be called to Washington in 1864 to become the North's chief quartermaster in charge of cavalry horses. But he and many other volunteer quartermasters got their feet in the door because they knew influential Republicans. This was particularly true for younger men. Simon Perkins Jr., only twenty-three years old when he served briefly with Ohio's 19th Infantry in the summer of 1861, had been working as a clerk in an uncle's Cleveland bank before the war. When Perkins managed to get an appointment as an assistant quartermaster of volunteers in February 1862, it was surely thanks to the influence of another uncle: Ohio's Governor David Tod. A personal connection to President Lincoln, of course, could be equally helpful. Samuel L. Brown, who would eventually become one of the top purchasers of animal feed for the Quartermaster's Department, joined the bureau in October 1861 only days after his father—a wealthy Chicago businessman who knew Lincoln—told the president that he would consider it "a particular favor" if his son could be given a spot in the army's pay bureau or another staff department.[48]

Nothing better indicates the power of Republican connections in the appointment of volunteer staff officers than the prominence of newspaper editors in the ranks of the wartime Quartermaster's Department. In Illinois, where Governor Yates clashed with U.S. Quartermaster Eddy over the control of procurement in late 1861, state officials managed to convince the War Department to send them a volunteer quartermaster. After Eddy was reassigned, the Springfield depot was handed to William Bailhache, a new volunteer quartermaster captain. Bailhache was already a familiar face in Springfield because he was the co-proprietor of the *Illinois State Journal,* the state capital's Republican newspaper. He was one of several Republican editors across the country who became volunteer supply officers. The man who became the chief quartermaster at Detroit for most of the war, George W. Lee, had founded a Republican newspaper at Howell, Michigan, during the birth of the party back in 1855. One of the top purchasing officers in wartime Cincinnati, David W. McClung, was another Republican editor. Soon after graduating from Miami (Ohio) University in 1854, McClung had helped to start a Republican newspaper in Butler County, Ohio. Like Bailhache, Lee and McClung were rewarded for their past work on building the infant Republican Party with relatively comfortable and well-paying positions as army quartermasters.[49]

Like regular quartermasters, volunteer assistant quartermasters had to file $10,000 bonds, with the signatures of two responsible sureties, before they could receive their commissions. Like regulars, the volunteers as U.S. disbursing officers had individual accounts with the Treasury for which they were personally liable. But veteran regulars could easily set themselves apart: nearly all of them had attended West Point between the 1820s and the 1850s, and many of them had worked together during the Mexican War or in the antebellum West. Unlike the volunteers, the veterans were familiar with the bureaucratic routines that Jesup had imposed over the course of the antebellum decades. Potentially, at least, the regulars' antebellum experiences gave them a special advantage when it came to filling the most powerful positions on the public side of the Northern military economy.

If the veterans of Jesup's bureau could claim special expertise in the business of military supply, however, they were also less likely than the volunteers to have good Republican credentials. Indeed, of all the differences between the regulars and volunteers, perhaps the most significant were political. Clearly, many volunteer staff officers—including all those newspaper editors—affiliated openly with the Republican Party. But the same could not be said of the regulars, whose politics tended to fall somewhere between neutral and Democratic. Many regular officers, who had struggled to climb the ladder of promotion during an era in which the Democrats were the leading party, were more sympathetic to the party that lost the 1860 election than to the one that won it. By the eve of the Civil War, perhaps three-quarters of those army officers who were inclined to take sides in national politics favored the Democrats.[50]

Because each individual entertained different political ideas, and because many officers subscribed (selectively, at least) to an ethic of nonpartisan professionalism, it is difficult to generalize about politics in the antebellum army officer corps. Certainly, young officers and their families cultivated relationships with party officials in their home states and in Washington: because congressmen controlled military academy nominations, they could not become West Point cadets in the first place without such relationships. Many West Point graduates maintained contacts with politicians over the years, if for no other reason than because Congress had the power to approve or block officers' promotions. And some officers, including top field generals, openly supported one party or another. After the Whig Party disintegrated in the early 1850s, ambitious officers tended to tilt toward the Democrats. At the same time, however, many officers saw themselves as professionals who had a responsibility to stay clear of the day-to-day squabbles of partisan politics. As the historian William Skelton has argued, few groups in antebellum America more closely approached the ideal of nonpartisan professionalism than the members of the army officer corps.[51]

In the end, the political orientations of most Quartermaster's Department officers and their peers lay somewhere between enthusiastic support for one party and strict neutrality. As the historian Samuel Watson has suggested, many antebellum officers were as much careerist as they were professional or partisan.[52] Many officers maintained contacts with party politicians not primarily for ideological reasons but because they wanted to enhance their own prospects for promotion. Robert Allen, who in 1846 found himself still a lowly lieutenant a full decade after graduating from West Point, undertook a major letter-writing campaign to land himself a spot as a captain in Jesup's bureau. At the beginning of that year, Allen wrote to President Polk to announce that, because he was "(laudably I hope) ambitious of realizing a little progress in my profession," he would very much like an appointment as assistant quartermaster. The junior officer reminded the president that his father-in-law, W. P. Peeble, had been one of Polk's "*first* and finest friends" in Maine (probably during the 1844 election). A few weeks later, Allen wrote directly to Quartermaster General Jesup to inform him that he had gathered letters of support from thirteen members of Congress (including most of the Democrats from the Indiana delegation) and thirty-two other eminent men. Such efforts were hardly unusual among young officers seeking promotions. But Allen's timing was unusually good: in May 1846, only weeks after he sent his letters to Polk and Jesup, the United States declared war on Mexico. Allen quickly received a spot as an assistant quartermaster. By the 1850s, when he served in San Francisco as chief quartermaster of the Department of the Pacific, Allen had become one of the top officers in Jesup's bureau.[53]

As the careerist efforts of Allen suggest, the veterans of the Quartermaster's Department—as they fought to rise through the ranks during an era of Democratic dominance in national politics—tended to gravitate toward a zone of political preference that was more Democratic than Whig or Republican. At the same time, many staff officers also hailed the ideal of nonpartisan professionalism, in part because this could enhance their own authority and independence. When senior quartermaster George Crosman wrote a private letter from Utah in 1858 to tell a friend his thoughts about the next presidential election, he identified Jefferson Davis as his top choice. This preference suggested that Crosman favored the Democrats over the Republicans: a former secretary of war in a Democratic administration, Davis was then serving in the U.S. Senate as a Democrat from Mississippi. But in favoring Davis—whom he described as a "noble . . . gentleman" and one of the "really good men of the country"—the Massachusetts-born Crosman was probably thinking of Davis not so much as a strident pro-slavery party man but rather as the secretary of war who had built support and prestige for the military in the early 1850s. Davis had treated the military as a professional organization that should not be used to build party or personal patronage machines.[54] Crosman and many of his peers tended to support Democrats not simply because they liked that party's ideological program but because they regarded conservative hawks like Davis as more likely than Republicans to maintain substantial military spending without treading on the traditional prerogatives of senior officers.

Crosman's top choice in 1853, of course, was not among the U.S. presidential candidates who split the vote in 1860. Instead, after many Southern states responded to the election of Republican candidate Abraham Lincoln by leaving the Union, Davis became president of the new Confederacy. The Quartermaster's Department, like the U.S. government as a whole, found itself in 1860 and 1861 in a state of crisis. Jesup, after a tenure of forty-two years, died in June 1860. His successor as quartermaster general, a veteran cavalry officer named Joseph E. Johnston, served for only a few weeks before April 1861, when he joined the Confederacy (for which he became a top field general)—becoming part of the 28 percent of the eleven hundred regular officers who sided with the South.[55] This meant that, when the Civil War started in April 1861, the North was without a chief supply officer. In June 1861, the Lincoln administration announced its choice as the new quartermaster general: a forty-five-year-old army engineer named Montgomery C. Meigs. During the Civil War and after, it was Meigs who led the organization that had been built and managed by Jesup for so many years. Although Meigs was an engineer rather than a veteran quartermaster, his nomination confirmed that well-established military institutions and officers would become powerful actors in the North's giant war economy in 1861–65.

Like many of the veteran officers he directed in the Quartermaster's Department starting in 1861, Meigs was a military professional who had managed to maintain ties with officials in Washington without becoming the creature of a particular party. After graduating from West Point in 1836 with a class rank of fifth out of forty-nine, Meigs joined the army's exclusive Corps of Engineers. He soon found himself surveying waterways in the Mississippi River valley with a young officer named Robert E. Lee. Unlike Lee, who in 1861 would side with his native state of Virginia, Meigs seems to have regarded himself first and foremost as a citizen of the nation. In 1840, a year before he married Louisa Rodgers, a daughter of a U.S. Navy commodore, Meigs wrote a remarkable letter to his father about politics. "You want me to join the Democratic party," Meigs wrote his father in May 1840 (half a year before a presidential election the Democrats would lose): "I have never thought much on the subject of politics. In the Army I consider it highly improper as it is for any employé of the Government to take an active part in party politics." Meigs told his father that, especially for a military officer, it was best to identify oneself not as a member of any particular party or locale but rather as an American. "I am a citizen of the United States," he wrote, "not of Connecticut where my grandfather lived or of Georgia where I was born or of Pennsylvania where I was educated. I should as soon think of boasting of being an Arch Streeter or a Chesnut Streeter as a Pennsylvanian or Georgian."[56] It was this young nationalist who, twenty-one years later, would become the chief overseer of the Northern procurement and logistics effort that would defeat Lee and the South and force them back into the national union.

By the 1850s, when he worked in Washington overseeing the construction of a new aqueduct for the city and additions to the U.S. Capitol and the main Post Office, Meigs had become more familiar with both procurement and politics than he had been in 1840. He enjoyed friendly relations with several party leaders, including the powerful Blair family of Missouri and Maryland, who by the start of the Civil War moved from the Democratic to the Republican fold. In the 1860 presidential contest, Meigs favored Democrat Stephen

Douglas over Lincoln.⁵⁷ But Meigs never became a party man. During the late 1850s, he fought with Secretary of War Floyd over construction contracts. After Meigs refused to allow Floyd to choose contractors, the secretary of war managed in 1860 to remove the engineer from the Washington projects and exile him to a remote army post on Dry Tortugas, a tiny island southwest of Florida. When Floyd was forced out of office at the end of that year, Meigs was vindicated. Soon after he returned to Washington in February 1861, the Lincoln administration sent him back to Florida, this time to investigate the problem of supplying its coastal forts. Only weeks after returning from this mission, Meigs became a candidate for the post of quartermaster general. There was no question that the North needed a highly competent man for the job. As the *New York Tribune* suggested in early June 1861, "There is no office connected with the Government of the United States that requires for its occupancy such varied accomplishments and qualifications" as the position of quartermaster general, which required considerable knowledge of transport, construction, machinery, and geography, as well as great integrity.⁵⁸ Only days after these words were published, Meigs was confirmed as the new chief of Jesup's old bureau.

Meigs proved to be a popular choice, even among those critics who believed that the existing American military bureaucracy was too arthritic and rule-bound to manage the war mobilization. Before Meigs was confirmed, President Lincoln himself (who had only recently come to know the engineer) declared, "I do not know one who combines the qualities of masculine intellect, learning and experience of the right sort, and physical power of labor and endurance as well as he." George Templeton Strong, a Wall Street lawyer who would serve during the war as treasurer of the U.S. Sanitary Commission (a nongovernmental aid organization not unlike the Red Cross), soon described Meigs as "an exceptional and refreshing specimen of sense and promptitude, unlike most of our high military officials. There's not a fibre of red tape in his constitution." In this judgment, Strong agreed with the *New York Tribune,* which by December 1861—as it called for mandatory retirement as a means to eliminate "dead-wood" in the military and other government departments—had concluded that, in Meigs, "A new and most efficient head, by the best fortune in the world, was given to the Quartermaster's Department."⁵⁹

Such tributes, along with Meigs's success in overseeing a giant procurement and logistics effort that allowed the North to win the Civil War, have led several historians—most notably Allan Nevins and Russell Weigley, writing a half-century ago—to identify the North's quartermaster general as one of the most important leaders in the entire conflict. This is surely correct. But by emphasizing Meigs's impressive personal accomplishments, it is easy to lose sight of the fact that he led a well-established organization staffed by experienced senior officers. Certainly, as the *New York Tribune* had suggested before Meigs was confirmed, it was critical that the North fill the post of quartermaster general with someone of high abilities and integrity. Meigs fit the bill, and his leadership was important. He did not, however, have to build an organization from the ground up. Instead, he managed a much-expanded version of the continental supply network that Jesup and his subordinates had created over the preceding decades. For oversight of the main procurement depots across the North—including Philadelphia, New York, Cincinnati, Saint Louis, and Washington—Meigs was able to rely upon the senior quartermasters who had been serving Jesup's bureau for years, most of them lately in the trans-Mississippi West. Indeed, one of Meigs's main challenges during the first months of the war was preventing Secretary of War Cameron and other Republican officials from taking procurement powers away from this existing corps of veteran quartermasters.

One of the most important struggles over the military economy in 1861, of course, was the contest over purchasing power between national authorities and the various states. When they managed to consolidate procurement authority by the end of the year, Meigs and his veteran quartermasters were pleased. But Meigs also fought during his first weeks on the job to protect his chief quartermasters from interference by Secretary of War Cameron or the White House. In June, before he had a chance to settle into his office, Meigs wrote his father to say that Cameron had been talking about removing some of the quartermasters' clerks or other civilian employees—perhaps with the object of replacing them with good Republicans. Meigs informed his father that "in this I think he will find difficulty for he has not without good cause a right to interfere between an officer responsible for large sums of money and the agent or clerk to whose honesty the officer trusts it." In the months that followed, despite pressure from Cameron and other Republicans, the quartermaster general defended this policy of deferring many responsibilities to his senior officers. In late 1861, Meigs fended off a recommendation from the White House concerning civilian employees at the New York City depot by proclaiming, "I never interfere between the Quartermaster and those whom he hires, and I trust that the President will not permit the recommendation of any person's friends to induce him to interfere."[60]

The question of who would control purchasing and hiring in the North's military economy, which remained open during the first months of the war, was one of enormous importance. Meigs, a West Point–trained military professional who had struggled with Secretary of War Floyd over contracts in the late 1850s, fought hard during the first weeks of the Civil War to keep party politicians from controlling the work of his powerful bureau. When one congressman wrote in July 1861 to recommend a constituent who hoped to furnish the army with harness or other cavalry equipment, Meigs told him that, since the constituent's prices were not competitive, he could not be considered. In August 1861, after Cameron and Lincoln wrote him to recommend that the Quartermaster's Department buy mules from certain individuals in Kentucky, Meigs resisted. Given that the favored parties in Kentucky were offering mules at a price eight dollars higher than the average cost his officers were then paying for such animals, Meigs responded, awarding contracts to the men in question "will be difficult to explain." In fact, continued Meigs, he would tell his quartermasters to make the purchase only if Lincoln and Cameron provided him with a direct written order to do so. "If political considerations connected with the situation of Kentucky induce the President and yourself to direct this purchase," he explained, "it should be plainly ordered not left to my discretion." Although the secretary of war had the authority to overrule him, by refusing to cooperate the quartermaster general made it uncomfortable for Republican officials to dictate procurement policy. Meigs quickly became skilled at deflecting requests concerning army employees or contracts from Congress or the White House. When Mary Lincoln echoed the previous request from Cameron and her husband by suggesting that the Quartermaster's Department buy horses from certain parties in Kentucky, Meigs informed the first lady that "your young friend" in Kentucky should contact the local purchasing officer, since such procurement decisions were made by the depot quartermasters around the country.[61]

Although the wartime Quartermaster's Department was hardly untouched by partisan pressures from Republican officials, by early 1862 the regulars had managed to consolidate a great deal of power over the military economy. One reason for this was the stubbornness of Meigs, as well as the willingness of new Secretary of War Edwin Stanton (who replaced Cameron in January 1862) to allow Meigs to run his bureau with a minimum of partisan interference. But another reason was that even before Meigs became its chief, the Quartermaster's Department was already the best-qualified organization in America to handle complex procurement and logistical problems. Long before 1861, under the leadership of Jesup, the Quartermaster's Department had grown into a continental bureaucracy. Although it was relatively small in size throughout the antebellum period, especially after the Mexican War it stood as one of the most important—and most expensive—divisions of the American state. Although

Fig. 2.2. Group Outside the Office of the Quartermaster General, Washington, D.C. Taken at the end of the Civil War in April 1865, this photograph portrays only a small part of the huge staff employed by the Washington office of the Union's largest military procurement bureau, located at Pennsylvania Avenue and Seventeenth Street. The uniformed man with the white beard at *center left* is Brig. Gen. Charles Thomas, the senior army quartermaster who oversaw the office when Quartermaster General Montgomery C. Meigs was absent. *Source:* Library of Congress, Prints and Photographs Division, reproduction number LC-DIG-cwpb-04250.

a new political party took control in Washington in 1861, and although a new Quartermaster General took office in the same year, it was Jesup's old bureau and its officers who served as the core of the management team grappling with the giant project of Northern war supply.

Although they certainly shared power with Republican-appointed volunteers and occasionally found themselves taking orders from the secretary of war or other Republican officials, regulars (including Meigs himself) still managed to dominate the North's procurement effort. This is evident not only in the surviving correspondence of quartermasters and other military and civilian leaders but also in the record of wartime disbursements of procurement dollars by the U.S. Treasury. Of the twenty quartermasters who received the most money from the Treasury during the war, thirteen were graduates of West Point. A fourteenth regular, Daniel Rucker, had joined the Quartermaster's

Department just after the Civil War. More notably, the bureau's top seven wartime disbursing officers (and nine of the top ten) were all regulars, most of whom had worked in Jesup's bureau during the 1850s and before. Despite their lack of Republican credentials, these regulars became the top managers of the war economy—the biggest government spending project of the century.

Across the North, most of Meigs's supply depot chiefs—and thus most of the men given authority over the largest sums of procurement money—were the same regular quartermasters who had overseen continental logistics in the great West in Jesup's bureau during the 1850s. Stewart Van Vliet and John Dickerson, two quartermasters who had actually gone to Utah in 1857 in connection with the army expedition, became leading wartime purchasing officers at the large depots in New York and Cincinnati. George Crosman, the Philadelphia depot chief for much of the war, was the same man who had helped to outfit the Utah expeditions from Saint Louis before going himself in 1858 to Utah (where he penned the letter endorsing Jefferson Davis for president). David Vinton, who received some $100 million in Treasury disbursements in his capacity as senior officer at the New York depot during the war, was a chief quartermaster in the Mexican War before spending the 1850s handling military supply from Saint Louis and San Antonio. Another top Mexican War quartermaster, Thomas Swords (the man who had gone from Fort Leavenworth to Hawaii and back in 1846–47), served during much of the war as depot chief at Cincinnati. Robert Allen, who oversaw Northern logistics efforts from Louisville and directed the wartime Saint Louis depot along with a young West Point graduate named William Myers, had been the chief supply officer at San Francisco for several years during the 1850s. The Washington depot chiefs, Daniel Rucker and Morris Miller, were also veterans of Jesup's bureau. In short, although some volunteers came to hold considerable purchasing authority in the wartime Quartermaster's Department, the regulars largely maintained their grip on military purse strings.

By the beginning of 1862, when the various Northern states had given up procurement authority and when Meigs and Stanton had established a leadership style in Washington that left veteran staff officers with considerable autonomy, a network of regular and volunteer quartermasters across the North had begun to work on the giant problem of supplying the Northern armies. Although the war effort would continue to expand after 1862, it was already clear that this conflict would demand a procurement project that was entirely unprecedented in size. By the end of 1861, the Northern states had already mobilized over a half-million volunteers, far more than the one hundred thousand or so who had enlisted during the Mexican War. Eventually,

Table 2.2 Twenty Leading Procurement Officers (U.S. Army Quartermaster's Department) during the Civil War

Officer	Life Span	West Point Class	Supply Depot Posts, 1861–1865	Approx. Wartime Disbursements
Vinton, David H.	1803–1873	1822	New York City	$102 million
Crosman, George H.	1798–1882	1823	Philadelphia	$101 million
Allen, Robert	1811–1886	1836	St. Louis; Louisville	$100 million
Myers, William	1830–1887	1852	St. Louis	$79 million
Swords, Thomas	1806–1886	1829	Louisville; Cincinnati	$64 million
Van Vliet, Stewart	1815–1901	1840	New York City	$53 million
Donaldson, James L.	1814–1885	1836	Baltimore; Nashville	$46 million
Moulton, Charles W.	18??–1888	n.a.	Cincinnati	$42 million
Rucker, Daniel H.	1812–1910	n.a.	Washington	$40 million
Dickerson, John H.	182?–1872	1847	Cincinnati	$35 million
McKim, William W.	18??–1895	n.a.	Boston; Cincinnati	$32 million
Brown, Samuel L.	1825–1912	n.a.	New York; Washington	$30 million
Ekin, James A.	1819–1891	n.a.	Indianapolis; Washington	$26 million
Boyd, Augustus	18??–1896	n.a.	Philadelphia	$20 million
Miller, Morris S.	1814–1870	1834	Washington	$19 million
Babbitt, Edwin B.	1804–1881	1826	San Francisco	$18 million
Jenkins, Walworth	1832–1874	1853	Louisville	$17 million
Holabird, Samuel B.	1826–1907	1849	New Orleans	$16 million
Robinson, Henry L.	1812–1901	n.a.	Washington	$16 million
Hodges, Henry C.	1831–1917	1851	Fort Leavenworth	$15 million

Sources: Some information about postings and other biographical information is provided in George W. Cullum, *Biographical Register of the Graduates of the U.S. Military Academy*, 3rd ed., vols. 1–2 (Boston: Houghton, Mifflin & Co., 1891); Francis B. Heitman, *Historical Register and Dictionary of the United States Army* (Washington, D.C.: GPO, 1903); Ezra J. Warner, *Generals in Blue: The Lives of the Union Commanders* (Baton Rouge: Louisiana State University Press, 1964); Roger D. Hunt and Jack R. Brown, *Brevet Brigadier Generals in Blue* (Gaithersburg, Md.: Olde Soldier Books, 1990).

Note: Treasury disbursements to individual officers are calculated from annual "Receipts and Expenditures" reports. See appendix C.

the Northern armies would consist of nearly 1 million men at any given time. The Quartermaster's Department was charged with outfitting, transporting, and sustaining those armies. Over the course of the four-year war, regular and volunteer quartermasters would work together to purchase over a billion dollars worth of military goods, warehouse and package them, and get them into the hands of the soldiers who required them. Although some supply officers traveled with the armies in the field to supervise their logistics, many others were stationed at depots on the home front, managing the project of procurement.

The sheer size of the Civil War procurement project forced Northerners to perform unprecedented feats of public administration. It was the officers of the Quartermaster's Department and other military staff bureaus who managed the bulk of this work. Civil War quartermasters faced onerous bureaucratic demands, even as they exerted considerable personal influence over the shape of the war economy. Over two-thirds of the 600-page official handbook of U.S. Army regulations in force during the Civil War was devoted to the standard forms and procedures to be used by supply officers. This book reproduced dozens of standard forms, including the nine separate forms each quartermaster had to send with his monthly report back to Washington and thirty-seven other forms that he might need for his more substantial quarterly reports. During the war, several other guides to supply work—some of them written by quartermasters themselves—were turned out by commercial presses to meet the needs of the hundreds of Northern supply officers. One of these guides, written by regular quartermaster Walworth Jenkins, listed 1,108 separate rules in a book of three hundred pages. Volunteer quartermaster Roeliff Brinkerhoff, whose 1865 guide to war supply contained an appendix with seventy pages of sample forms, introduced his book by claiming that "the duties devolving upon the officers of the Quartermaster's Department are by far the most varied and intricate of any of the staff corps, and require the highest business capacity."[62]

ALTHOUGH THE CIVIL WAR turned it into a giant bureaucracy, the Quartermaster's Department that oversaw Northern procurement and logistics in 1861–65 still resembled its much smaller incarnation from the antebellum era. During the decades before the war, Quartermaster General Jesup had built a continental supply network that depended on a combination of bureaucratic controls and daring personal efforts on the part of officers working hundreds of miles from home. Although the scale of the Civil War procurement project overshadowed anything seen before, Jesup's bureau had faced serious antebellum challenges, most notably in Mexico and the new American West. Even before the Civil War, Congress had been accustomed to spending more on the Quartermaster's Department than on any other part of the national government. If the wartime bureaucracy was unprecedented in scale, it was nevertheless still the direct descendant of Jesup's organization. Most importantly, despite the influx of volunteers after April 1861, it was many of the same chief quartermasters who had served under Jesup who supervised the North's main procurement depots during the four-year conflict. Although a new quartermaster general and a new Republican administration arrived in Washington in 1861, the institutions and officers of the old Quartermaster's Department became powerful directors of the Northern war economy.

Because the Northern procurement project was neither a Republican patronage machine nor the result of top-down direction by Washington officials such as Meigs and Stanton, its political and economic history is especially complex and intriguing. Over the course of four years, U.S. supply officers around the North worked to solve the problem of how to outfit and sustain an army of a million men. Because the scale of this project was so huge, the economic and political decisions they made involved vast sums of money and affected the lives of thousands of people. Like their fellow officers in other supply departments, quartermasters—both regulars and volunteers—wielded more power from 1861 to 1865 than they ever had before. During the Civil War, when they became major manufacturers, employers, and buyers, quartermasters around the North became important participants in the ongoing struggles over business, labor, and government in America.

CHAPTER 3

The Making of a Mixed Military Economy

In October 1861, President Lincoln received a letter from Sarah Jacobs, a resident of Saint Louis. In recent weeks, Jacobs had been working as a seamstress for the army-run clothing halls in that city. Now, however, these facilities were being scaled back and might soon be closed altogether. Such a change, Jacobs advised Lincoln, promised to add to the burdens of hundreds of "women of the medium class, and those of humble life," who had been working side by side in the army clothing halls while the Saint Louis economy suffered from a war-induced slump. Surely, Jacobs wrote, Lincoln would not want U.S. policy to discriminate against her already-beleaguered city. "[A]s a Western man," she suggested, "you must feel an earnest interest in the welfare of the West." Anticipating possible objections to her proposal, Jacobs rejected the idea that she and other seamstresses should simply go to work for private contractors receiving army orders. Instead, Jacobs insisted, the president should "[o]pen again the workrooms [and] furnish through your quartermaster materials to these women, which can be wrought by their hands . . . into clothing, better than it is being made through the mediumship of private contractors, whose only care is to make the most profit—by the least work."[1]

Although it was written only six months after the beginning of the conflict, Jacobs's letter highlighted important questions about the military economy that would trouble Northerners throughout the war. In her defense of "the welfare of the West," Jacobs hinted at the ongoing struggle by particular regions to secure a share of national military expenditures. Like Jacobs in Saint Louis, Northerners across the country realized that the geography of the army supply system—including the placement of major Quartermaster's Department purchasing depots and manufacturing halls—would have significant economic effects if the war were to become long and expensive. But the geographical distribution of military dollars was not the only important issue. For Jacobs, there was also an important distinction between public and private enterprise. Some Northerners, including not only war industry workers like Jacobs but also some senior military supply officers, argued that it was better to operate army-run facilities than to contract out to private firms. Throughout the war, Northerners grappled with this fundamental question of the proper balance of public and private enterprise in the war economy.

As Jacobs's letter suggests, the struggle over this question was particularly intense in the clothing industry, which was one of the most important of all parts of the Northern military economy. Garments were the most expensive single class of manufactured goods consumed by Civil War armies, and seamstresses comprised the largest group of military-industrial workers in the North.[2] As Jacobs realized, the "make or buy" decision in the procurement of uniforms had especially high stakes. Throughout the war, both seamstresses and quartermasters worked to expand and maintain public clothing halls across the North. Echoing the arguments of Jacobs in 1861, both groups claimed that public enterprise produced superior goods while enhancing social welfare.

Ultimately, the campaign to sustain public military enterprise in the North proved fruitful but not to the extent that Jacobs and other seamstresses would have liked. Through the end of the war, the army continued to employ directly thousands of seamstresses, who produced close to a quarter of the garments shipped to Union armies. But most orders for uniforms, despite the objections of seamstresses and some officers, went to private contractors. There were many reasons for this tilt toward private sources of supply. Top Northern officials, including Secretary of War Stanton and Quartermaster General Meigs, concluded that it was good policy to send orders to the North's large existing commercial clothing industry. At the depot level, even those quartermasters who championed public enterprise proved less willing than seamstresses to endorse the most radical condemnations of contractors. Finally, constant cash flow problems in wartime public finance encouraged quartermasters to cut public payrolls and contract out. Although the public side of the North's mixed military economy remained robust throughout the war, its scope was limited by significant ideological and structural constraints.

The Wartime Expansion of the Depot-Arsenal System

By the end of 1861, the various Northern states had ceded control over procurement to U.S. military organizations. This transfer meant that the war economy was more centralized by early 1862, when General Grant began to lead large

Union forces in major battles near the Tennessee River. But the consolidation of national authority over procurement did not mean that the North's supply system was fully centralized in Washington. Instead, the army relied upon an expanded version of an existing continental network of depots and arsenals. During the war, public military enterprise expanded along with the depot-arsenal system. Even as the depots bought huge quantities of goods from the private sector, old and new public facilities swelled. This wartime growth of public military enterprise was driven not only by antebellum institutional arrangements and wartime pork-barrel politics but also by the efforts of ambitious procurement officers who created their own operations instead of contracting out.

Several of the North's most important military-industrial establishments dated back to the nation's infancy. In the 1790s, the Federalists had established six navy yards, along with two armories for small-arms manufacture at Springfield, Massachusetts, and Harpers Ferry, Virginia. These major yards and armories, along with smaller arsenals dedicated to producing lower-cost items such as ammunition, together employed hundreds of male and female civilians in peacetime during the antebellum years.[3] In short, the regular U.S. military was accustomed to having public facilities fill much of its modest demand for finished goods. Although the enormous rise in military demand during the Civil War guaranteed that there would be more purchasing from the private sector, many procurement officers worked quickly in 1861 to expand existing public facilities and build new ones.

Many antebellum public facilities, including the navy yards and Ordnance Department operations, grew by a factor of ten during the first months of the Civil War. In 1860, the North's five navy yards (Philadelphia, Brooklyn, Washington, D.C., Charlestown, Mass., and Portsmouth, N.H.) had employed between 200 and 600 men each. By the middle of the Civil War, each of these yards counted between 2,500 and 5,000 civilian workers.[4] The scale of expansion at army ordnance plants was similar. The Springfield Armory, which had been the nation's leading small arms manufactory for decades, had just over 300 workers in 1860. By the second half of the war, when Springfield was shipping roughly 20,000 rifles a month, it had roughly 2,500 employees. The Ordnance Department also hired hundreds of new workers—many of them women, teenagers, and children—to make ammunition at its well-established arsenals. The Watervliet Arsenal (north of New York City), the Watertown Arsenal (outside Boston), the Frankford Arsenal (outside Philadelphia), and the Allegheny Arsenal (outside Pittsburgh) had employed only a few dozen people each during the 1850s. By the middle of the war, each of these public arsenals was busy with the efforts of 1,000 to 2,000 people.[5]

Even food and medicine for the Union armies was produced in public facilities. The Commissary Department, which purchased huge quantities of pork, hardtack, and coffee, also bought tons of flour for use in army-run bakeries. One of these bakeries, located just outside Washington in Alexandria, Virginia, employed 200 civilians to operate twenty ovens that turned out as many as 100,000 loaves a day.[6] In the Medical Department, Surgeon General William A. Hammond decided in late 1862 to establish public drug laboratories in Philadelphia and Astoria, Long Island. By 1864, each of the laboratories employed between 200 and 400 women, men, and children to make pills, fill bottles, and perform other tasks.[7]

The Quartermaster's Department ran important public manufacturing operations at sites all across the country. Many of these enterprises were located close to major quartermaster depots, the procurement and warehousing centers that served as the heart of the military supply network. Philadelphia, for instance, was both the site of the main eastern depot of the Quartermaster's Department before the war and the location of the Schuylkill Arsenal. Like the navy yards and army arsenals, this uniform-making facility had been operating since the early days of the republic; it had about five hundred workers in peacetime. By the middle of the war, the Schuylkill Arsenal employed roughly five thousand men and women, most seamstresses who picked up cut material and returned finished garments.[8] As Sarah Jacobs's letter to President Lincoln shows, there were large public clothing halls in Saint Louis by the summer of 1861; similar facilities, along with a tent-making operation, were also established in Cincinnati during the first months of the conflict.

Together, the North's public arsenals produced a sizable fraction of the total wartime output of several key military goods. In the small arms industry, the Springfield Armory loomed large. The most widely issued weapon in the Union armies was the .58 caliber rifle-musket, a model that the Springfield Armory itself had developed during the antebellum period. Of the nearly 1.5 million Springfield rifles acquired by the Ordnance Department during the war, over 54 percent were assembled at the public armory in Massachusetts. Overall, about one-quarter of the $63 million worth of all small arms procured by the North came from the Springfield Armory. Because $18 million out of this total amount went to pay for arms imported from Europe during the first half of the war, the Springfield Armory could lay claim to accounting for at least one-third of the domestic production of finished goods in the wartime small arms industry. With its twenty-five hundred workers, Springfield employed more people than any single private supplier. To be sure, the Springfield Armory, like prime contractors, relied upon subcontractors for parts and materials. At the same time, however, it assisted many prime contractors and subcontractors by providing them with samples and gauges necessary to meet the army's standards for uniformity and interchangeability of parts.⁹

Although the Springfield Armory may have been the best known of the public manufacturing facilities, government-run operations constituted major sources of supply in many parts of the war economy. The Ordnance Department's arsenals made a high proportion of small arms ammunition: of the nearly $20 million worth of ammunition used by the North, over 60 percent was manufactured by the men, women, and children employed by the Ordnance Department at its own facilities.¹⁰ At the U.S. navy yards, the 55 new steam-powered vessels built and launched in 1861–65 accounted for 30 percent of all steam vessels completed in the wartime North. And of the 192 new ship hulls constructed at Northern shipyards during the four-year conflict, the public navy yards built over a third.¹¹ In the Quartermaster's Department, roughly 20 percent of the more than $300 million worth of finished clothing and equipage acquired during the war came from army-run operations. During the last year of the conflict, for which detailed information is most readily available, the Quartermaster's Department produced in its own facilities about 30 percent of all the finished garments and 20 percent of all the tents it acquired.¹²

As the depot system expanded during the war to handle the huge appetites of the Union armies, so did civilian employment by the Quartermaster's Department. By the middle of the war, the large clothing halls at Philadelphia, Cincinnati and Saint Louis were being supplemented by smaller operations at Louisville; Quincy, Illinois; and Steubenville, Ohio. At the larger depots all around the country, quartermasters employed dozens of inspectors, laborers, watchmen, carpenters, teamsters, and clerks. Even junior quartermasters supervised huge payrolls. Sylvester Dunan, a volunteer quartermaster captain at the Baltimore depot, by 1864 had nearly 600 civilian employees and a monthly payroll of $28,000. Meanwhile several individual quartermasters at the Cincinnati depot employed as many as 1,500 civilians each. At the offices of the Quartermaster General in Washington, the clerical staff alone swelled from 16 to nearly 600 by the end of the war.¹³

As Union armies advanced farther into the South, Quartermaster's Department payrolls increased apace Quartermasters were responsible for paying huge numbers of workers on the U.S. Military Railroad, the army-run rail network formed in the upper South as the Union advanced. (By the end of the war, this system, with more than 2,300 miles of track, 400 engines, and 6,600 railroad cars, was costing the North $1.3 million a month.) At Nashville in late 1864, Quartermaster John C. Crane was responsible for paying nearly $375,000

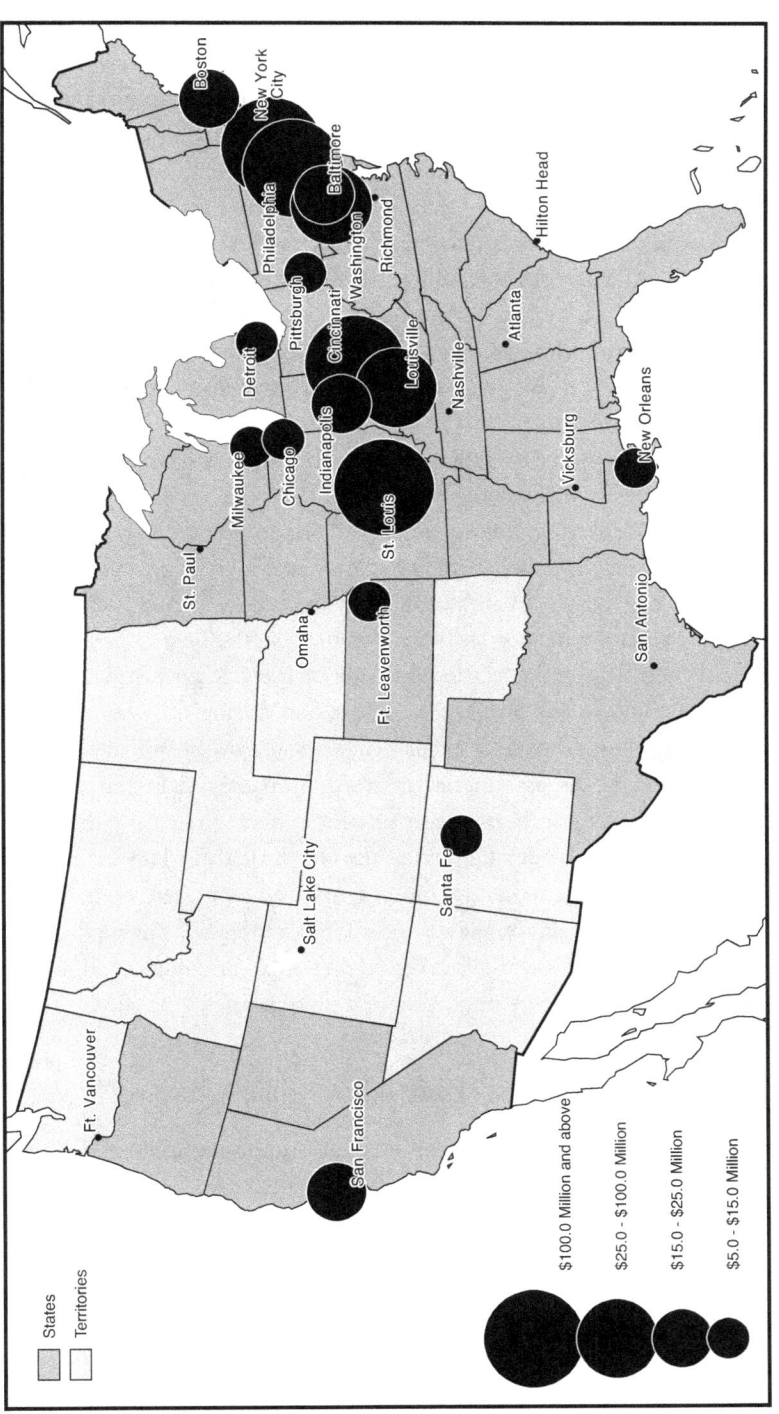

Fig. 3.1. Treasury Disbursements to Quartermaster's Department Depots, 1861–1865. *Sources*: U.S. Treasury "Receipts and Expenditures" reports, linked to information about supply officers' posts. See appendix C. *Credit*: Map by Patrick Jones, Cartographic Laboratory, Department of Geography and Earth Sciences, University of North Carolina at Charlotte.

in wages a month to some 10,000 civilian employees of the U.S. Military Railroad based in that city.¹⁴ At Chattanooga, U.S. quartermasters established a large cavalry depot to hold hundreds of horses and mules. Like its sister cavalry depots in Saint Louis and Washington, the Chattanooga establishment was run by over 1,000 civilian workers. Meanwhile, at City Point, Virginia, a major forward supply base for Grant's armies in the East, large army-run repair shops and warehouses employed as many as 1,600 civilians by the final part of the war. At these depots and in the field, the thousands of teamsters and laborers employed by quartermasters included large numbers of black men—including some who had escaped slavery only months before—as well as whites. By late 1863, as many as 25,000 African Americans worked for the army across the upper South.¹⁵

Together, these military enterprises across the country created a giant civilian work force that was paid directly by the army and navy. By 1865, the Quartermaster's Department alone employed an estimated 130,000 civilians; at the same time, the Navy Department had over 20,000 civilian workers, most of them at the navy yards. In all, the military bureaus at the end of the war counted approximately 160,000 workers. By contrast, the Post Office Department—easily the largest part of the civil side of the U.S. government—had between 20,000 and 30,000 employees before and during the war.¹⁶ By the standards of the times, the scale of civilian employment by the military bureaus was extraordinary: the largest American private companies, including railroads and textile manufacturers, almost never employed more than a few thousand workers at once during this era. Of course, the North also relied upon the work of tens of thousands of men and women employed by private contractors. But there was always a giant public side of military enterprise during the Civil War. All in all, the North's war effort rested on a mixed military economy, in which public enterprises were important and ubiquitous.

"Like the Dew of Heaven": Local Efforts to Secure Military-Industrial Money

Employing tens of thousands of people and spending millions of dollars a month, the North's depot-arsenal system was an economic powerhouse. As a national supply network managed by U.S. officers, it was largely outside the direct control of the various states and localities. Some state governors who had overseen their own procurement efforts during the first months of the war found that local businesses no longer received contracts under the national system. But states and localities continued to struggle for a share of jobs and contracts, even after the U.S. bureaus took over procurement at the end of 1861.

Fig. 3.2. "Filling Cartridges at the United States Arsenal, at Watertown, Massachusetts." Created during the first weeks of the Civil War, this image of operations at one of the Ordnance Department arsenals suggests the preponderance of women workers in many critical war industries. The image also points to the importance of military-run public enterprises, which supplied most of the Union army's small-arms ammunition and considerable fractions of its ships, rifles, tents, and uniforms. *Source:* Library of Congress, Prints and Photographs Division, reproduction number LC–USZ62-96445.

Although the Union was standing together to mount a national war effort, its various parts still demanded a share of the military-economic pie. Many states and cities found success, not only through their own efforts but also thanks to sympathetic supply officers. But there were losers as well as winners in the contest for war industry jobs and money. Although the Quartermaster's Department oversaw a decentralized supply network that could respond to local political demands for an equitable distribution of military money, this was still a national system, structured by longstanding institutional arrangements and

Fig. 3.3. U.S. Cavalry Stables at Chattanooga, Tennessee. Like similar establishments in Washington and Saint Louis, this large cavalry depot employed hundreds of civilians to help support the Union army's huge appetite for horses and mules. Built during the war, the giant new complexes of stables were among the most impressive physical manifestations of the unprecedented economic project managed by the Quartermaster's Department and other military supply bureaus. *Source:* MOLLUS Photograph Collection, U.S. Army Military History Institute, Carlisle Barracks, Pennsylvania.

officers, that could sometimes frustrate the best efforts of local politicians, businesses, and workers.

In her letter to Lincoln concerning the Saint Louis clothing halls, Sarah Jacobs had referred to the president's responsibility to protect "the welfare of the West." Similar concerns about the geographical distribution of military money were expressed across the North during the first part of the war. Savvy observers realized quickly that a major depot or arsenal might bring important economic benefits to a town or region. As early as 22 April 1861, just one week into the war, Congressman James K. Moorhead wrote directly to the Quartermaster's Department to suggest that Pittsburgh "affords the safest and best position for a purchasing and supply depot." In the weeks that followed, other cities around the North pressed Washington authorities with similar efforts to secure major war supply facilities. In July 1861, Congressman George H. Pendleton of Ohio wrote Secretary of War Cameron to remind him to be sure that the army distribute "a fair proportion of the contracts for army supplies to the Western part of the country, and especially to Cincinnati." The regional struggle for war dollars was under way.[17]

As Pendleton's letter to Cameron suggests, western states and cities were afraid that the business of war was being monopolized by New York and Pennsylvania, which contained major depots and arsenals. As river towns highly dependent on the Southern trade, Saint Louis and Cincinnati had been especially hard-hit by the onset of war. This made them especially eager to claim a sizable share of the jobs and contracts generated by the Union war mobilization. By July 1861, a Saint Louis newspaper was already complaining that the western states were being "overlooked" by army supply officials. "[T]he benefits to be derived from an expenditure of public moneys" during the war, the newspaper argued, "should be equalized as much as possible." Two weeks later, a Cincinnati paper complained of the "neglect of the West" in army procurement, which it called "a great and flagrant outrage." In December, Cincinnati's mayor met with Lincoln and its city council presented the War Department with a set of resolutions that described the economic "hard times" of the present winter and protested recent moves by the Quartermaster's Department to scale back local purchasing and public enterprise. "The paramount duty of our Government," the Cincinnati councilmen explained, "is that the protection and benefits it bestows, should descend like the dew of Heaven, equally upon all sections."[18]

Not only politicians but also diverse groups of men and women all around the North moved their feet and raised their voices to claim a share of U.S. military spending. In January 1862, hundreds of Cincinnati residents assembled at the city's Union Hall to protest a recent decline in local army spending.

According to these protestors and local newspapers, as many as twenty-five thousand local jobs, in a city still reeling from the loss of the Southern trade, might depend on the war economy. And yet the Quartermaster's Department had recently scaled back its employment of women in a public tent-making facility. "We do not wish to be sectional," explained one speaker at the Cincinnati meeting. "All we want is our proportion of public work."[19] Later that year, in August 1862, several Illinois manufacturers of woolen textiles wrote Governor Richard Yates to ask that they be given more army orders. "We believe," these businessmen told Yates, "that we can furnish as good materials and at reasonable rates east of the Alleghenies." By this time, Yates was already working on the behalf of potential suppliers in his state: a week earlier, he had written to Secretary of War Stanton to suggest that Illinois firms could fill contracts for uniforms "as expeditiously and cheaply as Eastern manufacturers."[20]

Such appeals for a wide distribution of procurement dollars found support from authorities in Washington, including Quartermaster General Meigs. "The policy of the Government and of this Department," Meigs informed General George McClellan in June 1861, "is to distribute its disbursements as much as possible, so as to equalize the compensation for the burdens of this war." To some degree, this commitment by Meigs lasted beyond the first months of the war, after U.S. authorities took over all contracting. In February 1862, Meigs assured Secretary of War Stanton that the Quartermaster's Department was trying to "distribute" the business of clothing and equipage manufacture "over the different parts of the country as far as possible." And as late as August 1862, just before Governor Yates's request for more clothing contracts for Illinois reached Stanton, Meigs was arranging to distribute among the Northern states the stockpiles of woolen textiles from the Philadelphia depot. Local tailors and seamstresses would then turn these materials into uniforms for the 300,000 men in the new nine-month militia.[21] In this case, war money still flowed according to the principles of federalism.

There were limits, however, to the equitable distribution of contracts and jobs. Although Meigs was indeed committed to a decentralized supply system, the state-by-state textile distribution program of August 1862 was a one-time event. After 1861, flows of war dollars were determined mainly by the contracting and employment activities of the U.S. depots and arsenals. Because the depot-arsenal system spread across the country, few regions were excluded completely from the war economy. But the structure of the national system, which established a hierarchy of major and minor depots and arsenals and took advantage of regional specialization, clearly favored some localities over others. After tailors in Wisconsin petitioned for more military work, Meigs responded by saying that future orders would go to "the lowest responsible bidders. The tailors of Wisconsin will have the opportunity to compete. It is the desire of this Department to distribute the work as much as possible compatible with the public interests and cheapness of supply."[22] In theory, anyone in the North could bid to fill quartermasters' orders. In practice, however, those located closest to the army's main depots—a group that did not include Wisconsin tailors—were best positioned to do the business of war.

Despite the early protests by residents of Cincinnati and Saint Louis, those cities were fully integrated into the North's war economy because they were home to large supply depots. Ironically, given the westerners' complaints of alleged favoritism toward the East, one of the biggest losers in the contest for military depots and dollars was a major eastern city, Boston. By October 1861, the Boston Board of Trade was already calling on the War Department to stop purchasing textiles and blankets in Europe; instead, the board argued, it should send orders to New England firms. Although this campaign failed to stop Meigs from spending $800,000 on woolens in Europe in the fall of 1861, its goal was achieved when Secretary of War Stanton ordered virtually all imports of military goods stopped in early 1862.[23]

But the Boston Board of Trade never managed to convince the War Department to locate a major depot in its city, despite considerable efforts. First in November 1861 and then twice more in March and May 1862, the board petitioned Meigs to establish a large depot in Boston. In the March petition, which was supported by Boston's congressmen, the board announced that "we insist upon a *just proportion of that supply* for the Northeastern States as a right which belongs to them, and one which should never be relinquished." Although they agreed to meet with board members, Meigs and Stanton ultimately brushed off these requests. In August 1862, Meigs responded to the complaints of Massachusetts governor John Andrew about the small scale of the Boston quartermaster office by explaining that it was best for the New York City depot to handle most purchases in New England.[24] Although many New England firms and their workers participated in the war economy as subcontractors, the lack of a major supply depot in the city impeded them from working directly for the Quartermaster's Department as prime contractors or employees.

As the case of Boston suggests, the national depot-arsenal system did not satisfy all of the calls across the North for an equitable distribution of procurement dollars. Nevertheless, the system never came close to centralizing all purchasing and employment in a single region or city. From the earliest weeks of the war, New York City and Philadelphia in the East were joined by Saint Louis and Cincinnati in the West as the North's leading supply depots.

As the conflict continued, the army also drew heavily on depots closer to the front, such as those in Washington, Louisville, Nashville, and New Orleans. In this way, large army purchases and payrolls were spread across much of the country. This decentralized military economy did not square so neatly with federalism as had the short-lived system of state-level procurement, but it did respond to local demands for a share of the business of war.

Military Enterprise and Welfare Work in the Wartime Clothing Industry

At first glance, the struggle in the North over the distribution of jobs and contracts might seem to suggest that the shape of the war economy was driven by negotiations between War Department officials in Washington and politicians and business leaders around the country. But this was not the whole story. In fact, depot-level army supply officers often led the way in shaping the procurement project. One important example of their influence is the development of the wartime clothing industry, one of the biggest in the North. Here, several key officers promoted a mixed military economy by championing public enterprise. As they expanded existing arsenals and created large new public facilities for the production of uniforms and tents, quartermasters argued that public enterprise had various advantages over contracting out. These advantages, the officers claimed, ranged from enhancing quality to promoting social welfare. Meanwhile, many of the seamstresses who actually produced the uniforms were pressing similar arguments. Together, officers and workers created a large public clothing industry during the first part of the war and resisted efforts to scale it back as the conflict went on.

Across the entire war economy, many procurement officers acted as public entrepreneurs. Accustomed to a peacetime military supply system that drew heavily on public facilities and fresh from a frontier West in which the army stood as the leading economic actor in many communities, these officers did not turn automatically to the private sector for their needs. At Nashville in 1864, depot quartermaster James Donaldson created his own shipyard, which employed as many as two hundred men to manufacture simple barges and repair other boats. At the same time, Donaldson spent thirty thousand dollars to turn an old Methodist printing shop into a regional U.S. printing facility that turned out over 5 million standard forms during the last year of the war.[25]

Some officers around the country saw their public enterprises as a means to bypass contractors who were inflating prices or even conspiring to fleece the government. James Moore, a volunteer quartermaster captain who supervised civilian carpenters in Washington, boasted at the end of the war that his shops

Fig. 3.4. Employees at a Government Horseshoeing Shop, Washington, D.C., Area. The dozens of blacksmiths and other civilian employees photographed outside this large army-run establishment were among the more than 100,000 civilians on Quartermaster's Department payrolls across the country by the latter part of the war. *Source*: MOLLUS Photograph Collection, U.S. Army Military History Institute, Carlisle Barracks, Pennsylvania.

made furniture for the army "at one half the price charged by merchants." In the Commissary Department, Louisville depot chief Henry Symonds concluded in 1864 that local pork packers were colluding to overcharge him. Symonds responded by creating his own pork-packing operation, which in a matter of a few weeks generated about 8 million pounds of salt pork, worth about $1 million.[26] These activities demonstrated a mentality shared by many officers: contractors in the private sector were often seen as rivals and not simply as cooperative partners.

This competitive mentality was also demonstrated in quartermasters' activities in the wartime clothing industry, where the North maintained what was truly a mixed military economy. The largest single public establishment in this industry was the old Schuylkill Arsenal in Philadelphia, where the number of employees jumped during April 1861 from a few hundred to around twenty-seven hundred. By midsummer, the Schuylkill Arsenal counted at least five thousand employees, most of them women doing "outwork": they picked up cut textiles from the arsenal, took the materials to their homes for sewing, and returned finished garments. In 1864, when Philadelphia depot chief George Crosman reported that he was paying 5,127 "sewing women" (most of them outworkers), the Schuylkill Arsenal also employed about seven hundred men, including at least fifty cutters of material and dozens of packers and laborers.[27] The scale of this arsenal's wartime expansion equaled that of the Ordnance Department facilities and the navy yards. It was all the more remarkable, however, because there were dozens of existing private firms in the clothing industry that were ready and willing to meet military demand.

The quartermasters' entrepreneurialism was especially evident in western cities, where they established new garment-making operations during the first months of the war. By late April 1861, only days after the start of the conflict, Quartermaster Justus McKinstry wrote from Saint Louis to Washington to ask for textile shipments so that he could have uniforms made "by the wives and daughters of volunteers in the service." By August 1861, McKinstry had succeeded in creating three large clothing halls that employed as many as fifteen hundred women. Meanwhile, in Cincinnati, depot chief John Dickerson (who four years before had served as quartermaster for the first Utah expedition) was setting up his own large public establishments. At the beginning of December 1861, one Cincinnati newspaper reporter found more than two thousand "poor women and girls" working at Dickerson's tent manufactory at Fourth and Main streets. The same reporter noted that in a different facility, dedicated to clothing manufacture, 'the roar of the sewing machines reminds one of a cotton manufactory in full operation."[28]

Energetic public entrepreneurs, some of these quartermasters claimed that their clothing halls were more efficient than anything offered by the private sector. In late 1861 Dickerson told a congressional committee that he had started his clothing manufactory as an experiment. He had expanded it, he claimed, after finding that he could beat contractors' prices by 10–25 percent. The superintendent of Dickerson's tent-making facility, West Point graduate E. F. Abbott, recalled in 1863 that the public operation saved money by shutting out colluding tent contractors, who had swarmed around the Cincinnati depot during the first part of the war. Such claims were not limited to a couple of West Point classmates. Quartermaster William McKim, a volunteer from Massachusetts who succeeded Dickerson in Cincinnati, reported to Washington in late 1864 that public manufacture of overcoats there was less expensive than contracting out. At the same time, junior quartermaster R. S. Hart in Saint Louis reported that the garments made in the public clothing halls there were "much superior to those furnished by private parties, or received from other depots both in quality of work and neatness of finish."[29]

But quartermasters' arguments for their public manufactories went beyond claims about cost and quality. Many officers also maintained that the public clothing halls served an important function as a public welfare measure. By keeping these halls open, quartermasters claimed, they could provide wages to hundreds of needy women who were being pushed into poverty by the economic downturn caused by the closing of Southern trade and the departure of many of their male relatives for military service. Such arguments were presented with particular force in Saint Louis during the winter of 1861–62. By the beginning of the new year, the Quartermaster's Department enjoyed a surplus of finished garments in its warehouses across the country, which led depot quartermasters to attempt to reduce civilian employment.[30] The situation now became a crisis, one that highlighted the importance of appeals to public welfare—in particular, the welfare of women on the home front—in the making of the North's mixed military economy.

By January 1862, former employees of the Saint Louis clothing halls were continuing to insist that they could not afford the loss of the small wages (about fifty cents a day) they had been earning by sewing uniforms. Their appeals impressed Samuel E. Meigs, a younger brother of the quartermaster general, who had recently arrived in Saint Louis as a volunteer quartermaster. On 5 January, he informed his brother that two of the three public clothing halls in the city were now shut down, "and the applications for work have been very pressing." He continued, "It is a terrible sight to see so many women, most of them appear to have seen much better days, almost crying for work, with starvation staring them in the face." The best policy, Samuel Meigs recommended, would be to re-open the clothing halls, even if this were not the most efficient course: "I can't but think that a disposition shown on the part of Uncle Sam, to 'pity the poor' here, will be of a sufficiently political value, to overbalance any additional cost, from the work being done here, instead of elsewhere." Frank P. Blair, the influential congressman from Saint Louis, had already reached a similar conclusion and was pressing the War Department on this issue. On 6 January, Quartermaster General Meigs promised Blair to maintain some public manufacturing in Saint Louis, in order to "distribute in some degree the relief which this employment gives to the working people of the cities in this time of distress."[31]

The initial concessions in this area made by Meigs and Saint Louis depot chief Robert Allen proved insufficient. On 1 February, Meigs told Allen that too much clothing had been purchased and issued over the past few months, dragging the North even further into deep debt.[32] Meanwhile, Allen was apparently trying to shut down the clothing halls, despite Meigs's promises to Blair the month before. The residents of Saint Louis, including hundreds of women who had been working on uniforms, refused to let this happen. On 3 February, Allen picked up one of the city's leading newspapers to discover that a major meeting had been planned to protest his policies. He probably received more appeals in person from seamstresses and their friends. The same day, Allen wrote Meigs to say that he wanted to abandon his plan to shut the last clothing hall. "I find that such a step will be attended to with much more distress than I had imagined," explained Allen. "The appeals of starving women are entitled to consideration and sympathy, and I am myself, fully impressed with it. They will work for half wages rather than not have work at all, and as they are now practiced in the particular trade of making clothing their work will compare favorably with that made in the East." Allen concluded by asking Meigs to send along textiles from the New York City and Philadelphia depots, so that he could put the Saint Louis seamstresses back to work.[33]

In the days that followed, Saint Louis residents continued to press the War Department to re-open the clothing halls. Two days after Allen wrote his letter to Meigs, a large crowd of men and women assembled at the city's Verandah Hall to draw up a petition claiming that, by closing the army facilities during a period of high unemployment "the government had done a great injustice to the working men and women of this city." This petition was soon sent to Congressman Blair. Meanwhile, Stanton received a report from his assistant Thomas Scott, then traveling in Saint Louis, recommending that the light cloth already on hand in Saint Louis be used to make summer garments for soldiers.

"[T]here is in this city a large number of women unemployed (many of them the wives of soldiers, with children)," Scott informed Stanton, "who have no means of support other than such employment, for their needles, as they can obtain." Days later, Blair presented Meigs with the Saint Louis workers' petition and urged him to order Allen to re-open the clothing halls.[34]

Now supported by workers, politicians, War Department officials, and even the depot quartermaster, the campaign to save the public clothing manufacture in Saint Louis had become irresistible. Before the end of February, Meigs not only reiterated to Blair his commitment "to distribute the public expenditure" but also ordered the New York and Philadelphia depots to send textiles to Saint Louis. The clothing halls remained open for the rest of the war. In late 1862, senior quartermaster Charles Thomas toured Saint Louis and found the halls to be well run. "It is certainly good policy to have such an establishment in this city," Thomas advised Meigs, "and it seems to employ many women who would without the work be in a destitute condition. . . . The people here claim a portion of the work as a right and probably they are correct in it." In the months that followed, the Saint Louis halls continued to employ hundreds of seamstresses. By 1864, two separate halls were employing 250 men and women within their walls and another 200 women listed on army payrolls as "working on clothing at home." During the last year of the war, army seamstresses in Saint Louis manufactured $3 million worth of finished garments.[35]

Elsewhere across the North, public clothing manufactories also flourished. In Cincinnati, where workers and local politicians also fought in early 1862 to secure jobs in war supply industries, the facilities created by Dickerson in 1861 also continued to operate on a large scale to the end of the war. In four years, Dickerson's operations produced nearly 300,000 tents and over 2.8 million items of clothing.[36] By 1862 and 1863, when it opened smaller clothing operations at the Louisville depot and in Steubenville, Ohio, and Quincy, Illinois, the Quartermaster's Department was clearly committed to maintaining a mixed public-private procurement strategy in this sector of the war economy. The modest Steubenville and Quincy facilities were clearly the products of patronage: Stanton ordered the former facility established in his home town, while Quincy was the home of U.S. Senator Orville H. Browning, a friend of President Lincoln.[37] The larger urban clothing establishments, however, were the product of a more complicated mixture of pressures, including military entrepreneurialism and local lobbying.

As the struggle in Saint Louis suggests, champions of the public clothing establishments emphasized the benefit of employment for female relatives of Union soldiers. Even the small Steubenville facility was designed, according to

Meigs, to provide income to the "destitute wives and families of soldiers who have gone into service from that Congressional district." At the giant Schuylkill Arsenal, Meigs told one applicant for employment, it was official policy "to employ as far as practicable, the wives and families of soldiers." This policy was formalized in July 1863, when Philadelphia depot chief George Crosman demanded that all army seamstresses prove that they were the "wives, mothers and sisters, or other near relatives" of Union soldiers. Many women with male relatives in the field supported such policies: earlier that year, Mary Morris and Elizabeth Moore of Philadelphia had written Stanton to say that "[t]he soldiers wifes orter be the first to get the worke." But especially in Philadelphia, where the clothing arsenal had been established for decades, such a policy displaced hundreds of longtime employees. Soon after Crosman issued his formal order, some two hundred army seamstresses assembled to criticize him. If the quartermaster would do less contracting out, they protested, there would be plenty of arsenal jobs for seamstresses, regardless of the military status of their male relatives.[38]

But many quartermasters regarded the public clothing establishments as something more than merely a work-welfare measure or an added bonus for families with soldiers in the field. As quartermasters' justifications in 1861 for the public enterprises in Cincinnati and Saint Louis suggest, some procurement officers also saw themselves as competing with the private sector. By the middle of the war, several officers claimed that the public facilities took profits out of the hands of contractors and placed money into the hands of deserving women. In 1863, Cincinnati chief quartermaster Thomas Swords recommended to Meigs that "[c]lothing should, as far as practicable, be manufactured by the [Quartermaster's] Department; by which means a better article can be secured, and employment given to the families of soldiers in the field at more remunerative prices than now paid by contractors." Officers at the Louisville depot, which had recently opened its own clothing operation, expressed nearly identical ideas. His own facility, boasted Louisville quartermaster Walworth Jenkins to Meigs in 1863, produced uniforms "[a]t a lower rate, than could have been procured elsewhere and at a great saving to the Government, besides giving employment and means of subsistence, to many of the wives, widows, sisters and daughters of our soldiers in the field, and who are certainly more deserving of the patronage of the Government, than the Army of Contractors, at home."[39] Supply officers were hardly the only Northerners to voice such ideas. Throughout the war, similar claims were made by the seamstresses who produced army uniforms, as they worked for the expansion of public work in their industry.

The Seamstresses' Campaign against Contracting

The clothing industry constituted one of the largest and most expensive parts of the entire Civil War economy. In addition to his pay, a Northern volunteer soldier had a clothing account that allowed him to draw up to forty-two dollars worth of clothing per year from his regimental quartermaster. The annual allowance included one or two caps and one hat, two coats or jackets, three flannel shirts, three pairs of trousers and three pairs of drawers, four pairs of stockings, and four pairs of shoes. One wool blanket, one waterproof blanket, and one overcoat were also part of the standard allowance, but these were supposed to last three years.[40] Although actual consumption varied from soldier to soldier, more than forty dollars per soldier per year added up quickly in a Northern army that reached 1 million men at once and in which over 2 million men eventually served. In four years of war, the Quartermaster's Department spent roughly $350 million on clothing and equipage (including items such as tents and knapsacks). This was twice as much as the Union armies spent on weapons.

Unlike the weapons industry, however, the ready-made clothing business was already selling millions of dollars worth of goods to Americans all over the country before the Civil War began. One of the nation's largest manufacturing sectors, the antebellum clothing industry included not only small tailors' shops but also large urban wholesaling firms accustomed to employing hundreds of seamstresses at once. During the Civil War, the size and maturity of this industry made it the site of a major struggle over the economics and politics of procurement. When quartermasters chose to make garments in army facilities instead of contracting out, they were bypassing existing capacities in the private sector. Seamstresses, known as some of the worst-paid workers in the antebellum American economy, came to see the army clothing establishments as a potential source of higher wages. Clothiers, many of whom were trying to make up for lost sales in the South, managed to secure contracts throughout the war despite the support among quartermasters and seamstresses for public enterprise. In this part of the Northern military economy, the army, labor, and business struggled throughout the war over the "make or buy" decision.

The size of the Quartermaster's Department wartime clothing establishments in Philadelphia and Cincinnati, which each employed as many as a few thousand seamstresses at once, was certainly impressive. But they were not much larger than leading clothing firms in the private sector. By 1860, when the men's clothing industry employed nearly 115,000 workers nationwide—only slightly fewer than the number working in the cotton textile or footwear industries—leading wholesale clothiers stood among the very largest enterprises in the United States. In New York City, the center of the industry, individual firms such as Lewis & Hanford, Devlin & Company, Wilde & Company, and Brooks Brothers each did close to $1 million in annual sales and each relied upon the labor of as many as 3,000 or 4,000 seamstresses during peak seasons. Among the nation's leading consumers of textiles, these enterprises often sold to customers all around the nation through branch offices or sales to local retailers.[41] It was these large wholesale clothiers, and not small tailors' shops, that would fill most of the North's contracts for uniforms.

Seamstresses, who were the bulk of the labor force for these wholesale clothiers, comprised one of the largest single groups of workers in industrializing America. Unlike their counterparts in the textile industry, however, seamstresses tended to work at home or in small workshops. Although wholesale clothiers might employ a few dozen sewing women on site, most of their employees were outworkers who took away cut material and returned finished garments. Even after the spread of sewing machines (foot- or hand-driven) in the early 1850s, outwork predominated. In many cases, wholesale clothiers reached their hundreds of outworkers through networks of subcontractors, who consolidated the business of distributing and collecting materials and garments. In antebellum New York City, many seamstresses were German and Irish immigrants. By working at home, these sewing women could combine some paid work with the unpaid labor of childcare or housework. The wages they earned, however, were meager. In urban labor markets full of women capable of using a needle or a sewing machine, the piece rates paid by clothiers for each finished garment stayed low. By 1860, the best-paid workers in the ready-made clothing industry were the skilled cutters of material—overwhelmingly male—who were paid an average of about fourteen dollars a week. Many full-time outworking seamstresses, on the other hand, were lucky to make four dollars a week.[42]

Like other workers in industrializing America, seamstresses used collective action both before and during the Civil War in an effort to protect their wages and autonomy against a rising tide of mechanization, division of labor, and labor market competition. During the antebellum era, male and female clothing industry workers, along with textile and footwear factory operatives, used strikes and petitions to advance their causes.[43] In the Civil War North, organized labor activity grew across the entire economy. As early as July 1861, New York City tailors met to denounce the "grinding system of wages" used by military contractors. In 1862 and 1863, there were at least three distinct waves of strikes at U.S. navy yards over extended winter working hours and proposed wage reductions.[44]

Especially during the second half of the war, when inflation began to eat into real wages, strikes became common. From Illinois coal workers to New York longshoremen, hundreds of Northern workers participated in strikes from 1863 to 1865. In several cases, as in the Pennsylvania coal fields and at the works of the Union's largest cannon supplier (Robert Parrott & Company) in Cold Spring, New York, troops were called out to break strikes. In April 1864, General William Rosecrans, dismayed by recent labor disputes in Saint Louis, issued a regional order outlawing strikes and the closed shop in any industry connected to the war effort. A similar order was issued by General Stephen Burbridge in Louisville.[45]

Although Northern seamstresses never engaged in major wartime strikes broken by troops, they did mount organized campaigns to lobby for higher rates of pay. Well before the war, seamstresses had been active in the labor movement and had been identified by middle-class reformers as victims of an unbridled capitalist economy.[46] From 1861 to 1865, as the historians Matthew Gallman and Rachel Seidman have shown in their studies of wartime Philadelphia, seamstresses became even more active and received even more attention from their fellow workers and the general public.[47] But it was not simply the stresses of war and the seamstresses' connections to a growing labor movement that made the struggle over the clothing industry so hotly contested. The seamstresses' campaigns were especially important and intensely political because the wartime clothing industry was part of the mixed military economy. In one of the most important political efforts of the entire war, seamstresses worked steadily—with mixed success—to encourage the North to reduce contracting and expand public enterprise.

During the first part of the war, while seamstresses in Cincinnati and Saint Louis fought to keep open the new army clothing halls in their cities, their counterparts in Philadelphia were also promoting arsenal production. In Philadelphia, the Schuylkill Arsenal had served for decades as the army's source of uniforms. When the Quartermaster's Department began to award some clothing contracts to private firms in the summer of 1861, some of the arsenal's longtime employees objected. In August, Hannah Rose and two hundred other Philadelphia women sent a petition to Secretary of War Cameron in which they emphasized their family connections to recently mobilized soldiers and called on the army to expand production at the arsenal. A month later, Martha Yeager and Anna Long led a meeting of army seamstresses at Philadelphia's Temperance Hall in which the sewing women drew up a petition for the eyes of Cameron and President Lincoln. Yeager, in an opening speech, likened contractors to parasites. "The person who would live off the labor of others, and thus grow rich at their expense," she told the assembly, "is an aristocrat" who deserved to be "loathed with contempt." Together, the Philadelphia seamstresses at this meeting resolved to "protest against, and vigorously and righteously denounce the infamous contract system, by which we are robbed of more than half our wages, while it puts large profits in the pockets of a few speculators and contractors, and by which we are impoverished."[48]

Unimpressed by the claims of depot chief Charles Thomas in the summer of 1861 that the Schuylkill Arsenal was already employing four times as many cutters as it had been in peacetime and was sending material out to some three thousand sewing women, Philadelphia seamstresses continued to insist that contracting out was immoral and harmful to the welfare of hundreds of Northern families. Later in the war, after Long—a widow with five children—lost her job at the Schuylkill Arsenal, she led 146 other dismissed seamstresses in petitioning Secretary of War Stanton to stop awarding contracts to private firms. They did not want to go to work for contractors, Long explained, because private firms "will not pay wages on which we can live." She demanded, "[W]hy should the government money be taken from the families of the poor to enrich the wealthy speculator without any gain to the government[?]"[49] For Long and other seamstresses, the issue was simple: public arsenals paid higher piece rates while allowing workers to deal directly with the state, while contracts allowed a few wholesale clothiers to profit through a subcontracting system that pushed the workers' pay down to a minimum.

Starting in 1863, when inflation began to be especially severe, workers around the North pushed for wage hikes that would respond to the rising cost of living. The tens of thousands of employees of the Quartermaster's Department joined workers in private industry in calling for more pay. In the spring of 1863, for instance, male workers at the Philadelphia depot petitioned local quartermasters for a 25 percent wage increase that would allow them to continue to support their families. These petitions were supported by U.S. congressmen from Philadelphia such as William D. Kelley and Leonard Myers, who asked Meigs and Stanton to approve the increases.[50] For much of 1863 and 1864, depot quartermasters across the North, including George Crosman in Philadelphia, David Vinton in New York, and Morris Miller in Washington, wrote Meigs to ask permission to raise the pay of discontented civilian employees. Aware of the recent inflation and the danger of losing some valuable workers to the private sector, Meigs approved several of these proposed increases.[51]

Like the military's male civilian employees, seamstresses on both sides of the mixed military economy became even more concerned with pay and employment at the public arsenals during the second half of the war. As before, many seamstresses and their supporters attacked the army's practice of contracting out as a boon to profiteers and a bane to workers. In New York City, which had no public clothing halls but was the center of the nation's ready-made clothing industry, sewing women became increasingly critical of contractors who let their workers suffer while reaping the profits of war. In a series of letters to the editor and articles published in 1863 by the *New York Sun* newspaper (whose editor, Moses Beach, supported women's labor organizations), seamstresses and their supporters called on the contractors and the national government to pay more to the women who produced the Union's uniforms. The author of one July 1863 letter suggested that, if the army was actually paying too little to contractors for them to pay decent piece rates for workers, "then let Government issue a few more 'greenbacks' and pay a higher price for army clothing"; on the other hand, if the contract prices were already high enough, the workers would need to create a "combination" to compel contractors to pay more. A month later, a correspondent identifying herself as "a working woman," complained of the long hours and low pay that were the fate of many women workers. This author also predicted that seamstresses would not sit back and accept the status quo. "May contractors take timely warning," she wrote, "and execute contracts with a view to paying more for labor. The war spirit against starvation prices is aroused in the sewing women."[52]

Seamstresses and their supporters began to form large associations that could wield economic and political power. By the end of 1863, New York City was home to a Working Women's Union and a Sewing Women's Protective and Benevolent Association; nearby, the Brooklyn Sewing Women's Beneficent Association had also been established. In Philadelphia, a Working Women's Relief Association and a Working Women's Association fought for seamstresses' interests.[53] At one organizational meeting for the Working Women's Union in New York, three hundred to four hundred women of various ages met in the Bowery district to discuss meager piece rates paid for military garments and tents. Even those who leased sewing machines (at a rate of eighteen cents a day), which allowed them to sew as many as eight shirts in a day, complained that they had trouble clearing three dollars a week. In some cases, piece rates for military garments had dropped over the course of the war, even as the cost of living rose. In a subsequent meeting, attended by men as well as women, the organization drafted a constitution in which it promised to work for higher wages and shorter hours, a wider range of occupational possibilities for women, and legal aid for workers with grievances against employers.[54]

During 1863, as prices continued to rise in advance of wages, Northern seamstresses and their supporters became even more critical of contracting and contractors. In November, the *New York Sun* printed a letter from a young woman who described how she and her mother, paid sixty-two cents for every cap they sewed, together were able to bring in only $2.50 a week. "Are we nothing but living machines," she asked, "to be driven at will for the accommodation of a set of heartless, yes I may say, soulless people?" By this time, *Fincher's Trades Review,* a Philadelphia weekly that stood as the leading labor publication in the country, had taken up the cause of the seamstresses and public enterprise. Leading Philadelphia clothing contractors such as Harkness Bros. and Kunkel, Hall & Company, the newspaper reported, paid only about eighty cents for all the sewing required to make a jacket; the army's Schuylkill Arsenal, on the other hand, paid over $1.12 for exactly the same work. "Who pockets the difference?" *Fincher's* asked. "Why should this amount be taken from poor seamstresses, and put in the pockets of a contractor? . . . [W]hy rob the poor to fatten contractors?" By the end of the year, *Fincher's* was calling on the army to manufacture all uniforms in its own facilities, "without the aid of contractors, who only step between the government and the seamstresses, to pocket half the earnings of the latter."⁵⁵

In Philadelphia, site of the army's largest clothing establishment, the fight against contracting—and for more public enterprise—came to a head in 1864. In mid-April, the city's Working Women's Relief Association held a major meeting at Sansom Street Hall. Led by T. W. Braidwood, a male reformer who headed the city's School of Design for Women, this meeting agreed to petition Philadelphia depot chief George Crosman to employ more seamstresses and raise piece rates at the Schuylkill Arsenal. A week later, Mary Pratt led a meeting of the same group, which resolved to call "upon the government to take the work of the arsenal into its own hands, that favoritism may no longer exist, and that the profit now going into the hands of contractors may inure to those who perform the labor." Such calls for the end of contracting continued in May in a series of meetings that brought together a broad coalition of middle-class reformers, leading male labor organizers, and seamstresses. In addition to attending these meetings, the nationally known labor editor Jonathan Fincher used his Philadelphia newspaper to broadcast the seamstresses' concerns and demands. By contracting out, *Fincher's Trades Review* claimed, the army effectively rewarded "a set of soulless sub-contractors" whose general practice was to "rent a cheap room in the suburbs, procure a lot of sewing machines, and employ young girls from 12 to 18 years of age, at just such prices as they choose to pay." The only solution to the problem of rock-bottom wages promoted by this system, *Fincher's* claimed, was for the army to demand that its suppliers pay "arsenal wages" and to expand the army's own public operations. By contracting out without a minimum wage, the newspaper concluded, "the Government becomes a party to the wrongs visited upon sewing-women."⁵⁶

One important result of the many meetings of Philadelphia seamstresses in the spring of 1864 was the drafting of a formal petition for higher wages and expanded arsenal production. At the beginning of June, the seamstresses presented depot quartermaster Crosman with a petition signed "Twenty Thousand Working Women of Philadelphia." Addressed to Secretary of War Stanton, the petition first noted that pay from the Schuylkill Arsenal was failing to keep pace with inflation. In fact, the document claimed, the arsenal's piece rates had actually declined by 30 percent over the past two years, even as the cost of living had increased by 75 percent. "What we need most," the petitioners told Stanton, "IS IMMEDIATE AID. You can give it; the power is lodged in you; issue an order to the Quartermaster General, authorizing or ordering him to increase the price of female labor until it shall approximate the price of living." But the petition did not simply call for higher arsenal wages. It also argued that any private contractors, who were "making immense fortunes off the Government," should be compelled to pay their workers the new arsenal rates. In any case, contracting out should be reduced drastically and public production expanded by a factor of four in order to have "millions of dollars annually saved to the nation." Under this plan, the public employment of seamstresses at Philadelphia would rise from about 5,000 to 20,000; the army would manufacture virtually all of the garments it required, instead of only a quarter of them, with its own establishments.⁵⁷

This major petition, produced by the organized efforts of one of the nation's largest groups of military-industrial workers and their allies, received considerable attention from the War Department during the summer of 1864. As he forwarded the document to Washington, Crosman endorsed its call for higher arsenal pay. The piece rates paid seamstresses, Crosman told Quartermaster General Meigs, "should be at once increased. . . . The tariff of prices formerly paid here, was reduced in August 1862, to make it correspond with prices then paid by Lt. Col. [David] Vinton in New York for similar work. It is now entirely too low." Nearly two years before, in other words, the Schuylkill Arsenal had actually cut wages to bring them closer to those being paid in the private sector; now, after months of high inflation, the arsenal was not paying enough.⁵⁸ Meigs's office responded by quickly polling the Cincinnati and Saint Louis depot chiefs about the piece rates paid to seamstresses at the army clothing halls and by contractors in those cities. The New York City depot, which did not have public manufactories but constantly signed contracts with clothiers, was also ordered to report on wage levels. Within a few days, Vinton responded by telling Meigs that seamstresses' pay in the private sector was "quite inadequate. . . . The competition from our contract system has the effect, I think, to keep these prices at the present low standard." Vinton reached this conclusion after speaking directly to leading New York contractors such as J. & W. Lyall, who said that if they were to pay their workers more than four to seven dollars a week, they could no longer compete for army contracts.[59]

The seamstresses' campaign reached unprecedented heights by the beginning of August 1864, when their leaders were accompanied by Congressman Kelley to meetings in Washington with President Lincoln and Stanton. The secretary of war, who had asked Meigs to submit a formal report on the seamstresses' petition in advance of these meetings, knew that top Quartermaster's Department officers supported the calls for higher pay at the public clothing establishments.[60] But these same officers, including Vinton in New York, had also reported that the design of the contracting system—which used competitive bidding to lower costs—was incompatible with the goal of raising pay for sewing women employed by contractors. For the seamstresses, the solution to this problem was to have the army create a public monopoly or enforce a minimum wage in the private sector. Some quartermasters, who were dissatisfied with contracting and enthusiastic about public enterprise, were sympathetic to these calls. In the end, however, Washington's response to the Philadelphia seamstresses' major petition of 1864 demonstrated that there were two sides to the North's mixed military economy.

Limits to Public Enterprise

Days after the delegation of Philadelphia seamstresses visited Washington, Stanton and Meigs settled on a response that answered only some of the petitioners' concerns. On 18 August 1864, Stanton approved the quartermaster general's recommendation to raise the pay of seamstresses working for army clothing operations by 20 percent. The War Department also agreed to employ 1,000 to 2,000 more women at the Schuylkill Arsenal. But Stanton rejected the idea of enforcing a minimum wage for contractors, who would continue to fill a large proportion of the army's needs for garments. As Stanton put it, "much difficulty has been found in adopting a system which will secure a fair rate of compensation to the operatives, without an unauthorized exercise of power over government contractors."[61] Although the seamstresses' campaign succeeded in raising wages and employment at the Philadelphia arsenal, it fell short of achieving a radical reorganization of the military clothing industry.

Unwilling to stop contracting for huge quantities of garments and unwilling to call upon prime contractors to pay minimum wages, the War Department had done little to address the complaints of seamstresses working on the private side of the military economy.

Sewing women in New York City, which lacked an army-run clothing establishment but was full of contract work, were especially disgusted with Stanton's decision. By the beginning of September 1864, organizations such as the Working Women's Protective Union had begun to gather signatures for a new petition to Stanton, asking him "in the name of justice and humanity" to enforce a minimum wage. The New York seamstresses began by emphasizing their patriotism and their connections to soldiers in the field. "[W]e do not ask charity," they assured the secretary of war; "we come to you as American women, many of whom have sacrificed the dearest treasure of our hearts on the altar of freedom." They continued, "We . . . respectfully ask you (if it comes within your province) to so modify the contract system as to make it obligatory upon all contractors to pay government prices." For the New York seamstresses who worked for army contractors, "government prices" would represent a big raise. By the end of September, the *New York Sun* and other newspapers were starting to publish comparative piece-rate data (perhaps taken from the Quartermaster's Department recent internal study), which showed that seamstresses working for the Schuylkill Arsenal received about 20–50 percent more per garment than did those employed by contractors.[62]

Throughout the remainder of the war, sewing women and their allies continued to call on Washington to either compel contractors to pay a minimum wage or stop contracting altogether. In doing so, they attacked contractors as exploitative profiteers. In late September 1864, the *New York Sun* referred to "those fiends in the shape of men—Government Contractors—who, by their inhuman conduct, are driving ten thousand working women into the very jaws of hell." The *Sun* also seconded a recent editorial from the *Philadelphia Evening Bulletin*, which had argued that the dynamic of subcontracting and low wages on the private side of the military garment industry was "literally shameful; there is no other word to express it, and no community is justifiable in seeing the poor amongst it crushed in this frightful way. It is the clear duty of government so to control its contractors that they shall be forced to do some kind of justice to those whom they employ." If this was impossible, the *Evening Bulletin* continued, "let the whole contract system be swept away, and let the Government employ its sewers directly, as now done at the arsenals, so that no middle profit shall be squeezed out of the poor workwomen." The clothing contractors, the newspaper concluded, "who are thus grinding the poor to death, are squandering their excessive profits in every species of foreign luxury. . . . The whole matter is too infamous to be tolerated."⁶³

In January 1865, another delegation of Philadelphia seamstresses, led by Martha Yeager of the Schuylkill Arsenal, went to Washington to meet with President Lincoln. Again, they succeeded in winning limited concessions in the form of piece-rate hikes and more arsenal employment. But again, the War Department refused to stop contracting out; the Philadelphia seamstresses were left repeating their old demand that Washington force contractors to pay arsenal rates. In February 1865, a leading Cincinnati newspaper complained that "[i]t is one of the most pitiable and shameful things of the times that the clothing contractors, who make enormous profits upon the work done for them, are enabled to oppress the poor women who do the sewing, and constrain them to work at prices, most justly styled 'starvation.'" Later that month, Cincinnati sewing women petitioned President Lincoln to stop contracting and expand public enterprise in their city. "We are unable to sustain life for the price offered by contractors," they claimed, "who fatten on their contracts by grinding immense profits out of the labor of their operators." The solution, they told the president, was to have "the Government, through the proper officer of the Quartermaster's Department, issue the work required directly to us."⁶⁴

Despite their vigor and their considerable achievements, the well-organized campaigns by seamstresses in several Northern cities never convinced the army to stop signing large contracts for garments or to compel its contractors to pay arsenal rates. One reason for this was that some top officials in Washington— including Stanton and Meigs—were unwilling to support the seamstresses' demands for radical reforms in this large part of the military economy. But the limits on the extent of public enterprise in the North did not come solely from top officials' sympathies for contractors. Problems with war finance, as well as ideological differences between workers and War Department officials, worked to limit the size of army-run operations.

There were always two sides to the North's mixed military economy. Even in major war industries with large existing public facilities, such as small arms, shipbuilding, and clothing, the North awarded tens of millions of dollars in contracts to private firms. The contracting began on a large scale during the first weeks of the war and continued to its end. In the case of uniforms, Meigs decided soon after taking office that, instead of trying to rely solely on the Philadelphia arsenal, his bureau would use wholesale clothiers as a major source of supply. In July 1861, Meigs told Philadelphia depot chief Charles Thomas that the Schuylkill Arsenal "should be kept up and worked to its full capacity, serving as an important source of supply and also as a model, to whose products

... supplies should conform." However, Meigs continued, "it is thought best to resort for the great additional demand to the contract system which will bring into use and give employment to the many large establishments well organized by present capital and enterprise where business is cut off by the rebellion [and] where operatives are suffering."[65] Contracting out, the quartermaster general believed, would simultaneously aid industries then suffering from the end of the Southern trade while allowing the army to save money by tapping the experience and underutilized existing capacities of private firms.

In comparison to some of his most entrepreneurial depot chiefs, the quartermaster general was more circumspect about the ability of the army to outdo private firms in terms of cost. Not all officers of the Quartermaster's Department were of the same mind about the "make or buy" question. At one extreme were officers such as Saint Louis transport chief Lewis B. Parsons, the former Ohio & Mississippi Railroad executive, who tried to limit outright purchases of boats in favor of short-term contracts and charters. "[P]rivate enterprise," Parsons declared in 1863, "will always perform the same service cheaper than the government can." By contrast, many of the quartermasters who established new clothing halls and other army-run operations claimed that they could match or outperform the private sector. From Washington, Meigs took a middle road. He approved large public enterprises but also encouraged contracting. One benefit of contracting, his July 1861 letter on clothing procurement suggested, was that it could help business to rebound from a slump caused by the war's outbreak. Wary of having the government spend huge sums on buildings or capital equipment that it would have to unload at war's end, Meigs refused to let the public operations expand endlessly. More than some of his depot chiefs, the quartermaster general believed that contracting could save money. In February 1864, after Senator Henry Wilson forwarded suggestions that the army buy feed for horses and mules by dispatching agents to deal directly with farmers, Meigs responded by saying that the marketing of forage for the Union armies "is now done by private individuals under the keen competition of commercial rivals, and is probably done as cheaply as it could be done by special government agents."[66]

Ultimately, the extent of public enterprise in the North was limited both by Washington officials' sympathies for business and by the workings of the wartime financial system. Throughout the war, quartermasters had trouble getting enough cash from the Treasury to pay their employees on time. Although supply officers also had problems securing funds for contracts, the particulars of military finance made it easier to pay off large contracts than to pay the wages of dozens of individual workers. This problem, which existed across the entire war economy, was evident in the clothing industry, where quartermasters were compelled to cut back on army-run operations because of financial difficulties. If it had been easier for them to secure cash to pay seamstresses and other army employees, supply officers would have made the public side of the military economy even larger than it turned out to be.

Throughout the war, quartermasters were constantly behind in their payments to their creditors, including contractors and employees. Although many seamstresses and other employees lobbied throughout the war for more arsenal employment and less contracting out, they were often dissatisfied with the slowness of pay at the arsenals. Quartermasters were frequently beset by complaints, if not outright strikes, brought by their employees to protest late pay. In April 1863, Saint Louis depot chief Robert Allen complained to Meigs that the "women in the Manufacturing Halls have not been paid for three months; and I have an army of teamsters constantly besetting my office clamorous for pay." In 1863 and 1864, when he oversaw huge public payrolls at depots in Nashville and Louisville, Quartermaster James Donaldson saw dozens of artisans and laborers organize strikes to protest late pay. In Philadelphia in September 1864, Quartermaster Alexander J. Perry informed Meigs that the depot did not have enough cash on hand to pay its several thousand seamstresses. "The necessary cash disbursements at the arsenal are about two thousand dollars per day," Perry reported. He needed cash from the Treasury immediately, Perry explained, so "that we shall not be compelled to suspend labor at the arsenal or create confusion." A month later, seamstresses at the Saint Louis clothing halls sent the secretary of war a formal petition complaining that they had been paid for only one month's work in the last four months.[67] Although seamstresses preferred arsenal work over contracting, this was not because the arsenals were smoothly run operations that always paid them in full and on time.

Constant shortages of funds forced supply officers to delay payment to contractors, as well as workers. Significantly, however, the financial system made it easier to settle debts with contractors. This was because contractors did not require small change and could more easily forgo being paid in cash. Starting in the spring of 1862, after Congress passed new legislation, the War Department began to pay many of its contractors with short-term, interest-bearing notes called "certificates of indebtedness." Issued in amounts of one thousand dollars or more, these certificates could be converted to cash easily by well-established business firms. They were not helpful, however, to employees who earned as little as twenty or thirty dollars a month. Meigs himself recognized this problem as early as November 1862, when he urged Stanton to press the Treasury for more cash disbursements, which would help to "restore the credit of the Quarter Master's Department, with the working people." Meigs was pushed to make such requests by quartermasters across the country, who found it easier to get Treasury funds in the form of certificates and complained to Washington that their large payrolls could only be settled with greenbacks. Typical was the complaint of Memphis depot quartermaster Asher Eddy in September 1864, when he told Meigs, "My employees have not been paid for three months. . . . Certificates of indebtedness are of no use whatever to pay the employees with."[68]

In the end, the scarcity of cash had significant effects upon the shape of the military economy. On the one hand, the genuine concern of Meigs and his depot chiefs about their employees' welfare, a concern that was tied up with their interest in avoiding labor protests and strikes, caused the Quartermaster's Department to limit any cash outlays for nonpayroll purposes. In some cases, at least, this limited the bureau's procurement options in ways that tended to increase the prices it paid for supplies. Meigs himself admitted as much in early 1864, in correspondence with Quartermaster Samuel L. Brown, who was then acting as the Union's chief buyer of forage in the East. "The attempt to supply you with money for your purchases," Meigs told Brown, "has caused great suffering among laborers, teamsters, and other poor men employed by the Quartermaster's Department." In the future, therefore, Brown would have to eschew cash purchases, even though it was clear that this would be more expensive. "The Department is aware that forage must cost more if bought payable in certificates," Meigs explained to Brown, "but as forage can be bought for certificates, and labor cannot, and as the Treasury is unable to supply promptly money enough for both purposes, the Quartermaster's Department is obliged to limit its requisitions."[69]

Because such internal accounting adjustments failed to generate enough greenbacks for Quartermaster's Department employees, the cash shortage also served to promote contracting over arsenal production. This constraint on public enterprise became especially clear in the case of the Cincinnati clothing halls during the latter part of the war. In August 1864, President Lincoln received another letter from a seamstress who wanted him to protect the public clothing halls and stop contracting out. The author of this letter, Jane Hasler, informed Lincoln that "[t]he work the soldiers' wives have been getting from the government shop of Cincinnati has been taken from them, and given to a class of *wealthy men, mostly Jews,* who are re-letting it to them, at starvation prices." Although it was more openly anti-Semitic than most seamstresses' appeals, Hasler's letter otherwise included a typical combination of references to the immorality of profiteering contractors and the dire plight of the sewing women whose male relatives were serving the nation in the field. Contracting out, Hasler argued, "is taking the very bread from out of the soldiers wives and childrens [mouths] and mostly giving it to a class of men who are not willing to own they are citizens of the United States but who are making enormous sums off of the government if I am rightly informed." The solution to this crisis, Hasler told the president, was for the government to "return the work to us": that is, to stop contracting and instead re-expand the public clothing halls, which paid higher wages.[70]

The response to Hasler's letter, which the White House forwarded to the War Department, demonstrated that the scale of public enterprise in the North was constrained by financial difficulties. On 1 September 1864, only days after Hasler's letter was received in Washington, Meigs ordered new Cincinnati depot chief William McKim to look into Hasler's complaints and report "with as little delay as possible." In his report on the matter, Cincinnati quartermaster Charles W. Moulton explained that he had become responsible for the procurement of clothing at that depot in the summer of 1863. At that time, Cincinnati depot chief Thomas Swords had ordered "that Union people *only* be employed, and that particular attention be given to employing those who were dependent upon our soldiers in the Army. I was also instructed to *increase* the manufacture of clothing," to the extent that this was possible. Moulton then began to do so and found that the quality of garments produced by the army-run operations was quite high. "I found however," Moulton recalled, "that I could not carry on this business at all unless I was punctual in paying these employees every week, as they were mostly dependent upon their labor for their living." Because Moulton had difficulty securing enough cash from the Treasury in a timely fashion, he proceeded over time to sign more contracts with private firms and cut production at the army-run halls. In some cases, Moulton went so far as to contract out for the production of garments and tents from textiles previously purchased by the army and designated originally for use in the government-run Cincinnati establishments.[71]

Moulton's report made it clear that the constant cash scarcity had led him to contract out for hundreds of thousands of dollars worth of garments that he might have otherwise acquired from the army-run facilities in Cincinnati. Back in Washington, the quartermaster general suggested that the army's cash flow problems alone could be reason enough to contract out. In October 1864, after reading the reports generated in response to Jane Hasler's letter, Meigs informed Stanton that, in order to continue "the practice of giving out clothing to be made up by the wives and relatives of soldiers," more cash from the Treasury would be required. "Unless the Treasury can do this promptly," Meigs continued, "I am of the opinion that it will be better to give out the work to contractors entirely. They probably do not pay quite so much to the working people, but, borrowing money from the banks and capitalists, they are generally able to pay their working people with reasonable punctuality." For the North's top supply official, the slowness of Treasury cash disbursements had become reason enough to scale down some of the country's biggest public enterprises and send more orders to the private sector.[72]

Although this episode demonstrates that the structure of Union finance tended to promote buying over making, the seamstresses and supply officers who campaigned for more arsenal production were never entirely defeated. Despite Meigs's comments in late 1864 about the financial advantages of contracting out, the continuing pressure applied by seamstresses and some depot officers caused the War Department to maintain the public clothing halls in Cincinnati and elsewhere. Indeed, at the beginning of 1865, Quartermaster Alexander Perry—then chief of the clothing and equipage division at Meigs's office in Washington—informed Cincinnati depot chief William McKim that he should procure significant quantities of garments from army-run operations. "I am directed by the Qr. Mr. Gen'l to advise you," Perry wrote, "that it is believed to be good policy to encourage the manufacture of clothing by the wives and relatives of soldiers in Cincinnati. . . . This would relieve much distress and the U.S. will get the garments without an intermediate contractor's profit." One important advantage of army-run production, Perry added, was that it eliminated "the whole controversy as to the government interfering between contractors and their work people."[73] That is, more public production meant that there would be less need to worry about trying to enforce a minimum wage for contractors. Remarkably, this order from the very top of the procurement system did not differ much in its rhetoric and logic from some of the most strident petitions of the seamstresses. At the end of the war, as at the beginning, many army supply officers openly endorsed public enterprise as a means to boost the welfare of soldiers' families while reducing private profit taking in the military economy.

THE NORTH'S MIXED MILITARY economy grew out of antebellum procurement practices, which had long combined contracting and arsenal production. But it was also the product of intense wartime struggles over the "make or buy" decision. Many cities and regions fought for army-run manufacturing facilities because they saw such operations as an important source of jobs and military dollars. War industry workers, including seamstresses, mounted organized efforts throughout the conflict to expand public facilities, which tended to pay higher wages than contractors. Some procurement officers, including depot quartermasters, quickly created new public enterprises and expanded existing ones. Although these champions of public enterprise did not always agree, their arguments often overlapped. Both quartermasters and seamstresses fought to limit contracting by claiming that public production enhanced the public welfare, especially that of soldiers' families. At the same time, they both claimed that contracting allowed a few unscrupulous profiteers to benefit from war, while most Northerners suffered. This ideological defense of public enterprise, not only the existing institutional arrangements or efficiency calculations, kept the output of army-run clothing manufactories high enough to account for roughly a quarter of all garments issued to the Union armies during the war. The large public side of the North's mixed military economy was very much a political creation, but it was ordinary workers and army supply bureaucrats, rather than party politicians, who generated much of the pressure that maintained it.

Although the army frequently chose to make instead of buy, in the end it bought more than it made. Faced with a well-developed commercial and industrial economy full of private firms eager to take war orders, many War Department officials and supply officers saw contracting as essential. Even for some officers who were concerned about the legitimacy of wartime profit taking, contracting appeared in many cases to be less costly or at least easier to handle than army-run operations with giant civilian payrolls. But when they decided to buy instead of make, Northern procurement officers did not cease to be concerned about the politics of the military economy. Exactly how to buy, supply officers discovered, was a difficult question entailing high stakes. On the private side of the North's war economy, no less than on the public side, officers and civilians struggled over the ethics of procurement and the flows of military dollars.

CHAPTER 4

The Trouble with Contracting

Although public enterprises were important in many areas of the North's military economy, in the end most procurement dollars went to contractors. From 1861 to 1865, the Quartermaster's Department and other military supply bureaus signed thousands of agreements with hundreds of prime contractors all over the country. But contracting, while routine and essential to the Union's war effort, never ceased to be troublesome—in more ways than one. The laws and regulations governing contracting created significant administrative costs for military procurement officers and their suppliers. For both the army and its contractors, the financial aspects of the war supply business proved especially difficult. For all concerned, the bureaucratic institutions that ballooned after 1861 to handle the business of war made that business complicated.

But the trouble with contracting went beyond the costs associated with large-scale government regulation and finance. In the eyes of many Northerners, contracting seemed to create undesirable economic outcomes. It was not only that the act of buying instead of making limited the wartime state's influence over wages paid to war workers on the home front, as seamstresses and others constantly argued. Beyond this, there was the problem of individual contractors receiving huge payments—and, potentially, huge profits—even as hundreds of thousands of their countrymen risked life and limb, often with little compensation.

This stark inequality bothered not only common soldiers and war workers but also many of the Union's top procurement officers. To these officers, it seemed that, in addition to the sheer scale of the war supply business, its unique administrative and financial complexities worked to push the bulk of the business of war and the potential wealth it represented into the hands of a small group of unusually large dealers. Having grown up in a political culture that was generally hostile to the concentration of economic power, supply officers—no less than many other Northerners—were disturbed by this evident trend. In response, some of them waged a campaign to discard the formal contracting procedures that seemed to be preventing small producers from selling directly to the giant wartime state. In at least one large part of the North's military economy, the one that provided the army with horses and mules and their feed, this campaign resulted in a partial victory for the critics of contracting. On the whole, however, the North continued to supply its armies by negotiating formal contracts, often with merchants and manufacturers whose business operations were on a scale that put them well outside the circle of small producers long idealized by many Americans. In the end, the war carried an economic logic of its own, which could overwhelm the intentions of even the most influential managers of the North's procurement project.

The Life of a Civil War Contract

In 1862, after the various Northern states had given up independent procurement authority to the U.S. bureaus and the chaos of the initial war emergency had been tamed, the Quartermaster's Department settled into a contracting routine. In the bureau's national depot system, there were three main procurement depots: Philadelphia, New York City, and Cincinnati. By the summer of 1862, the chief quartermasters at each of these three depots were following the standing orders of Quartermaster General Meigs to accumulate and maintain enough goods to outfit 100,000 new troops and maintain another 200,000 already in the field. In October of the following year, after President Lincoln called for 300,000 more troops, Meigs raised the depots' target for supplying new troops to 150,000 soldiers each. This decentralized system served as the foundation of supply for the North's million-man army.[1] Meanwhile, Saint Louis, Washington, and Louisville emerged as three more leading procurement depots, where quartermasters bought and warehoused huge numbers of horses and mules, tons of grain and hay, and other goods. Together, the six largest depots handled the bulk of Northern contracting (see fig. 3.1).[2]

Civil War contracting was governed by federal law and military regulations, which had developed over several decades and had been revised as late as 1861. Under normal circumstances, U.S. procurement officers were required to circulate advertisements for proposals. The laws and regulations did allow that, in times of "public exigency," when goods were required immediately, the advertising requirement could be dropped and the officer could make an "open market" purchase, as private parties might.³ But Quartermaster General Meigs encouraged his depot chiefs to advertise whenever possible. In June 1861, when virtually any purchase might have been seen to fall under the "exigency" created by the new war emergency, Meigs reminded the senior depot quartermaster in Philadelphia that recent congressional statutes required advance notice. "[M]ake public advertisement and contract to lowest bidder whenever time permits," Meigs told Colonel Charles Thomas. "[O]ne two or three days even of advertisement," he continued, "are better than none."⁴

Advertising, which was supposed to increase competition by encouraging multiple bids, occurred at the beginning of the contracting process. In their newspaper advertisements, quartermasters called on potential suppliers to submit physical samples of the goods they proposed to supply, along with a per-unit price. After receiving the sealed proposals, a quartermaster typically required all bidders to gather in his office for the opening of the bids. In June 1861, after Thomas placed an unusually large advertisement for a variety of supplies, he collected a total of 425 bids, including as many as 50 for a single class of goods. By the time these bids were opened and contracts awarded at Thomas's Philadelphia offices, six hours had passed. Even relatively narrow advertisements, which were more typical, commonly attracted many potential contractors. By the middle of the war, depot chief Thomas Swords in Cincinnati was often surrounded by close to twenty bidders as he opened and read their sealed proposals. Typically, a quartermaster's inspectors attended this ceremony so that bids accompanied by unacceptable samples could be rejected immediately. Bidders who could not furnish or promise acceptable financial sureties or bonds were also rejected. At this point, the regulations required that the contract be awarded to the "lowest responsible bidder," whose "responsibility" was measured by the procurement officer on the basis of the bidder's proffered bonds and evident capacity to fill the contract.⁵

In practice, this process could be complicated. Because many advertisements allowed proposals that offered to furnish a fraction of the total quantity required, the full orders described in a single advertisement were often divided among several contractors. In one ordinary advertisement for eleven hundred cavalry horses (which were to be six to eight years old, fifteen to sixteen hands high, and dark colors) in April 1863, for example, Indianapolis depot chief quartermaster James Ekin stated that bidders must offer to furnish a minimum of one hundred animals. Ekin collected proposals from a total of eleven bidders. Upon opening the sealed bids in the presence of the bidders, the quartermaster found that the proposed prices per animal ranged from $106.45 to $127.00. But the low bidders, Foudray & Evans, had staggered their bids along an ascending price scale by lots of 100 animals, so Ekin gave them a contract for 100 horses at $106.45 and 100 more at $106.95. Ekin then signed a different contract for the remaining 900 horses with J. M. Moorhead, who had bid the single price of $107.50 for all 1,100 animals. During the war, such divided orders occurred hundreds of times at supply depots across the North.[6]

The path from bid opening to contract signing could also be complicated by the inspection of samples. Although most Quartermaster's Department depots offered prospective contractors both written specifications and material models for the goods required, there could be considerable variations in the quality of contractors' samples. John Dickerson, chief quartermaster at the large clothing and equipage depot in Cincinnati, used the bid-opening ceremony to conduct a public analysis of samples. Together with his chief inspectors, Dickerson weighed the samples, pulled at them to gauge the strength of materials, and treated them with acid to test colorfastness.[7]

At this stage of bid opening and reviewing of samples, depot quartermasters could find any number of reasons to reject a proposal. The process could frustrate prospective contractors. In one small transaction at the Philadelphia depot in May 1863, for instance, quartermasters called for proposals to furnish eleven hundred wagon covers made of ten-ounce "army standard cotton or linen duck." When the eleven bids were opened and the samples viewed, however, the Philadelphia inspectors decided that linen duck would not be suitable after all. In the end, the entire $13,000 contract went to Henry Simons, a leading Philadelphia military contractor who had promised covers made of cotton duck. The several bidders who had underbid Simons with proposals for linen covers doubtless wondered why the army had wasted their time by mentioning linen duck in the original advertisement.[8]

If a prospective supplier succeeded in winning a contract, he hurried to make the necessary arrangements—which might include purchases, the hiring of facilities and labor, or subcontracts—required to complete the order and begin deliveries to the depot. The speed at which he had to work depended on the terms named in the contract. In the case of Ekin's April 1863 horse transaction, the advertisement and contract made it clear that a third of the lot was to be delivered every ten days, meaning that the whole order had to be filled within one month.[9] Other contracts might demand even faster deliveries, especially if the Quartermaster General's office in Washington or the staff of a commanding general in the field had made a special request for quick action. Most routine contracts for clothing and equipage, on the other hand, called for a delivery schedule that could easily stretch over two or three months.[10]

It might be assumed that, once the final delivery was completed and the goods passed inspection, a contractor could simply collect a cash payment and calculate his profits. But nothing could be further from the truth. Payment for Civil War contracts was only occasionally made directly in specie or greenbacks. Normally, for contractors and procurement officers alike, the financial side of the Civil War supply business revolved around paper instruments known as quartermasters' vouchers and certificates of indebtedness.

Despite their importance in the contracting business, these vouchers and certificates are not among the best known of the new financial instruments introduced during the Civil War. Since the Mexican War, the United States had relied upon the "Independent Treasury" system, which required a strict separation of national government funds from banks and required U.S. officers to use silver or gold—or "specie"—for all transactions. This system was quickly overwhelmed by the Civil War. Although eastern banks lent the government $140 million in the fall of 1861, by the end of the year many banks suspended specie payments. In February 1862, Congress authorized the issue of $150 million in the national paper money that would become known as "greenbacks." At the same time, the Treasury issued $500 million in new 6 percent bonds. These bonds, most of which would be sold to Northerners through the remarkable sales network of Philadelphia banker Jay Cooke, were the famous "five-twenties," so named because they matured in twenty years but could be redeemed by the government in as little as five years. As the war went on, Congress authorized the issue of more greenbacks and bonds; it also created the first federal income tax and a new system of national banks.[11]

The greenbacks and bonds, along with the income tax and national banks, lasted long after 1865 and have been well remembered as important parts of Civil War history. Equally important for Northern contractors during the war itself, however, were shorter-lived instruments such as procurement officers' vouchers and Treasury-issued certificates of indebtedness. During the early months of the conflict, senior quartermasters quickly found themselves short of funds to pay their enormous debts to suppliers. In October 1861, Cincinnati depot chief quartermaster John Dickerson explained to Quartermaster General Meigs in Washington that he had put up a sign reading "No Money" in his office "to save myself the trouble of answering 'Have you any money' fifty times a day or oftener." Dickerson was hardly alone. Soon after taking over the Saint Louis depot in the fall of 1861, Quartermaster Robert Allen was forced to tell leading contractors such as Philadelphia wagonmakers Wilson, Childs & Company that he could not pay them for the goods they had delivered. At the beginning of 1862, Allen informed Meigs, "I am carrying on the current business almost entirely on credit."[12]

Quartermasters continued to be short of funds throughout the war, frequently forcing suppliers to wait several months for their pay. For contractors, who had to pay workers and their own subcontractors and suppliers, this could mean serious cash flow problems. But quartermasters and contractors quickly found a workable (if imperfect) solution to these difficulties. When they received deliveries, quartermasters issued the contractor an official voucher, confirming that the goods had been accepted and that pay from the Treasury would be forthcoming. By the beginning of the war's second year, if not before, quartermasters issued these vouchers as a matter of course. Testifying in October 1862, Allen explained that it was the "constant practice" of quartermasters to give vouchers to contractors awaiting payment. Contractors could then raise money with these vouchers by exchanging them for cash (at a discount off their face value) with a broker or bank. A year into the war, Allen noted that "vouchers have become so much a matter of currency that they pass from hand to hand by endorsement."[13]

Evidence of the use of these vouchers may be found in the records of Whitaker & Sons, a Philadelphia woolens mill that filled some Quartermaster's Department contracts for blankets. As the company delivered blankets to the Cincinnati depot during the last months of the war, Quartermaster William McKim issued them a separate voucher for each lot. Like many depot quartermasters, McKim withheld 10 percent of the value of each delivery until the contract was complete. By the eve of their final delivery, Whitaker & Sons had already received vouchers for nearly 90 percent of the total value of the contract. Like many contractors, Whitaker & Sons used these vouchers to raise cash to pay their workers and suppliers. During the two months before their blanket contract was completed in April 1865, the company sold (to a bank or broker) nearly all of McKim's vouchers at discounts of 5–7 percent below their face value. Before the final delivery, Whitaker & Sons had used the vouchers to generate $45,000 in cash on a contract with a total nominal value of $55,000.[14]

Contractors took a great interest in the discounts on quartermasters' vouchers, which could vary from day to day, depending on expectations about financial and military developments in Washington and in the field. The steeper the discount, contractors knew, the less profitable the contract. Although future discounts were impossible to predict, prospective suppliers factored voucher discounts into their bids, inflating the prices paid by the army. Such was the cost to the United States of failing to offer immediate cash payments. For many American merchants and bankers, who were accustomed to transactions involving credit and who routinely handled and discounted a bewildering array of commercial paper instruments, the quartermasters' vouchers cannot have been especially troublesome in and of themselves.[15] Nevertheless, the value of vouchers could change quickly in the North's unstable wartime financial markets. At certain times and places, discounts on vouchers were steep. In February 1862 in Saint Louis, for example, Assistant Secretary of War Thomas A. Scott (the Pennsylvania Railroad Company executive turned war organizer) reported back to Stanton that bidders on army contracts were adding as much as 15 percent to their prices because they saw quartermasters' vouchers being discounted at that rate in local markets. Three years later, when the flow of Treasury funds was more stable, the Quartermaster's Department found that the use of vouchers instead of cash was adding 8 percent to the price of horse and mule feed delivered to eastern depots.[16]

Contractors realized that the discounts on quartermasters' vouchers varied not only by time and place but also according to the identity of the individual officer who issued the paper. This meant that, even within the Quartermaster's Department, each individual disbursing officer effectively had his own credit rating in the private sector, just as he had his own account with the Treasury. Differences in individual officers' discounts might come from a variety of sources, such as the directness of his connection to the Treasury (with the chiefs of the major depots generally having the most direct access to large disbursements), his foresight in requisitioning sufficient funds, or his payment practices. In 1863, the vouchers of Cincinnati chief quartermaster John Dickerson were often discounted at the relatively low rate of 2 or 2.5 percent, while those of some more junior quartermasters in the region—including Francis Hurtt, a volunteer captain in Cincinnati who was apparently unpopular with local farmers—faced discounts at least 1.5 points higher. When a Cincinnati newspaper reported in January 1864 that discounts of under 4 percent were enjoyed by quartermasters' "vouchers of recent local issue, and from under the hand of prompt officers," it suggested the extent to which contractors and financial markets drew distinctions among individual officers.[17]

The financial complexities associated with contracting continued when quartermasters' vouchers were exchanged for final payment at the depot, which occurred when the contracting officer had received enough Treasury disbursements to his account to cover the debt—a date that often fell at least two to three months after final delivery. Even now, however, most final payments were not made in greenbacks. Instead, they came in the form of certificates of indebtedness, a new Treasury instrument created especially for military contracting.

"A very important financial measure adopted by Congress last week," the *New York Tribune* advised its readers on 4 March 1862, "has not attracted the attention it deserves." The *Tribune* was referring to the act passed by Congress three days before, which authorized the Treasury to issue—without any set limits—new "certificates of indebtedness," special one-year notes that carried 6 percent interest. Issued in amounts of one thousand dollars or more, these certificates were designed to serve as payment for military contracts. The *Tribune* explained why these certificates promised to ease contractors' financial burdens: "Capitalists have been shy of lending money upon simply approved claims and large shaves have been paid by public creditors to obtain facilities for carrying out their contracts, but with these certificates they can go into the money market to make negotiations upon the footing of the most favored borrowers."[18] More than the vouchers issued by individual supply officers, in other words, these Treasury certificates inspired the kind of confidence that allowed contractors to raise cash easily.

Unlike vouchers, certificates were not issued until after the completion of a contract. But they were still useful—and indeed quickly became the most important medium of payment for Civil War contracts—because they flowed much more quickly and easily from Treasury to depot than did scarce specie or greenbacks. Thomas Swords, the Cincinnati depot chief, paid out his entire first batch of certificates within two days after receiving them in early April 1862. He immediately wrote Meigs to ask for another $1 million in certificates. "The public creditors prefer to take certificates," Swords reported, "rather than wait under the uncertainty of their accounts being paid in other funds."[19] From the spring of 1862 to the end of the war, U.S. supply officers paid most of their debts to contractors in certificates. Treasury disbursements to depot quartermasters typically consisted of a small amount in greenbacks or other "currency" (with which the officers could pay employees or small contractors) and a larger amount of certificates. When Cincinnati depot quartermaster John Dickerson submitted his monthly request for funds in May 1863, for example, he asked for $750,000 in currency and $3 million in certificates. A year later, a routine disbursement to Saint Louis depot chief William Myers consisted of $2.8 million in certificates and only $85,000 in currency.[20]

Although they were less celebrated than greenbacks and five-twenties, certificates of indebtedness comprised a significant fraction of the North's war debt. From March 1862 through June 1865, the Treasury issued a total of $507.6 million in certificates of indebtedness, slightly more than its issues of greenbacks over the same period. Certificates accounted for 13 percent of the total value of all Northern financial instruments issued during the war (including greenbacks, bonds, and other Treasury notes), but because they were paid one year after the dates of issue, certificates absorbed fully 31 percent of all U.S.

funds used to redeem public debt during the war itself. By the last year of the war, the value of outstanding certificates (which were constantly being issued and redeemed) stood at approximately $250 million. Given that the entire annual U.S. budget just before the Civil War had come to only about $65 million, this was a staggering sum. And yet, as short-term debt, it disappeared quickly: by 1866, when greenbacks and five-twenties were still ubiquitous, virtually all the certificates had been retired.[21]

As they did with vouchers, contractors used certificates to raise cash to pay workers and suppliers. This meant that certificates quickly found their way to banks. Even if a contractor held his vouchers until he could exchange them for certificates, he probably could not afford to wait a full year for the certificates to mature. Instead, he would sell the certificates for cash at a discount. (When they were first issued in the spring of 1862, certificates were discounted in New York by 3% or 4%; over the next three years, the discount was often higher. Quartermasters' vouchers, which could usually be exchanged for certificates within two or three months, were subjected to discounts at least 1% above those for certificates.)[22] Thus, in many cases, it was banks rather than contractors who awaited final payment. As Henry D. Cooke, brother of the North's champion bond salesman, explained at the end of the war, "certificates are mostly held by banks." The Cooke brothers' banking house had been buying millions of dollars worth of vouchers and certificates since 1862, in some cases at the urging of the Treasury itself. Across the North, banks held considerable stocks of vouchers and certificates. As early as January 1863, banks in New York state held at least $17.5 million worth of certificates, while their counterparts in Massachusetts held $7.7 million worth.[23] By the end of the war, the First National Bank in Nashville alone held more than $1 million worth of vouchers.[24]

As the use of vouchers and certificates suggests, a Civil War contract was not a simple transaction. To be sure, doing business in the peacetime commercial economy could be equally complex. But the unique difficulties associated with contracting could discourage some firms from doing business with the government. The less comfortable a given firm with large credit transactions and banking, for instance, the less that firm would likely be attracted to the contracting business. No less than major long-distance commercial transactions in the peacetime economy, war contracts could require significant amounts of knowledge, credit, and time. Although dozens of small firms still managed to do business directly with the military, the administrative and financial burdens associated with war supply tended to push much of the business of prime contracting toward those firms most comfortable with large-scale production, distribution, and finance.

Civil War Supply as a Big Business

When students of American history think about the timing of the emergence of big business in America, they often point to the Gilded Age. It was not until the 1870s and 1880s, after all, that truly giant industrial concerns were established by men like John D. Rockefeller, Andrew Carnegie, and Gustavus Swift. There is nothing wrong with such a chronology of American economic development—so long as it does not obscure the considerable transformations wrought by industrialization and commercialization during the earlier part of the nineteenth century. Although the Union's military purchasing system may have favored large-scale firms by placing bulk orders and negotiating large formal contracts, it rarely had to invent them. By 1861, the North was already home to many factories full of expensive machinery operated by dozens or hundreds of workers. Equally important, its economy contained large wholesaling firms that were accustomed to distributing large quantities of goods and undertaking the financial operations required to do so. For the most part, it was these large-scale manufacturers and merchants—unusually big businesses in their day—that signed and filled the prime contracts that sustained the Union armies. Although many Northerners inside and outside of the military establishment preferred a war economy in which small farmers or manufacturers could deal directly with the national state, in practice the supply system depended largely on sizable factories and wholesalers.[25]

In some important military supply industries, a handful of large-scale firms had come to dominate their trades even before the Civil War. In these parts of the supply system, procurement officers saw little reason to advertise in newspapers for bids, because there were so few firms that had the capacity to fill a major contract. In the case of horseshoes, for example, the Quartermaster's Department turned to the nation's leading mass producer of these items, Henry Burden & Company of Troy, New York. A pioneering inventor of horseshoe-making machines, the Scottish-born Henry Burden during the antebellum decades had built a large facility that used a sixty-foot water wheel, one of the largest in the world; eventually, his business employed as many as fifteen hundred men to produce a variety of iron products.[26] During the Civil War, Burden & Company "had a practical monopoly of the manufacture of horse and mule shoes," according to Quartermaster General Meigs. "[N]o other establishment," he informed Secretary of War Stanton at war's end, "has had the capacity of machinery or capital to supply the enormous quantities required for the use of the Army."[27] Before the end of 1861, Meigs's office in Washington had taken the unusual step of sending regular orders for tens of thousands of horse and mule shoes directly to Burden & Company, instead of having the depots procure them through the normal contracting process.[28] By the middle of the war, there were several potential rivals to Burden & Company, including the Gosnold Mills of New Bedford, Massachusetts, and Stone, Chisholm & Jones of Cleveland.[29] But Meigs's comments suggest that the vast majority of orders for horse and mule shoes—on which the army spent close to $5 million in four years—went to Burden & Company, the industry leader.

Another "practical monopoly" in the North, which did even more business with the army in dollar terms, was the Union India Rubber Company. In this case, the near-monopoly came from patent rights, rather than the firm's unique capacities. During the first months of the war, a few of the Northern states had purchased rubber blankets and ponchos for their soldiers, even though they had not been standard issue in the regular army to that point. But after Secretary of War Cameron ordered the Quartermaster's Department in November 1861 to begin procuring waterproof rubber ponchos or blankets for each Union soldier, the supply of these items, which were priced initially at about $2.50 apiece, became a truly big business.[30]

This trade was quickly dominated by a single firm, the Union India Rubber Company of New York and Connecticut, which owned exclusive rights for the manufacture of rubber goods for military purposes. Because the Civil War ended two months before the end of the seven-year extension in 1858 of the famous Goodyear patent, Union India Rubber was able to maintain a lucrative near-monopoly for the entire war. The company used the courts to maintain its dominance. In the summer of 1864, the Quartermaster's Department split some orders between Union India Rubber and a firm called Providence India Rubber, based in Rhode Island. On 26 July 1864, for example, Philadelphia depot chief George Crosman contracted with the president of Union India Rubber for 15,000 rubber blankets at $4.81 each, while the president of Providence India Rubber got an order for 50,000 at $4.00.[31] But later that year, Union India Rubber won a suit in federal court that forced the rival to give up military contracting and hand over any profits from military sales.[32] All in all, together with its subsidiary Phoenix India Rubber, Union India Rubber signed over $5.7 million worth of prime contracts during the 1862–64 calendar years, which accounted for fully two-thirds of the value of all contracts for goods made of rubber and gutta-percha (a natural rubber substitute) during that period.

Although the near-monopolies enjoyed by Burden & Company and Union India Rubber were exceptional, there were several important military supply industries in the North in which only a handful of firms acted as significant prime contractors. One of these was the gunpowder trade. In 1860, when most black powder was used not by the military but rather in mining and construction, U.S. census officials counted fifty-eight powder-making "establishments" across the country. However, just four firms—the E. I. Du Pont de Nemours Company in Delaware, the Hazard Powder Company in Connecticut, Laflin, Smith & Boies in New York, and the Oriental Powder Company in Maine—accounted for 69 percent of the industry's $3.2 million in annual output. During the Civil War, the Union's army and navy supply officials were able to rely upon these firms and a couple of others for their gunpowder needs. Du Pont and Hazard, which each employed about two hundred men at their mills, led the way by selling the army and navy close to $1 million worth of powder during each year of the war. As in the case of horseshoes, the military ordered from these suppliers directly at a given price, instead of trying to stimulate competition by advertising regularly for bids. Competition was further constrained because leading powder companies colluded during the war to fix prices. In this case, pre-existing concentration led easily into an even more concentrated war supply industry.[33]

A similar pattern prevailed in the field of heavy ordnance. To be sure, the North's heavy ordnance industry was never as concentrated as that of the South, where a single large firm—the Tredegar Iron Works in Richmond—had an effective near-monopoly.[34] But in the North, of the $17.6 million worth of big guns and projectiles purchased by the army's Ordnance Department between 1861 and 1866, two-thirds came from just four foundries: Robert P. Parrott & Company in Cold Spring, New York; the Fort Pitt Works outside Pittsburgh, Pennsylvania; Cyrus Alger & Company in Boston; and Hotchkiss & Sons in Connecticut. Long before the beginning of the Civil War, these leading suppliers were already well established in the business of heavy ordnance manufacture (see table B.2, in app. B).

The Fort Pitt foundry, for example, had started to fill military contracts in the 1820s; during the 1840s and 1850s, it was at the center of important technological changes in the production of the biggest guns (those with a bore of eight inches or more). In 1847, army lieutenant Thomas J. Rodman, an ordnance officer, had patented a new cannon-making process in which iron guns were cast hollow and cooled from the inside out, making them stronger. Between 1849 and 1860, the Fort Pitt Works experimented successfully with the Rodman method, which the War Department endorsed on the eve of the Civil War as its new standard for big smooth-bore guns. When war broke out in early 1861, Fort Pitt proprietor Charles Knap—who had purchased the rights to the Rodman patent—hired dozens of new workers and invested $240,000 on new equipment, including furnaces and steam-driven cranes. After having produced about $125,000 worth of cannon and other goods in 1860, Knap's company sold nearly $3 million worth of guns and projectiles to the Ordnance Department during the Civil War. Although several other Northern foundries made big smooth-bore guns with the Rodman method during the war, the Fort Pitt Works used its antebellum experience to remain the leader in this field.[35]

An even larger gun-making operation, which supplied the army with nearly $5 million worth of heavy ordnance, was the foundry near West Point overseen by Robert Parrott, a former army officer. Although he eventually produced some giant weapons, Parrott specialized at first in the production of smaller rifled field guns, distinctive for their wrought-iron reinforcing band. Rifled guns used elongated projectiles that could spin like small-arms ammunition, increasing their accuracy. To turn out more than fifteen hundred rifled iron guns and some 3 million projectiles during the war, Parrott tripled his prewar work force of 350 men. Further east, the Alger family's well-established South Boston Iron Works (which had been filling military contracts since the 1830s) and Hotchkiss & Sons in Connecticut each employed several hundred men to make guns and projectiles. Hotchkiss, the North's fourth-leading supplier of heavy ordnance in dollar terms, specialized in the production of elongated projectiles for rifled guns with three-inch bores, such as those built by Parrott's West Point Foundry and the Phoenix Iron Company, located outside Philadelphia. Although the army did buy heavy ordnance from other firms, it was the four leaders that kept the Union armies supplied with most of the artillery and projectiles they used.[36]

Unlike the gunpowder and heavy ordnance industries, the American wheeled vehicle industry was not particularly concentrated before the Civil War. In this part of the North's military economy, it became evident that the army's preference for large quantities of standardized goods could promote concentration in less concentrated industries. Although there were over 3,000 manufacturers of wagons and carts in America in 1860, most of the 40,000 or 50,000 wagons and ambulances purchased during the Civil War by the army came from a handful of large, industrialized firms in Philadelphia and Cincinnati.[37] To be sure, the Quartermaster's Department did purchase vehicles directly from as many as a few dozen firms all across the North, and dozens of other firms participated in this business as subcontractors providing spokes, hubs, hinges, and other parts.[38] Nevertheless, there can be no question that, in this sector, the business of military supply was remarkably concentrated. Army contract registers indicate that roughly half of the vehicles purchased by the Union were supplied by just three or four large manufacturers: Henry Simons & Company and Wilson, Childs & Company, both of Philadelphia, and the linked concerns of J. C. C. Holenshade and Jacob W. Holenshade in Cincinnati (see table B.3, in app. B).

In contrast to the vast majority of wheelwright shops in the United States, these leading wagon contractors were practicing large-scale industrial production before the beginning of the war. Significantly, all of them had also supplied wagons to the army during the 1850s. In Philadelphia, Henry Simons's company was using steam-driven saws and employing as many as 250 men even before the Civil War; by the fall of 1861, his work force had expanded by a hundred and the company—with the help of subcontractors for parts—was turning out as many as a dozen wagons per day. Wilson, Childs & Company, a similar Philadelphia firm, had sold hundreds of wagons to Southern plantations during the antebellum era; even before the Civil War, it had a six-acre facility with steam-powered machinery and a large lumber yard.[39] In Cincinnati, the other great center of wagon and ambulance manufacture, the Holenshades and other prospective contractors solicited large army orders from the beginning of the Civil War. Employing between three hundred and six hundred men at once, J. C. C. Holenshade's wagon-making operations became the North's leading supplier of vehicles.[40] Although army inspectors found that they did not necessarily outperform smaller suppliers in terms of meeting standard specifications,[41] these big prime contractors with industrialized production processes were best positioned to respond to military demand. With some fifteen thousand wagons in the field that required regular repair and replacement, the Union army was drawn to those suppliers who were closest to achieving the higher uniformity and lower costs associated with large-scale production.

Even in war supply industries in which prime contracts were spread somewhat more widely, the bulk of the goods used by Union forces came from unusually large mechanized factories and commercial firms. In the case of small arms, many of the North's top contractors—including Colt, Sharps, Remington, Burnside Rifle, the Providence Tool Company, and Alfred Jenks & Son—employed between five hundred and two thousand men each during the war years, making them unusually large manufacturing operations for this era (see table B.1, in app. B).[42] In the footwear industry, several leading contractors, such as T. & E. Batcheller and Kimball & Robinson in Massachusetts, owned factories that before 1861 were already employing several hundred men and women and steam-driven pegging machines to finish up to one thousand or two thousand pairs of boots and shoes per day. Another top contractor, Benedict, Hall & Company of New York City, was also a leading wholesaler and manufacturer during the 1850s. Although it was possible for less well-established firms to become leading army suppliers, in most cases they did so by creating highly mechanized operations. John Mundell and James Harmer in Philadelphia, for instance, apparently became leading contractors because they acquired McKay stitching machines (introduced in 1862) or other machine-sewing technologies—a good investment when most of the Quartermaster's Department depots in the East demanded sewed footwear. In Cincinnati, where the army depot bought mostly machine-pegged boots and shoes, the local wholesalers A. Simpkinson & Company and R. M. Pomeroy & Company supplied goods that came mainly from the New England factories (see table B.4, in app. B).[43]

As the success of footwear wholesalers in the military supply business suggests, the ranks of top Northern contractors were not limited to manufacturers. Instead, purely commercial firms or those that combined mercantile and manufacturing operations were among the very biggest of all prime contractors in dollar terms. Although this situation bothered the many Northerners who believed that the military economy should be built upon direct transactions between the army and small producers, it was not surprising to those who were familiar with the structure of American business in the mid-nineteenth century. During the antebellum years, many parts of the American economy had been coordinated largely by mercantile firms, which not only distributed goods but also supplied much of the credit required by manufacturers.[44] In the North's military economy, distribution and finance, as well as production, remained critical. Although small and medium-sized manufacturing firms served throughout the war as contractors and subcontractors, the North's leading suppliers in dollar terms were a handful of wholesaling enterprises that coordinated the work of thousands of women and men. In the field of military clothing and equipage, one of the biggest in the war economy, some leading suppliers became general-purpose contractors, providing several different sorts of goods. It was these giant clothing and equipage contractors, even more than railroads or arms manufacturers, who ranked as the North's biggest operators in the business of war supply.

One field of military contracting in which big mercantile firms predominated was the supply of woolen textiles and blankets, for which the wartime Quartermaster's Department spent over $50 million directly, not counting its much larger indirect consumption in the form of finished garments. Although the army signed prime contracts worth hundreds of thousands of dollars with medium-sized Philadelphia-area textile manufacturers such as Whitaker & Sons, James Lord Jr., and John Dobson, who employed only a few dozen people each, its main suppliers of woolen textiles were firms that had long specialized in the mercantile end of the business.[45] Several of these

Table 4.1 Leading Civil War Contractors with the U.S. Army Ordnance and Quartermaster's Departments

Supplier Name(s)	Location(s)	Item(s) Supplied	Nominal Value of Goods Supplied (in millions of dollars)
John T. Martin/Martin Bros./ John E. Hanford	New York, NY	clothing, textiles	14.8
C. W. & J. H. Freeland & Co./ Alvin Rose	Boston, MA	clothing, tents, blankets	8.7
Albert Jewett & Co./ George Chapin	New York, NY	textiles, tents, shoes	8.2
Theodore Polhemus Jr./Fox & Polhemus/T. J. Van Wyck	New York, NY	tents, cotton textiles	7.7
Benjamin Bullock's Sons	Philadelphia, PA	woolen textiles, blankets	7.6
Hanford & Browning/ Browning, Button & Co./ W. B. Button	New York, NY	clothing	6.7
A. & C. Slade/Slade, Smith, & Colby	Philadelphia, PA/ New York, NY	textiles, tents	6.5
Union India Rubber Co./ H. D. Hadden/Phoenix India Rubber Co.	New York, NY/ Naugatuck, CT	rubber blankets	5.7
John Boylan & James B. Boylan	Newark, NJ	clothing, textiles	5.7
Lewis, Boardman, & Wharton	Philadelphia, PA	textiles, clothing, tents	5.0
Robert P. Parrott	Cold Spring, NY	heavy ordnance	4.7
Colt's Patent Fire Arms Co.	Hartford, CT	small arms	4.7
W. C. Houston	Philadelphia, PA	textiles	4.6
Henry S. McComb	Wilmington, DE	tents, knapsacks, shoes	4.1
Naylor & Co.	New York, NY	imported small arms	3.8
Joseph Lee	New York, NY	clothing, tents	3.5
George W. Jones/Jones Bros.	Cincinnati, OH	textiles, blankets	3.4
Thomas F. Carhart/ George Opdyke	New York, NY	clothing	3.2
E. Tracy & Co.	Philadelphia, PA	clothing, haversacks	3.2
E. Remington & Sons	Ilion, NY	small arms	2.9
Charles Knap/Fort Pitt Foundry	Pittsburgh, PA	heavy ordnance	2.8
H. Boker & Co.	New York, NY	imported small arms	2.8
E. I. Du Pont de Nemours	Wilmington, DE	gunpowder	2.7
Cyrus Alger/South Boston Iron Works	Boston, MA	heavy ordnance	2.6
Sharps Rifle Manufacturing Co.	Hartford, CT	small arms	2.4

Sources: This table combines data from sources that are not strictly comparable. Data on small arms and heavy ordnance purchases from 1861 to 1867 are provided in a published source: "Ordnance Department," House Exec. Doc. 99, 40th Congr., 2nd Sess., ser. 1338. Figures for Quartermaster's Department contractors are from 1861–64 contract registers, most of which are unpublished and held in RG 92, National Archives. See appendices B and C. Du Pont's wartime sales to the army are provided in Harold B. Hancock and Norman B. Wilkinson, "A Manufacturer in Wartime: Du Pont, 1860–1865," *Business History Review* 40 (1966): 213–36.

leading Civil War contractors—including Benjamin Bullock's Sons & Company, W. C. Houston, Tredick Stokes & Company, and Joseph B. Hughes—had been central players in the Philadelphia wool and woolens trade during the 1840s and 1850s, when many of them had done at least a small military contracting business with the Quartermaster's Department.[46] Houston, for instance, had been a member of leading Philadelphia wool dealers H. Robinson & Company during the antebellum years; in 1861, he came out of retirement to fill more than $4.6 million worth of prime contracts for woolen textiles.[47] In the end, this veteran Philadelphia merchant signed contracts with a total dollar value that was roughly equal to that of the total sales of each of the North's leading weapons contractors, Parrott and Colt (see table B.5, in app. B).

The dominant firm in the supply of woolen textiles to the Union was undoubtedly Benjamin Bullock's Sons & Co., also of Philadelphia. Filling over $7.6 million worth of prime contracts for textiles and blankets, the Bullocks ranked among the army's top five suppliers in dollar terms. The operations of this firm, which had historically engaged in both distribution and production, suggest the impossibility of making absolute distinctions between merchants and manufacturers in the North's military economy. Before the firm's English-born founder Benjamin Bullock died in 1859, he and his sons had spent two decades selling wool that came from sheep raised in Ohio and other parts of the Midwest, but they also sold textiles manufactured in the Philadelphia region.[48] At the beginning of the Civil War, Bullock's sons probably used their well-established purchasing networks in the Midwest to buy large quantities of raw wool suitable for military goods. They also moved to buy up much of the output of manufacturers in and around Philadelphia. By the winter of 1862–63, they controlled the output of at least six woolen textile mills in the region; by the end of the war, this figure had reportedly reached fourteen mills, with a total of three thousand employees. Meanwhile, the Bullocks operated their own factory, located just west of Philadelphia in the town of Conshohocken, where some of its two hundred employees worked a night shift under the light of gas lamps. In 1864 alone, this factory used 675,000 pounds of wool in the production of textiles for their army contracts.[49]

The only army contractors in the North who saw even more military dollars than the Bullocks were the top suppliers of finished clothing and equipage. Although the Quartermaster's Department produced roughly a quarter of the garments and tents consumed by the Union armies in its own public enterprises, this left tens of millions of dollars in annual orders from private firms. Like the footwear trade, the ready-made clothing industry was already so competitive by 1861 that the army was able to purchase from a variety of well-established wholesaling merchant-manufacturers in several cities. In Cincinnati, two multifirm wartime alliances called Heidelbach, Kuhn & Rindskopf and Mack, Stadler, & Glaser each filled over $1 million in prime contracts during the first half of the war.[50] In Philadelphia, the well-established wholesale clothiers clustered in the city's sixth ward also filled large contracts. One of the city's top clothing contractors was Rockhill & Wilson, a thirty-year-old company that expanded during the war to employ as many as eight hundred men and twenty-eight hundred women to make uniforms.[51] In Boston, the leading contractor was the wholesale clothing firm led by brothers Charles W. and James H. Freeland, who had relocated to the city in 1860 after a decade in Worcester. Overcoming the absence of a major purchasing depot in Boston, C. W. Freeland & Company succeeded in winning more than $8.6 million in prime contracts for tents, blankets, and clothing before the end of 1864 (see table B.6, in app. B).[52]

The biggest of the clothing contractors, based in New York City, were associated enterprises led by John E. Hanford, William C. Browning, and John T. Martin, which together filled well over $20 million worth of prime contracts during the Civil War—easily qualifying them as the North's very biggest war suppliers. Hanford & Browning, established under that name in 1858, was the direct descendant of Lewis & Hanford, long one of the largest wholesale clothiers in New York City. In 1854, when it was known as Hanford & Company and had branches in Saint Louis, New Orleans, and Richmond, the company took in the thirty-eight-year-old John T. Martin as a partner. Martin, who had made a small fortune in the clothing business in Saint Louis during the 1830s and 1840s, invested $50,000 in Hanford & Company. Now living in a $30,000 mansion in Brooklyn Heights, Martin also had over $100,000 invested in his old clothing business, now run by his brothers.

During the Panic of 1857, Hanford & Company was forced into bankruptcy. But it emerged quickly in 1858 as Hanford & Browning, having gained a new partner in the person of William C. Browning, a twenty-five-year-old, second-generation New York clothier. Hit hard by the loss of its Southern trade at the onset of the war, the company moved into large-scale contracting by winning a $1.25 million order from the Quartermaster's Department in May 1861. In 1862, John E. Hanford left Browning to join Martin. By making extensive use of subcontracting for outwork, as well as some large in-house operations with dozens of sewing machines, Martin and Hanford employed directly and indirectly hundreds of men and several thousand seamstresses in the New York area. By the end of the war, the prime contracts in Martin's name alone accounted for over a fifth of all army purchases of garments. Meanwhile,

Browning took in new partners—including Worthington Button, who signed many contracts—and continued to fill roughly $1.5 million in clothing contracts per year through the end of the war.[53]

The ranks of the North's top army suppliers were filled out by a handful of firms who came, by the second half of the war, to serve as something like general contractors for clothing and equipage. C. W. Freeland & Company, the leading Boston wholesale clothiers, who also filled large contracts for tents, blankets, and textiles, was among these companies. Cotton textile merchants, who were among the biggest economic actors in the country during this period, also served as top military contractors.[54] Fox & Polhemus of New York City, among the nation's top distributors of cotton duck (the heavy canvas used in sails and tents), became the North's leading prime contractors for finished tents, as well as for duck used by the army's own tent-making facility in Cincinnati. In 1864 alone, Fox & Polhemus signed nearly $4.2 million worth of prime contracts for tents. Meanwhile, huge quantities of tents, textiles, and garments were supplied by two firms descended from Slade, Smith & Company, a Philadelphia firm whose partners had spent the 1850s marketing the products of large New England textile mills. By 1863, Jarvis and Calvin Slade had moved to New York, where they became large suppliers to the army of tents, duck, woolen textiles, and knapsacks. Back in Philadelphia, J. Frailey Smith invested $100,000 in a new partnership called Lewis, Boardman & Wharton; this company, which represented at least ten New England cotton mills, proceeded to fill large contracts for tents and clothing, as well as textiles. Together, these two descendants of Slade, Smith & Company—both acting as merchant-manufacturers to become something like general-purpose suppliers to the Quartermaster's Department—filled roughly $11.5 million worth of prime contracts during the war.[55]

One of the most important large-scale general contractors with the Quartermaster's Department was Henry S. McComb of Wilmington, Delaware, a former apprentice to a leather currier who had risen fast in the leather manufacturing business since entering it in the mid-1840s. McComb, who was just thirty-five years old when the Civil War began, had filled some small contracts for leather with Philadelphia depot quartermasters during the Mexican War and afterward. In 1855, he had inherited a small fortune upon the death in a carriage accident of his father-in-law, Charles Bush, a partner in Bush & Lobdell, the successful Wilmington manufacturer of railroad car wheels.[56] During the Civil War, when he filled over $4 million worth of prime contracts, McComb ranked as one of the most diversified and well connected of all the North's top suppliers. Before the end of 1861, he was already employing several hundred men and women in and around Wilmington to make tents and haversacks, above and beyond the predictable jump in his sales of leather to the army's Philadelphia depot.[57] Acquiring the necessary textiles and other materials and distributing them through many subcontractors to be sewn into finished goods in the homes of local residents, McComb became the North's top supplier of haversacks and knapsacks. In 1863, he ranked as the second-leading prime contractor for tents; he also continued to supply leather for the army's footwear manufactory in Philadelphia and filled some small contracts for shoes. To complete these contracts, McComb's wartime account books show, he engaged in large transactions with leading military contractors in Philadelphia, including the textile merchant firms of Tredick, Stokes & Company; Slade, Smith, & Company; and Lewis, Boardman, & Wharton.[58]

It would be easy, taking the American economy of the early twentieth century as a point of comparison, to characterize even the operations of top Northern contractors such as McComb as relatively small business enterprises and to conclude that most of the North's military supply sectors were highly competitive and decentralized. To many Americans living in the 1860s, however, the Civil War economy seemed remarkably concentrated. The top contractors in the North, which did over $1 million in annual sales to the army, qualified as some of the biggest businesses in the country during this era.

While some small shops and medium-sized manufacturers participated fully in the war economy as subcontractors or prime contractors, the army acquired most of what it needed by dealing directly with well-established big manufacturers, commercial firms accustomed to large-scale distribution and finance, and a few ambitious upstarts who were willing and able to undertake the heavy financial risks that ranked among the most troublesome aspects of the contracting business. While most Northern households saw less than five hundred dollars in income per year, top prime contractors routinely engaged in multiple transactions amounting to hundreds of thousands of dollars each. It was these leading prime contractors, more than the subcontractors they employed and much more than the average citizens of the North, who stood to lose or gain the most from the business of war. And although many Northerners believed that gains were certain once a contract was awarded, it is clear that prime contractors faced some real risks, which served to limit the participation of many smaller firms in the first place. Despite the risks that contractors faced, however, the large profits that many of them collected during the national emergency could only appear to many Northerners—including army supply officers—as nothing less than stunning.

Risks and Returns

In November 1861, when the debts of the North to its suppliers were beginning to mount, Quartermaster General Meigs responded to the requests for immediate Treasury funds from New York City depot chief David Vinton. "I know that injustice is done to well-deserving contractors, who had the right to expect cash" in return for their deliveries, he wrote. However, Meigs continued, "[m]any other injustices are the result of this war, and great as it is, it is one of the least, so long as there are found merchants, manufacturers, or capitalists who will take the risk of supplying this Department with clothing or other indispensable stores for the defence of the country, we must exert ourselves to obtain them."[59] As Meigs suggested, the Northern war effort would rely heavily on those firms willing and able to do business with the government in the face of the uncertainties associated with the war—especially those involving finance.

However large its potential rewards, the business of military supply did entail special demands and risks. Many of these risks were connected to the often substantial gap in time between contract award and final payment. Indeed, many potential suppliers apparently preferred to act as subcontractors, rather than prime contractors, because they did not want to wait an indefinite amount of time for pay in uncertain forms. William H. Stevens, a veteran Cincinnati livery stable owner, told a congressional committee in early 1862 that he had filled a few contracts himself early in the war but "had got out of money" while waiting for quartermasters to pay him. "Many will sell to contractors quicker than to the government," Stevens explained, "because their capital becomes absorbed by waiting upon the government." Another leading horse contractor, Charles Elleard of Saint Louis, testified that, during the first year of the war, "many persons preferred to sell to me for $105" per horse, even knowing that his contract price with the army was $119, because they knew that the army would not pay them promptly in cash.[60]

For many suppliers, the key to success or failure was betting correctly on the future military demand for certain goods. Those contractors or subcontractors who invested heavily in key commodities in 1861 or 1862, well before the end of the war and the peak of wartime inflation, might make considerable gains. To the entrepreneurs themselves, their profits could be justified by the risks they had taken under conditions of great uncertainty. As leading Philadelphia tent contractor William Cozens put it at the end of the war in describing his early purchases of cotton and linen duck, "Whilst others timidly hung back until the article was needed, I went boldly into the market and bought all that was for sale or would be for sale for months. Had this enormous stock fallen upon my hands, my disaster would have been great. It rose, and who will say that the gain was not fairly earned by the risk encountered?"[61] Many of Cozens's fellow Northerners would question this logic. But he correctly identified one important source of risk for himself and his fellow leading contractors: the chance that the war would end suddenly, leaving them holding huge stocks of goods no longer wanted.

The inspection process at the depots, in which some lots of goods or even entire deliveries might be rejected, also raised potential problems for prime contractors. Depending on their severity, inspectors could have an enormous influence over a contractor's profits. If corrupt, negligent, or simply too generous, they could do great harm to the Union. In some cases, inspectors were found to be insufficiently strict. In an internal report completed in October 1863, the War Department found that inspection standards at the central Quartermaster's Department depots were so uneven as to allow "favored contractors nominally the *lowest bidders,* to obtain in reality, the highest price."[62] On the other hand, from the point of view of many contractors, inspectors were often much too careful, making the cost of doing business with the government higher than normal. As its accounts were being audited by a government committee in early 1862, Boker & Company of New York City, a leading importer of small arms still waiting for $1 million in outstanding pay, argued in a letter to the War Department that, "in a business with the government, where we are subjected to the errors of judgment or caprices of inspectors, we are entitled to a larger margin of profits than when dealing with regular and well established firms."[63]

Even successful contractors found that rejection upon delivery was a real and costly possibility. Experienced gunpowder suppliers, including the industry-leading E. I. Du Pont de Nemours & Company, had certain lots of goods rejected by stringent Ordnance Department officers and inspectors throughout the war.[64] Many of the firms that would become the North's leading contractors, it seemed, had to persevere through significant rejections by inspectors. In mid-July 1861, Henry S. McComb received news from the Philadelphia depot that 210 of the last 221 tents delivered under his contract had been rejected; this did not prevent him, however, from winning many more large orders for tents later on.[65]

Apparently, many suppliers moved along a learning curve, even as they proceeded to fill individual contracts. In November 1864, William Whitaker & Sons signed a contract with Cincinnati chief quartermaster William W. McKim for 15,100 blankets. In the case of one early delivery, which reached Cincinnati in early December, 423 blankets out of a lot of 500 were rejected. According to the company's internal records, "Most of these blankets were rejected for being greasey. Some few were narrow and others had thin places in them." Despite its early problems with "greasey" blankets, Whitaker & Sons succeeded in overcoming inspection problems over time. Although fully 40 percent of the first 4,300 blankets it delivered were rejected by the Cincinnati depot, only 3 percent of the final 10,600 blankets failed inspection.⁶⁶

Even when the delivered goods passed inspection, prime contractors could become frustrated with the depots' inspection and payment rules. Kunkel, Hall & Company, a Philadelphia-based major supplier of clothing and textiles, complained to the Cincinnati depot in early 1864 about the standard practice among quartermasters of withholding 10 percent from the vouchers they issued, which was designed to encourage the timely completion of a contract. Kunkel, Hall & Company argued that, because all suppliers were already filing bonds with each contract to protect the army from nondelivery, the 10 percent rule was overkill. At that moment, the firm explained, they held $1 million worth of prime contracts, "which were taken at the very lowest prices and on which we make but a slight profit. The withholding of ten per cent from the vouchers—which you will issue to us—on these contracts, will practically compel us to loan the United States near one hundred thousand dollars until the completion of each contract."⁶⁷

Although some contractors may have exaggerated their difficulties in their correspondence with public officials, there can be little doubt that they risked substantial losses because of unexpected delays in payment or steep discounts on Treasury paper. In February 1862, a friend of the leading New York City clothing house of Seligman & Company wrote to Secretary of War Stanton that the firm was now waiting for roughly $1 million in pay, some of which was as much as four months overdue. Faced with the prospect of trading quartermasters' vouchers for cash at a discount of over 6 percent, he claimed, Seligman & Company was "pressed to the verge of failure."⁶⁸ A week earlier, Stanton had received an even more desperate plea from F. B. Loomis of New London, Connecticut, who had delivered $18,000 worth of woolen textiles to the Cincinnati depot in November 1861. Only half of his pay was in cash, Loomis told Stanton; the rest was in vouchers, which he could not use to pay workers. "My help who are poor," he explained "want their pay and must have it in order to maintain their families. My whole capital including what I have borrowed has been handed over to our Govt. in the shape of Army Goods." Loomis concluded, "I beg of you. . . . For God's sake help me at once."⁶⁹

If contractors had been paid entirely in gold, waiting several months to exchange vouchers for final payments would have been less of a burden.

Because they were paid in greenbacks and certificates of indebtedness, prime contractors had to consider the costs associated with inflation, which could cause the prices for labor and materials to rise before they had completed a given order (see app. A), and the discounts on vouchers and certificates. Naturally, contractors tried to figure these hidden costs into their initial bids.

As early as October 1861, the *New York Tribune* observed that "no one can afford to deal with the Government without securing a large margin to cover the delay and expense of getting paid."[70] Tyler, Stone & Company of Philadelphia, the army's leading prime contractor for coal, complained to Quartermaster General Meigs in early 1863 that they had recently been compelled to sell a batch of vouchers at a discount of 8 percent. If it were not paid faster, the company warned, it would be compelled to raise prices.[71] Similar dissatisfaction was expressed by Wall Street bankers Read, Drexel & Company, who told Stanton in June 1862 that they were having trouble with some $117,000 worth of Saint Louis depot vouchers they had "purchased in the usual course of business and forwarded to the Treasury Department for payment." The bankers had been unable to exchange the vouchers immediately for certificates because the Treasury had not yet advanced sufficient funds to the account of Saint Louis depot chief Robert Allen. "This delay causes us no little inconvenience," complained Read, Drexel & Company, adding, "we think it but proper to call your attention to the injury to the credit of the different Quartermasters that will result from the delay in payment of their checks."[72]

Because calculating the potential costs of inflation and discounting was an uncertain science, it was possible for contractors to underestimate these factors and take real losses. Benjamin Bullock's Sons & Company, the Union's leading supplier of woolen textiles, signed a major contract in August 1863 to deliver 1 million yards of kersey at the price of $0.795 a yard. This contract contained a special clause that allowed the United States to place a subsequent order for a half-million more yards of kersey at the same price. When the Quartermaster's Department exercised this clause in November, Pennsylvania state treasurer Henry D. Moore wrote the War Department on behalf of his friends. Rising prices for materials and labor, Moore explained, had caused the Bullocks to take heavy losses even on the initial contract. Given that the Bullocks ranked "among the most loyal and patriotic citizens of our state" and had been filling government orders faithfully since the first weeks of the war, Moore argued, the War Department should let them out of the contract's special clause "in the spirit of right and justice!"[73]

Supply officers believed that inflation could cause contractors to suffer real losses. Days after Moore wrote his letter to the War Department, across the country in Cincinnati, Quartermaster David McClung reported to Washington that, because of unexpected price increases in recent months, "[a]ll my grain contracts have been filled at a heavy loss to the contractors. The corn of L. & M. Stone [a leading forage contractor based in Cincinnati] cost them about 20 [cents per bushel] more than I paid them."[74]

For many of the North's leading suppliers, the risks of contracting seemed greatest in 1864, when the combination of growing inflation and major Union military setbacks in June at Cold Harbor and Kennesaw Mountain shook public confidence in the Treasury, even as the Treasury itself was having trouble filling War Department requisitions. By July, certificates of indebtedness were selling at only 93 percent of their par value, prompting the North's bond trader extraordinaire Jay Cooke to advise the Treasury to cut back on new issues, guarantee that all vouchers would be paid within sixty days, and purchase some of those certificates currently on the market.[75] The same month, several of the North's top contractors, including the Bullocks and Lewis, Boardman & Wharton, wrote Philadelphia chief quartermaster George Crosman to protest "the very serious loss we are daily forced to sustain on the vouchers issued by your department. . . . When we agreed to furnish the articles the vouchers issued by your department found ready sale from 1 to 1½ pr. ct. discount; now they can scarcely be sold at all, even at 10 pr. ct. discount, which, as you are aware would be a ruinous loss to many of us." Beyond this, the contractors explained, they were all losing money simply "on the *cost* of all the materials out of which we are furnishing the contracts." Some kind of immediate relief was necessary, the leading contractors argued, "both as an act of justice" to the North's suppliers and in order for the government to avoid huge price hikes on future contracts.[76]

During the crisis of the summer and fall of 1864, the Philadelphia suppliers were joined by many contractors across the country in concluding that it might not pay to do business with the government unless they changed the terms of their agreements drastically. According to a July letter to Meigs from Kunkel, Hall & Company, which sold to both the Philadelphia and the Cincinnati depots, Crosman's vouchers were virtually impossible to sell, even at a 10–12 percent discount, but "the Cincinnati vouchers cannot be sold at all, and we find it utterly out of the question to borrow money at any price, with Cincinnati vouchers as collateral." Pressing the War Department to get Treasury funds immediately, Kunkel, Hall & Company informed the Quartermaster General that, because of the recent severe depreciation of vouchers and certificates, "We in common with other manufacturers are suffering very heavily from losses" on a daily basis.[77] Such complaints came not only in letters from contractors to the government, but in businessmen's discussions among themselves. Days after Kunkel, Hall & Company wrote to Meigs, Augustus G. Hazard, one of the North's leading suppliers of gunpowder, wrote to his peers at Du Pont to suggest that they try to insist in all future contracts that they be paid in cash, since "we may and probably will lose from 8 to 10 per cent on certificates."[78] In August, William Brand—a New York City contractor who was a major supplier of tents and cotton duck—proposed to the Quartermaster's Department to sell 1.25 million yards of duck at 78 cents a yard in greenbacks, but only if this price were adjusted at the time of final payment to account for any variation in the gold-greenback ratio since the contract date.[79]

However dangerous it seemed at the time, the contractors' crisis of the summer and fall of 1864 was eventually contained, and those who weathered it were rewarded. By the end of the year, after President Lincoln was re-elected and General Sherman delivered the "Christmas gift" of Savannah, Georgia, the Union's war effort seemed on much surer footing than it had just a few months before. The worst of the inflation was now past, and Treasury bond prices were climbing back up. After the Union declared victory in April 1865, the leading contractors who constantly held large volumes of government paper—along with the many banks who did likewise—had little to fear. Indeed, many of them could now be confident of considerable profits. This was certainly true of John T. Martin, who was the North's biggest prime contractor. After the war, Martin's fellow New York wholesale clothier (and wartime contractor) William C. Browning recalled that "[t]he unsettled conditions, due to the prolongation of the war and the depreciation of the certificates with which the government paid, made the business [of prime contracting] one of many hazards." At war's end, however, Browning explained, Martin and a few others among "the larger and more responsible dealers, having faith in the government, reaped their reward in the reestablishment of credit and the corresponding appreciation of government certificates from seventy or eighty cents to par."[80] For leading contractors such as Martin, the large potential losses that had come from holding vouchers and certificates in 1864 were transformed the following year into big profits.

Even before the Union victory, many contractors succeeded in turning profits. Because detailed business records from the 1860s are rare, the exact extent of leading contractors' profits or losses is impossible to determine. Moreover, contractors may have taken in large sums without achieving extraordinary rates of return on their investments. By 1862, the North had begun an economic boom, as output and employment pulled out of the hard times that had been caused first by the Panic of 1857 and then by the closing off of the Southern trade in 1861. For those few Northerners fortunate enough to have large funds available, it was not terribly difficult to achieve real wartime returns of 5 or 6 percent a year (or more) by investing in government bonds or railroad stocks.[81] A leading army contractor who sunk $400,000 into war supply operations in early 1863 and achieved a net profit of $20,000 by the end of the year would have made a small fortune by the standards of the day but not an exceptional rate of return on his investment. Nevertheless, for many Northerners, wartime profits on this scale were disturbing.

As they followed the progress of the war in the newspapers, Northerners on the home front saw not only reports of recent battles and new casualties but also detailed income tax lists. These income tax data suggested that leading military suppliers managed to do well, despite the risks associated with contracting. The Treasury's Bureau of Internal Revenue, created by congressional legislation in 1862, was charged with collecting the nation's first federal income tax. Under the 1862 rules, annual incomes between $600 and $10,000 were taxed at the rate of 3 percent, while incomes above $10,000 were taxed at 5 percent. (The rates were raised in 1864 to create a three-tier system with rates of 5%, 7.5%, and 10%.)[82] By 1864, publishers in cities across the North had begun to print lists of local taxpayers and their incomes, in pamphlet form and in the pages of the daily newspapers.

Surveying the declared personal incomes of leading military contractors for the second half of the war, many Northerners may have concluded that the business of war supply was quite profitable. To be sure, many of those contractors had enjoyed high incomes and considerable wealth before 1861; furthermore, those firms that reached the top ranks of suppliers by doing repeat business with the army were probably more successful than the average contractor. Still, the tax lists suggest that the gains from large-scale contracting could be considerable. Many of the North's leading contractors easily reached the top tax bracket of annual incomes over $10,000 a year, giving them some of the highest incomes in the country. H. G. Hadden of New York, president of the Union India Rubber Company, was taxed on a personal income of $17,000 in 1863–64. The following year, John Childs and William Wilson, the senior partners in one of Philadelphia's two giant wagon-making concerns, had incomes of nearly $23,000 each. In Cincinnati, leading wagon contractor J. W. Holenshade took in $14,000 in 1863–64 and more than $25,000 the following year. John Simpkinson, the Cincinnati footwear wholesaler whose company did over $1.7 million in annual sales (about half to the military) in 1864, reported personal incomes of $16,000 and $29,000 during the last two years of the war.[83]

Leading suppliers of textiles and clothing and equipage, who did a larger business with the army in dollar terms than any other contractors, reported even higher wartime incomes In Philadelphia, top woolen textile contractors such as W. C. Houston and the Bullock brothers reported individual incomes for the last year of the war of close to $100,000 each, placing them at the very top of the city's tax lists. Some Philadelphia clothing contractors, including Joseph F. Page, William Anspach, and M. Hall Stanton, reported 1864–65 incomes of over $60,000. Three partners in the firm of Lewis, Boardman, & Wharton—which had led the group of top contractors in 1864 who complained of steep losses due to inflation and discounting—managed to take in nearly $110,000 in combined personal income during the war's last year. Henry S. McComb, the general war contractor extraordinaire from Wilmington, Delaware, averaged nearly $50,000 a year in declared personal income during the war years. As early as May 1863, in fact, McComb was able to buy $50,000 worth of the Treasury's five-twenty bonds. In contrast, for the first year after the war, McComb's income dipped below $5,000. A similar pattern held in the case of James H. Freeland, senior partner in C. W. Freeland & Company, the giant clothing and textile contractor in Boston. Freeland's personal income for the last year of the war was over $150,000, about twenty times that of his income the following year.[84]

Such information about the personal incomes of top Northern contractors does not show conclusively that the military supply business was exceptionally profitable. It does suggest, however, that many major suppliers succeeded in negotiating the risks associated with contracting well enough to reap sums of money that seemed immense to most Northerners. These rewards were not guaranteed and might have been erased altogether if either the Union armies or the U.S. Treasury had been less competent or less fortunate. In the eyes of many Northerners, however, the business of military supply seemed to enrich a small number of men, even as hundreds of thousands of less fortunate citizens risked their very lives to defend the Union. Seamstresses, as they attacked contractors and championed public enterprise, regularly condemned this evident injustice. Many army supply officers, as well, were troubled by the distribution of procurement dollars, which they saw as being skewed by a formal contracting system that worked against small farmers and other small-scale producers. By the second half of the war, in one of the very largest sectors of the Northern military economy, some of these officers sought to break free of the formal contracting system that was mandated by Congress.

The Campaign against Contracting

Northern procurement officers negotiated thousands of formal written contracts during the war. Required by federal law and military regulations, formal contracts were supposed to encourage competition and transparency in the procurement process. In many cases, however, supply officers concluded that contracting was unsatisfactory. Formal contracting, which required advance advertising and considerable paperwork, was slow and troublesome. By advertising the army's purchasing plans, formal contracting could encourage speculation and collusion that could inflate prices. And by encouraging supply officers to make a small number of bulk purchases, formal contracting could discourage small dealers from selling directly to the army. All of these problems with contracting troubled Northern quartermasters. By the middle of the war, some of them concluded that formal contracting should be scrapped altogether in certain large sectors of the war economy. Working against the letter and spirit of federal procurement law, quartermasters bought millions of dollars worth of horses and grain without signing formal contracts. They did so not simply to promote their own autonomy but also to shape the military economy in ways that they believed might favor small farmers and dealers while discouraging mercantile intermediaries they referred to as speculators and middlemen.

Even before quartermasters consciously turned away from formal contracting in markets for agricultural goods, many military transactions were made without advance advertising and official contracts. One important part of the North's war economy that dispensed with regular contracting was the field of railroad transportation. Led by Thomas A. Scott, a vice president of the Pennsylvania Railroad Company who joined the War Department as an assistant secretary, the army negotiated standard rates with the railroads. In July 1861, Scott provided the Quartermaster's Department with a rate schedule to be observed "as a general basis" in paying the railroads for troop and military freight transport. Railroads that had received federal land grants, such as the Illinois Central, were paid at rates one-third lower than these standard levels. In February 1862, soon after Congress authorized the War Department to take over any private railroad for military purposes, Secretary of War Stanton met with railroad directors to draft a new but similar set of standard rates. Although the army went on to run its own large U.S. Military Railroad networks in the upper South, existing Northern railroads remained private and were paid for their military work according to the standard rate schedule.[85]

The rate schedule system used for railroads was unique. But all sorts of other goods and services were also purchased by the army without formal contracts. According to standing laws and regulations, advanced advertising and contracting could be bypassed in times of "public exigency." During such emergencies, procurement officers were allowed to make "open market" purchases, such as those that might be made by private individuals or firms. Of course, such purchases were common in wartime, when some requisitions by commanding generals came on short notice. During the first months of the war, when the Union was scrambling to assemble large armies, open-market buying was especially common. When asked by a congressional committee during the winter of 1861–62 to describe their recent purchasing practices, top depot quartermasters admitted that they had often ordered goods directly from suppliers. David Vinton, the New York City depot chief, testified that he bought over 350,000 pairs of boots and shoes (worth roughly $700,000) in 1861 without advertising or drawing up official contracts. "I am aware," Vinton admitted, "that, as far as possible, everything should be contracted for; but such a course has been impracticable with myself at this depot."[86]

For Vinton and other quartermasters, the emergency of the initial war mobilization justified the use of open-market transactions. By the second year of the war, however, Washington called upon procurement officers across the North to confine such purchases to true emergencies. On 2 June 1862, Congress passed a new law requiring supply officers to make all contracts in writing and send copies of their advertisements and lists of bids to Washington. Although the War Department interpreted this law in a way that allowed some open-market purchasing to continue, it increasingly demanded formal contracting.[87] When the Philadelphia depot chief, George Crosman, asked permission in April 1863 to order wool textiles directly from firms that had supplied them successfully in the past, Quartermaster General Meigs scolded him. Only "public exigency," Meigs reminded Crosman, could override the requirement that he advertise for proposals and draw up formal contracts. "If it costs more to advertise," continued Meigs, "the law of the land has indicated the choice of the people. They prefer equal chance to all to compete, even if the cost be greater." To do otherwise, Meigs explained, tended to lead to public complaints about "monopoly" or preferential treatment for "favorite contractors." Meigs concluded his scathing rejection of Crosman's proposal by asking the Philadelphia quartermaster, "Why . . . do you embarrass the Dep't by recommending such propositions—clearly illegal?"[88]

While Meigs defended formal contracting, many depot quartermasters came to believe that it hampered their ability to best serve the interests of the North. These officers argued that, if open-market purchasing could be used for routine purchases as well as emergencies, they could create a military economy that was more efficient and more just. Crosman himself had suggested as much in March 1863, in a letter to former Secretary of War Simon Cameron. After Cameron recommended a certain supplier of footwear, Crosman explained that he could not order directly from any firm, however low their prices or fine their products. "The law and regulations, and my instructions," explained Crosman, "require that *all* purchases of army supplies must be on public advertisement, and I cannot, therefore, purchase in open market, even at a less price, except in an *emergency,* when immediate delivery is required."

Crosman and other depot quartermasters believed that these formal contracting requirements made their jobs more difficult and cost the North money. In his letter to Cameron, Crosman defended his ideas by explaining that, if supply officers were not compelled to contract, they "could, without publishing to the world the wants of the government, make purchases in the open market, and often save large sums of money to the Treasury." Crosman then suggested that, if the North would not give its procurement officers sufficient authority to best serve the public interest, it might be better off nationalizing all war industry: "Either all materials and supplies for the Army and Navy should be manufactured in establishments owned by the public and under the supervision of government officers, or else, the purchasing officers ought to be empowered to use their own discretion in obtaining these supplies; and held accountable for an honest and faithful execution of their duties."[89]

Besides arguing that procurement officers should be given more autonomy, Crosman's letter also pointed out that advertising could actually inflate prices by telegraphing the army's impending purchases. Many quartermasters believed that, when particularly large orders were advertised in advance, certain unpatriotic parties in the private sector would try to corner short-term markets or forge collusive agreements in an attempt to force the army to buy at high rates. Crosman and other top quartermasters thought that such schemes, inevitable in capitalist markets that were full of speculation, could be circumvented by the intelligent use of open-market purchasing. At the end of 1861, the Cincinnati depot chief, John Dickerson, admitted that he deliberately avoided advertising for the full quantity of goods he required. "I have made but few contracts for anything," he told a congressional committee. "If I want thirty thousand suits, I advertise for say five thousand suits. The lowest responsible bidder gets the amount called for by the advertisement, and the remainder wanted is distributed among the clothing houses which do the best work, at the same prices at which the award was made."[90] In other words, during the early months of the war Dickerson was using formal contracting merely as a mechanism for setting competitive prices. Eschewing the winner-take-all dynamic inherent in the contracting process, he preferred to spread orders among several competent firms.

Although Dickerson and other quartermasters were soon pressed by the War Department to make formal contracts for all of their regular purchases of clothing and equipage, open-market methods were used through much of the war to acquire grain and hay. Quartermasters regarded grain dealers as particularly prone to collusion and speculation. As late as September 1864, when he was supervising the purchase of 1.8 million bushels of grain a month for the Nashville and Saint Louis depots alone, Quartermaster Robert Allen wrote Meigs from Louisville to say that he routinely avoided contracting. "My course hitherto (which was suggested by you)," Allen reminded the quartermaster general, "had been to advertise for a small amount, and having thus established a rate purchase largely at that rate, through confidential agents. In this manner the Government never appeared to be largely in the market, and advance in prices was avoided." Allen worried that recent congressional legislation, which reiterated the statutory requirement of formal contracting, might force him to pay higher prices. "Now, if I advertise for eighteen hundred thousand bushels of grain, it will advance in price at least ten or fifteen per cent." Despite his defense of formal contracting in the case of textiles and clothing and equipage, the quartermaster general apparently approved more secretive purchasing schemes in the case of forage. Soon after receiving Allen's letter, Meigs urged War Department lawyers to interpret the new congressional legislation in a way that would support Allen's customary methods.[91]

Quartermasters across the North regarded markets as dangerous animals that were easily roused. They saw open-market purchasing, which allowed quicker and smaller transactions and hid the actual needs of the army, as an important method for keeping markets quiet. In August 1862, Quartermaster General Meigs himself confided to Secretary of War Stanton that, although he had "desired to procure a reasonable increase of supply without too much exciting the market," recent Union troop movements had signaled large impending purchases, driving up prices. The following year, David McClung, a volunteer quartermaster in Cincinnati, complained, "I have been compelled to buy forage on a scant, excited and constantly rising market. . . . I have to resort to every expedient to get any supply, and not to excite the market to ruinous prices." McClung's counterpart in Saint Louis, Quartermaster Edmund Chapman, reported in 1864 that any requirement to use formal contracting for grain procurement "will render it necessary to advertise for very large amounts, which will have a tendency to agitate the markets and raise the price." Chapman, who had purchased nearly $12 million in forage during the 1863–64 fiscal year, recommended that purchasing in smaller batches would keep prices lower.⁹²

One of the North's leading practitioners of quiet open-market purchases was Samuel L. Brown, a volunteer quartermaster from Chicago who rose to become the chief buyer of forage for Union armies in the East. During the second half of the war, Brown spent nearly $30 million for grain and hay, nearly all of it without formal contracting. Even more than his colleagues in the West, Brown was wary of wartime markets and regarded secrecy as essential. In late 1863, he went so far as to ask Washington to avoid announcing the disbursements being sent him at the office of the assistant treasurer in New York. Announcing the transfers, he explained, would "put the whole flock of speculators and dealers in information of my being in the market which for very prudential reasons I do not wish to do." Brown's open-market transactions were so secretive and so huge that they worried the quartermaster general. In March 1864, when Brown was requesting disbursements of $2 million a month in cash, Meigs warned him, "Your purchases without public notice go to the limits of the law, which requires contracts. . . . Can an emergency last three months?" But Brown continued to buy millions of bushels of forage without advertising. Later in 1864, he explained that, if he were to publicize his wants in advance, "dealers would become aware of the necessities of the Govt., combinations would be formed, and prices advanced largely; but by purchasing quietly from day to day no one is aware of the vast amounts used, and consequently prices do not advance."⁹³ By eschewing formal contracting, Brown and others believed, they could save the government money.

Another virtue of open-market purchasing, in the eyes of many supply officers, was that it could make it easier for them to buy from small dealers. Formal contracting, they believed, worked in several ways to encourage bulk orders that discouraged smaller firms from competing for prime contracts. From advertising to opening bids and samples to filing the final paperwork, contracting demanded a great deal of time and energy on the part of the officers themselves and their staffs. "It was a herculean task," David Vinton, the New York depot chief, complained less than a year into the war, "to arrange and collate the bids and decide who was the lowest bidder in many instances." When Congress passed its June 1862 legislation that seemed to demand that all purchases be made with formal contracts, procurement officers complained immediately that they typically made hundreds of small purchases a month. If forced to contract in all cases, they would have to consolidate these orders into a few bulk purchases from a small number of firms. The quartermaster general himself warned that a strict construction of the new legislation would lead the North toward a procurement system such as those that "prevailed in some monarchical governments, where great contractors, commanding millions of capital, make general contracts to furnish all supplies for the government under periodical lettings." Although the War Department quickly issued a general order informing officers that the new legislation would not in fact preclude open-market purchasing, quartermasters' concerns about the consolidation of purchasing continued.[94]

For quartermasters and many civilians, open-market purchasing seemed to be especially superior to formal contracting in the procurement of horses and mules—one of the most expensive parts of the entire war economy. During the conflict, the army purchased a minimum of roughly 640,000 horses and 300,000 mules, or close to one million animals in all.[95] Together they cost the army well over $100 million.

Although thousands of Northern farmers owned small numbers of horses and mules, the giant scale of military demand together with the formal contracting system evidently pushed the army to buy animals in bulk from men with particularly deep pockets, good bank credit, or high tolerance for risk. In one typical horse contract, signed in April 1863, Quartermaster Stewart Van Vliet in New York City advertised for 2,000 horses, stipulating that bidders must propose to supply a minimum of 500 animals. After receiving a total of twenty-two bids, Van Vliet awarded the entire order at a price of $111.75 per animal to the low bidder, Asa D. Wood of Brooklyn.[96] To fill the contract, Wood would have to spend roughly $200,000 in a matter of a few weeks—not an operation that was easily contemplated by average farmers.

As Van Vliet's 1863 contract with Asa Wood suggests, depot quartermasters rarely purchased directly from small farmers in counties dozens of miles away. Pushed by the contracting system to consolidate their orders, quartermasters came to rely heavily on a small number of suppliers who used extensive subcontracting networks to tap stocks of animals located far and wide, especially in leading horse-producing states such as Pennsylvania, Ohio, Kentucky, Indiana, and Illinois. Although quartermasters continued to buy from smaller dealers and farmers throughout the war, by 1863 a handful of big contractors were supplying a large fraction of the Union's horses and mules. By the autumn of that year, for example, Edward A. Smith of Cincinnati and his associates had already sold nearly fifty thousand animals to the army, for roughly $4.5 million. To finance these operations, Smith had borrowed some $2.5 million from the Commercial Bank of Cincinnati, along with nearly $2 million from other midwestern banks. Smith and his partners used this money to buy animals from dozens of farmers and dealers scattered around eastern Indiana, western Ohio, and northern Kentucky. His net profits on these army orders, Smith claimed, were about $70,000 (a margin of under 2%). One of the North's leading contractors during the first half of the war, Smith was no ordinary farmer.[97]

Smith's peers among the North's leading horse and mule suppliers also did business with the army on a huge scale. William Calder, the leading horse and mule contractor in Pennsylvania, was a forty-year-old stagecoach company operator and investor from Harrisburg who already had considerable land holdings when the war began. But to fill military contracts, Calder went far beyond his own farms. During the first year of the war, when he delivered about 8,000 animals to the army, he relied on subcontractors across Pennsylvania and in other states. "I employed everybody to fill my contracts," he told a congressional committee. As the war went on, so did Calder's large operations. According to a short biography published after the war, Calder claimed to have sold the Union a total of 42,000 horses and 67,000 mules—which would have cost about $13 million.[98] Even if these figures were somewhat exaggerated, they suggest the giant scale of operations of the North's top horse contractors. In Saint Louis, Charles M. Elleard and Robert W. Peay formed a wartime partnership that sold the army about 30,000 animals—worth over $3 million—by the spring of 1864.[99] A handful of other dealers, including J. M. Moorhead of Indianapolis and John Spicer of Chicago, also supplied close to $2 million worth of animals for the Union armies by purchasing from small farmers and stock dealers across the Midwest.[100]

The evident inequities in the distribution of prime contracts for animals troubled many Northerners, who believed that buying these animals with formal contracts was a mistake. In 1863, a farmer from Illinois articulated this view in a letter to a Saint Louis newspaper, in which he noted that Saint Louis quartermasters had advertised recently for several hundred horses or mules at once. If they would simply advertise for smaller lots, he advised, they would attract more bidders and higher quality animals. Most farmers and small dealers could not contemplate signing a contract for hundreds of animals. But they might well be interested in selling the army ten or twenty. Under the present system, they could participate only by serving as subcontractors.[101] Another critic of contracting for horses was James M. Hopkins of Lancaster, Pennsylvania. In May 1862, Hopkins wrote to Secretary of War Stanton to "do away with the contract system." Contracting, explained Hopkins, favored "professional horse jockeys, whose whole aim is to victimize the government, as well as the Farmers."[102] Over a year later, Hopkins wrote Stanton again, informing him that many farmers were selling horses at prices as low as $80 to contractors who sold large numbers of horses to the army at a contract price of $145. In both letters, Hopkins argued that the army should create a more extensive purchasing network of its own, which could buy directly from smaller dealers. After Stanton referred this letter to the Quartermaster's Department, Assistant Quartermaster General Charles Thomas replied that he agreed with much of what Hopkins had to say. But because the laws demanded contracting, Thomas continued, the army's hands were tied.[103]

In early 1864, Quartermaster General Meigs responded directly to the complaints about the army's failure to buy agricultural products directly from farmers. Meigs's comments were inspired by a communication from U.S. Senator Henry Wilson of Massachusetts, the chairman of the Senate's committee on military affairs, who had forwarded a letter from a Northerner who suggested "that the United States should purchase its forage directly from the farmers, and thus avoid all the expense of intermediate agents." Meigs offered two distinct replies to this suggestion. First, he challenged the assumption that it was always more expensive to the army to buy from "intermediate agents" rather than original producers. To establish a more extensive public purchasing network, Meigs argued, would entail considerable costs.

Like many of his fellow procurement officers, however, Meigs did not simply endorse the contracting system. Instead, his February 1864 letter to Wilson expressed considerable anxiety about the consolidation of the business of war in the North. The critics of contracting, Meigs admitted, were right to complain that a small number of firms were dominating the war supply business. The quartermaster general explained that the depot system, which provided several major purchasing sites across the country, was designed to combat the problem of consolidation. "The evil of dealing too much with large dealers," he explained, "I have attempted to mitigate by establishing as many depots for contracting for supplies as possible. I fear that there is now too great a disposition to run into centralization, to require too many of the contracts to be made at the centre of political influence and intrigue" in Washington. "France and Russia," Meigs continued, "are centralized governments; ours should, I think, distribute its action as widely as possible."[104]

But Meigs knew that even the relatively decentralized depot system did not eliminate the "evil" of a few large-scale firms dominating the prime contracting business. Like many of his officers at depots across the country, the quartermaster general believed that the formal contracting system tended to promote consolidation. Congressional statutes that required advertising and awards to low bidders, Meigs explained to Senator Wilson, served to "compel the Government to purchase in a manner which tends to throw the business into the hands of large dealers." For the quartermaster general, this concentration was regrettable. Nevertheless, he was not willing to create the giant purchasing network that some Northerners proposed as the best way for the army to bypass commercial intermediaries in the market for agricultural goods. "It is not possible," Meigs wrote, " . . . to employ and trust the army of agents needed to purchase direct from farmers. . . . *In medio tutissimus ibis*. The subject is surrounded with difficulties."

As Meigs knew, the large purchasing operations of the leading horse and mule contractors provided a valuable service to the Union that could not be replaced by the army without considerable trouble and expense. Nevertheless, Northerners had become increasingly troubled by the consolidation of prime contracting, especially in the supply of agricultural goods. Small farmers, who still accounted for much of the North's population and who were traditionally regarded as among the most virtuous citizens of the republic, were finding it virtually impossible to do business directly with the army. Within the military itself, officers proposed to counter this problem by using more open-market purchasing. Among the advocates of this solution was General George Stoneman, head of the Cavalry Bureau that the War Department created in July 1863. Stoneman, whose bureau was authorized to oversee cavalry horse procurement by the Quartermaster's Department, recommended in October 1863 that "the contract system be dispensed with as far as consistent with the public interest, and that most of the purchases be made in open market and from first hands."[105]

Like Stoneman, many quartermasters believed that it was the legally mandated formal contracting system itself, and not simply economies of scale or special competence or favoritism, that pushed the business of military supply toward a few big operators. In October 1863, just as Stoneman made his own recommendation, the Cincinnati depot chief Thomas Swords filed an annual report that also promoted open-market purchasing as a way to better distribute army orders. "The law at present," Swords reported, "throws the furnishing of supplies entirely in the hands of large capitalists, thus creating a monopoly; for instance, five thousand horses, a million bushels of grain, or ten thousand tons of hay (no unusual quantities) are required; it is impossible to make contracts with the growers for what small quantities each may have." However, continued Swords, "If officers could go into open market or give notice that they would purchase any quantity of a required article at a certain price, it would be an inducement for the farmers to bring in their produce themselves, as they could get better prices, and the Government save what now goes into the pockets of middle-men or speculators."[106]

In expressing a preference for dealing directly with farmers and other small dealers, military officers shared with many other Northerners an ideal vision of the military economy in which the nation at war would circumvent intermediaries, avoid concentrations of wealth and power, and rely upon virtuous small producers. Swords's 1863 recommendations, for instance, anticipated the words of Congressman Robert C. Schenck of Ohio, who explained in June 1864 that new legislation establishing more formal administrative divisions in the Quartermaster's Department in Washington was *not* designed to concentrate purchasing authority. "We wish to drive off professional contractors," Schenck declared on the House floor, "and to get contracts where the farmers and producers of goods can enter fairly into competition with others, and where they will not be subjected to the overshadowing and grasping power of any central company that may be built up at the seat of Government."[107]

Inspired in part by complaints from inside and outside the military about the dominance of "large dealers," in 1864 the War Department reformed its practices in the procurement of horses and mules. Most importantly, it authorized the use of open-market purchasing as a routine practice for buying animals. Remarkable because it clashed with standing laws and regulations, this was one of the most dramatic manifestations of the power of ethical concerns to alter the shape of the North's military economy. James Ekin, who had served as the chief quartermaster at Indianapolis since the fall of 1861, moved to Washington in December 1863 to become the bureau's new horse procurement chief. In the spring of 1864, the secretary of war authorized Ekin to order supply officers at any newly established depots to use open-market transactions. Despite the statutory requirements, in other words, the War Department would no longer require formal contracting in this part of the war economy. By August, Quartermaster General Meigs received a report by an inspector touring the Midwest that suggested formal contracting was now quite rare. "Almost universally," the inspector wrote, "I found the Quartermasters purchasing supplies in open market; as far as my examination went this method was cheaper for the government than purchasing upon contracts."[108]

The policy shift in the procurement of horses and mules was reflected in a significant drop between 1863 and 1864 in the fraction of purchases made with formal contracts. Certainly, open-market purchases were widely used before 1864, even in nonemergency transactions. In 1863, not much more than $16 million worth of formal contracts for horses and mules were recorded in Quartermaster's Department registers. This was slightly more than half of the total value of all the animals purchased in that year. Across the North during that year, several quartermasters began the contracting process by advertising

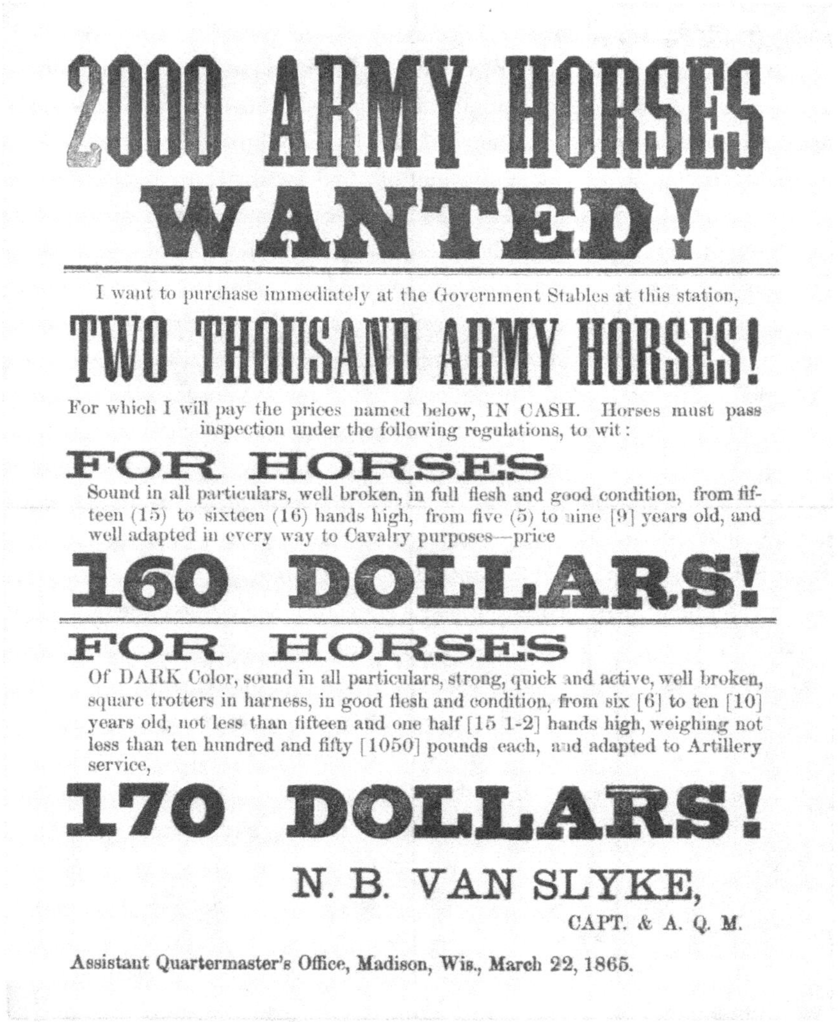

2000 ARMY HORSES WANTED!

I want to purchase immediately at the Government Stables at this station,

TWO THOUSAND ARMY HORSES!

For which I will pay the prices named below, IN CASH. Horses must pass inspection under the following regulations, to wit:

FOR HORSES

Sound in all particulars, well broken, in full flesh and good condition, from fifteen (15) to sixteen (16) hands high, from five (5) to nine [9] years old, and well adapted in every way to Cavalry purposes—price

160 DOLLARS!

FOR HORSES

Of DARK Color, sound in all particulars, strong, quick and active, well broken, square trotters in harness, in good flesh and condition, from six [6] to ten [10] years old, not less than fifteen and one half [15 1-2] hands high, weighing not less than ten hundred and fifty [1050] pounds each, and adapted to Artillery service,

170 DOLLARS!

N. B. VAN SLYKE,
CAPT. & A. Q. M.

Assistant Quartermaster's Office, Madison, Wis., March 22, 1865.

Fig. 4.1. "2000 Army Horses Wanted!" In this broadside from the last weeks of the war, Assistant Quartermaster Napoleon Bonaparte Van Slyke of the Madison, Wisconsin, depot advertises his requirements for horses. Notably, he offers to purchase immediately and in cash. By the latter part of the war, the Quartermaster's Department was moving increasingly toward this kind of direct, "open market" purchasing, which some officers saw as a way to bypass big contractors and deal more directly with farmers. *Source:* Chicago Historical Society.

and receiving eight or ten bids but ended up buying in the open market after deciding that all the bids were too high.[109] But in 1864, both the efforts to contract and actual formal contracts declined considerably. Just above $6 million worth of horse and mule contracts were recorded in 1864, which represented a decline of over 60 percent from the previous year. This was not because the army stopped buying animals but because fewer purchases were made with formal contracts. Whereas roughly half of purchases had employed contracts in 1863, under 20 percent did so in 1864. In this sector, more than any other part of the war economy, the supply officers who complained about constraints brought by contracting had succeeded in avoiding them.

Just as seamstresses and quartermasters used appeals to virtue and justice to defend public enterprise in the case of manufacturing, critics of the acquisition of agricultural goods inside and outside the army also succeeded in altering procurement policy. It is far from certain, however, that the turn to open-market buying in 1864 created radical changes in the distribution of money and power in the agricultural goods markets. Although quartermasters could more easily buy small lots of animals, it is not clear how many smaller farmers and dealers could actually undersell the big prime contractors. Because open-market purchases, unlike formal contracts, were not carefully recorded, the pattern of purchasing during the latter part of the war is difficult to reconstruct. There is evidence that some of the same partnerships that dominated prime contracting also received some open-market orders. When leading horse contractors such as Edward Smith and William Calder testified during the war about their operations, for example, they consistently indicated that only a fraction of their sales to the army came under formal contracts. Of Smith's estimated $4.5 million in sales to the Union through the autumn of 1863, for instance, only about $1 million were recorded in Quartermaster's Department registers of contracts. Even if it is allowed that some of this gap came from animals Smith supplied under subcontracts for other recorded prime contracts, it seems likely that he and other leading contractors filled many open-market orders. While formal contracts may have indeed favored large dealers, the results of the mainly open-market purchasing system that prevailed late in the war probably disappointed those who had predicted it would create a democratic war economy in which small farmers found it easy to do business directly with the army.

WHATEVER THEIR ULTIMATE EFFECTS, the unique developments in the Union's effort to supply its forces with horses and mules indicated Northerners' deep concerns about the workings of the military economy. Formal contracting itself, many supply officers and other Northerners believed, actually promoted the consolidation of army orders into the hands of a few firms. Some Northerners, including quartermasters, believed that procurement policy and practice could be adjusted in ways that would limit such concentration. Under the existing legal environment, however, such changes were not easy to make. Even when they were accomplished, as in the field of agricultural goods, the changes never allowed the army to bypass commercial intermediaries completely. Even those supply officers who articulated their commitment to public enterprise and a producer-friendly military economy found themselves buying many of the goods they needed from a handful of leading contractors. The system of formal contracting did favor bulk orders, but even before 1861, many Northern industries were already more concentrated than many Northerners would have preferred.

Some Northerners were troubled by the structure of the war supply system not simply because they objected to the concentration of economic activity per se but also because some suppliers seemed to be mere intermediaries who came between the wartime state and the actual producers of goods and services. Within the war supply system, seamstresses and quartermasters, among others, complained frequently of profiteering middlemen in the military economy. In this, they were joined by many members of the Northern public. In the halls of Congress, the pages of newspapers and magazines, and thousands of public and private conversations during the war, Northerners subjected their war economy to a thoroughgoing critique. Although the contracting system survived these attacks, Northerners' discontent with the military supply effort was deep enough throughout the war to lead them to pursue major investigations and punishments.

CHAPTER 5

The Middleman on Trial

For the men and women who worked directly for the North's military supply system—including seamstresses and quartermasters—the procurement project was never simply a problem in maximizing economic efficiency but also had significant political and ethical dimensions. Many Northerners, even those outside the supply system, shared this understanding of the business of war. In a maelstrom of accusations and investigations running from the very beginning of the war through the end, members of Congress, the press, and ordinary Northern citizens identified dozens of government officers and contractors as perpetrators of illegal or immoral behavior in the military economy.

Given that the North depended on profit-taking contractors for most of its supplies throughout the war, it would be easy to conclude that this copious criticism amounted to all bark and no bite. Alternatively, it is possible to suggest, as many historians have done, that after the first few chaotic months of the war, honest administrators such as Secretary of War Stanton and Quartermaster General Meigs succeeded in taming the worst problems, so that Northerners' cries against illegitimate activities in the military economy became softer over time.[1] Such interpretations, however, underestimate the depth of Northerners' discontent with the dynamics of the economic mobilization for war and their willingness to pursue significant reforms. The well-known congressional investigations into war contracting in 1861 and 1862 were not the only important official response to this discontent. Indeed, the widespread attacks on improprieties in the war economy were manifested in much more striking fashion, long after the administrative confusion of the first months of the war had passed, by a series of military courts-martial of supply officers and contractors.

In all of these protests, investigations, and trials, a common theme emerged: many Northerners were concerned above all with the problem of profit-taking intermediaries, or "middlemen," who came between the wartime state and the nation's small farmers and producers. More profound than a series of complaints against individual criminal acts, Northerners' campaign against middlemen raised difficult questions about the legitimacy of commercial profit taking during a national emergency. Like the seamstresses' struggles against contracting out and quartermasters' efforts to bypass intermediaries in military markets for agricultural goods, this broader challenge never overturned the basic structure of the supply system. Nevertheless, as several military supply officers and contractors came to understand, the campaign against the middleman was fully capable of punishing those who defied it.

Early Complaints and Investigations

"While the many are asking what they can do for the Republic," the influential *New York Tribune* warned in early May 1861, "a few are seeking to make the Republic minister to their own selfish ends through jobs, contracts, and all the underhanded machinery of peculation." Over the following weeks, such criticisms spread across the North. In early June, the illustrated humor magazine *Vanity Fair* published a poem entitled "Army Contractors," which imagined that the gates of heaven would soon feature a sign declaring "No Jobbing Contractors Admitted." By the autumn of 1861, when most Northern states were transferring procurement authority to U.S. officials, cynicism about the military economy was rampant. In late October, the *New York Tribune* reported that "cheating the Government in its contracts is so common as to be presumed nearly universal." The situation was so bad, the *Tribune* suggested, that many potential suppliers were avoiding military contracts to preserve their reputations. "Men of fair repute," Horace Greeley's paper informed its readers, "shrink from the imputation of contracting as though it were akin to forgery or theft."[2]

The widespread discontent in the North with early supply efforts led to linguistic innovations. The word *shoddy* was originally a technical term used to describe the scattered pieces of wool that were by-products of textile production; these leftovers were sometimes recycled by mixing them into new materials to add softness and reduce costs. During the summer of 1861, some of the uniforms delivered under contracts made by the various states had been found to be thin and fragile because their materials included excessive amounts of shoddy-like filler. As the historian Stuart Brandes has shown, by

August 1861 the *New York Herald*, a popular anti-Republican newspaper, had begun to use the terms *shoddy* and *shoddy aristocracy* to refer to substandard military goods and contractors. Soon, the *Herald* would come to use the word *shoddy* to denounce Republican administrations at the state and national levels. In September, *Vanity Fair* ran a poem entitled "Song of the Shoddy," in which a quartermaster in the field discovered a shipment of bad uniforms. These garments, according to the poem, contained "Coats too large and coats too little / Coats not fitting anybody / Jackets, overcoats, and trowsers / Made of cheap and shameful Shoddy." The poem concluded by evoking the image of a guilty contractor being named by the press and then forced to stand in public in the tattered clothes of his own making. For the rest of the war, many Northerners would continue to use the word *shoddy* to refer to all kinds of alleged rottenness in the military economy.³

Besides using existing words in new ways, Northerners also made use of an older political and economic vocabulary that had been developing across the Atlantic world for decades. On the Civil War home front, the term *middleman*, in particular, became used widely to denounce illegitimate activity in the military economy. For decades, farmers and workers in Europe and the Americas had been attacking middlemen—who might include bankers, merchants, and lawyers—as economic parasites who enriched themselves by creating artificial barriers between honest producers and consumers. During and immediately after the American and French revolutions at the end of the eighteenth century, profiteering military contractors had been placed among the ranks of the nonproducers. Especially in the United States, the celebration of producers and denunciation of parasitical middlemen became a central part of mainstream political rhetoric. The Jacksonian Democrats, who dominated national politics in antebellum America, regularly praised farmers and laborers while demonizing middlemen. In the years leading up to the Civil War, producerist ideas—which came in a variety of strains, ranging from radical to conservative—remained at the heart of the popular political economy endorsed by most Americans.⁴

By promoting a public spirit of patriotism and shared sacrifice, the Civil War mobilization amplified Northerners' concerns about economic intermediaries and their allegedly illegitimate profits. The *New York Tribune*, whose weekly edition of nearly 190,000 copies made it the most widely read newspaper in the country at this time, set the tone in this campaign with an influential editorial printed in late May 1861, only six weeks into the war. "In these venal times, when men are not ashamed to live by levying black-mail upon legislation," the *Tribune* wrote, "we can conceive how it is possible for peculators to get between the Treasury and the Army, and, while seeming to be impelled by patriotic impulses, can bend all their energies to making money out of the existing war, by standing as middlemen between a hard-pressed government and its self-sacrificing defenders." Before the war, the offending intermediary might have been a grasping merchant who took excessive profits by coming between a farmer and the families who actually consumed his products; now, the middleman was standing between the wartime state and its soldiers. Such a view of the military economy was convincing to many Northerners who grew up in Jacksonian America, including many of the procurement officers who managed the supply effort. In August 1861, soon after he took office, Quartermaster General Meigs urged the Philadelphia depot to make every effort to contract directly with those firms who actually manufactured wagons. Allowing a prime contractor to subcontract for finished goods, Meigs declared, was "injurious" to the "reputation" of the government. "The Agents of the United States," he continued, "should tolerate no middleman but should deal directly with the manufacturer or the workmen."[5]

In the context of the national war emergency, the well-established American critique of parasitical intermediaries became sharper. In its anti-middleman editorial of May 1861, which was reprinted by other papers across the North, the *Tribune* had declared that "[s]omething must be done to stop these outrages upon our brave troops. Vultures that prey upon the hearts of the dead on the battle field," the newspaper cried, "are human compared with monsters who furnish rotten blankets and rotten meat to the living in the camp." One possible response to the "outrages" of monstrous middlemen, the *Tribune* suggested, would be to subject contractors to trials by courts-martial and penalties including capital punishment—measures that had been employed in Europe during the Napoleonic Wars. This idea, put forward during the very first days of the war, was quickly endorsed by people across the North. Before the end of 1861, the idea of applying martial law to contractors and executing corrupt procurement officers and suppliers was endorsed by the *Chicago Tribune* newspaper, *Scientific American* magazine, and U.S. Senator Orville H. Browning of Illinois, an influential friend of President Lincoln. By the first weeks of 1862, the state legislatures of New York and Ohio had passed resolutions calling on Congress to pass new laws to prevent and punish corruption and fraud in the military supply system.[6] These calls for new formal punishments went unfulfilled—for the moment.

Although many special legislative committees formed by the various Northern states criticized state-level procurement transactions during the first year of the war, it was investigations at the national level that most captivated

Northerners and achieved the most important effects. By the time most of the state legislative committees were filing their reports in early 1862, Northerners were already familiar with a congressional committee on military contracts, whose activities were covered closely in the press. Known officially as the Select Committee on Government Contracts of the U.S. House of Representatives, this body was widely known as the Van Wyck committee, after the name of its first chairman, Representative Charles H. Van Wyck of New York. Formed in July 1861, the Van Wyck committee spent the next year touring the North, taking testimony from some five hundred witnesses. Its major reports, issued in December 1861 and August 1862, ran to a total of more than twenty-seven hundred pages of commentary and testimony. Many of the individual people and specific transactions targeted by the committee have stood ever since as illustrations of the North's chaotic mobilization during the first months of the Civil War.[7] Less well remembered, however, is the committee's broader critique of illegitimate profit in wartime, which used the figure of the middleman as the symbol of all that was wrong with the military economy.

In the East, the Van Wyck committee's investigations concentrated on New York City, where special procurement agents appointed directly by cabinet officials had accumulated astonishingly large commissions. In these cases, the Van Wyck committee criticized both Secretary of War Cameron and Navy Secretary Gideon Welles for bypassing existing military procurement bureaus. Early in the war, Welles had appointed his brother-in-law George D. Morgan—a wholesale grocer from Westchester County who was also a cousin of New York's Governor Edwin D. Morgan—as a special agent to charter and purchase vessels. Collecting a 2.5 percent commission on most of these transactions, Morgan made roughly $90,000 in under five months. Morgan defended his work as an economical solution to the North's needs. But others disagreed. In a letter to Van Wyck committee member Elihu B. Washburne of Illinois, the Wall Street lawyer Dexter Hawkins condemned Morgan as one of the "swindlers who stand between the ship owner and the Navy Department." Hawkins continued acerbically, "If Mr. Welles must have his relations supported by Government, let an almshouse be opened for them where they may be supported at a reasonable price." Hawkins had little trouble convincing the Van Wyck committee, which in its first report called Morgan's activities "reprehensible" and "demoralizing" to the Union. "No citizen," the committee declared, "can justify any such attempt to convert the public necessities into an occasion for making private gain." Morgan should clearly have been employed "at a fair fixed salary," the committee continued, instead of being allowed to take commissions at a rate of $300,000 a year. The "family arrangement" between brothers-in-law Welles and Morgan, it concluded, brought "great discredit upon the public service." Although Welles would survive this scandal, the congressional investigation had quickly shaken the leader of the Navy Department.[8]

The Van Wyck committee's investigations in late 1861 did even greater damage to the War Department. Like Welles, Secretary of War Cameron had authorized a close acquaintance to serve as a special purchasing agent in New York, even though the army maintained regular supply depots in that city. Cameron's special agent, the newspaper editor and publisher Alexander Cummings, claimed that his purchases were necessary during the emergency of the first months of the war when "more despatch was required than the official routine would permit." But the Van Wyck committee concluded that the transactions undertaken by Cummings should have been left to the "regular and responsible agents" of the United States—that is, the veteran quartermaster and commissary officers at the New York depots—instead of the "irresponsible personal friends of parties holding position under the government." Cameron's decision to give special procurement powers to his friend Cummings, the committee declared, promoted "a system of public policy [that] must lead inevitably to personal favoritism at the public expense, the corruption of public morals, and a ruinous profligacy in the expenditure of public treasure."[9] Unlike Welles, Cameron was unable to survive the investigations of the Van Wyck committee. In January 1862, less than a month after the appearance of the committee's first report, President Lincoln asked for Cameron's resignation and named him ambassador to Russia.

The Van Wyck committee did not confine its activities to New York and Washington nor its targets to members of Lincoln's cabinet. Traveling hundreds of miles across the North to take testimony from hundreds of witnesses in 1861 and 1862, the congressmen became especially concerned with developments in Saint Louis, headquarters for the army's Western Department. As a civil war raged between pro-Union and pro-Confederate factions within the state of Missouri, the military and political efforts of General John Frémont, the commander of the Western Department, were watched closely across the North. In late August 1861, a month after he arrived in Saint Louis, Frémont had declared martial law across the state. At the same time, he suggested that Confederate partisans would risk having U.S. forces confiscate their property—including property in the form of human slaves. Although it was overturned quickly by President Lincoln, Frémont's confiscation order found support among so-called Radical Republicans—those most sympathetic to abolition. At the same time, Frémont alienated the Blair family, the powerful

Missouri-Maryland clan that had formerly supported him. By mid-September, when Frémont ordered the arrest of Saint Louis Congressman Frank Blair Jr. for alleged "insubordination," the split was complete. The political tension still prevailed when the Van Wyck committee arrived in Saint Louis in October to investigate alleged procurement problems.[10]

Preceded in Saint Louis by investigators from the executive branch, the Van Wyck committee proceeded to confirm existing suspicions of procurement troubles in that city. In September, Lincoln had sent Postmaster General Montgomery Blair (Frank Jr.'s brother) and Quartermaster General Meigs to investigate Frémont's handling of his department. Secretary of War Cameron and Adjutant General Lorenzo Thomas had followed in early October, after which they wrote a brief report—soon leaked to the press—condemning Frémont's administration of the Western Department. But it was mainly thanks to the Van Wyck committee that Northerners quickly came to regard Saint Louis as a center of mismanagement and corruption in their military economy. Just after he arrived in that city, committee member Elihu Washburne sent Lincoln a series of enthusiastic telegraphs to say that Frémont's department was "in the most deplorable condition imaginable" and that "the disclosures of corruption[,] extravagance and peculation are utterly astounding." Washburne and the other committee members were encouraged to reach this conclusion by Frank Blair Jr., who wrote his fellow congressmen to say that Frémont had "ordered some of the largest and most remunerative contracts to be given without bidding or competition to the gang of Californians by whom he is surrounded." Blair blamed not only Frémont but also U.S. quartermaster Justus McKinstry, who had been Saint Louis depot chief since 1860. Just five months earlier, Blair had recommended the quartermaster for a promotion. But now Blair identified McKinstry as a man "cunning without being able," who influenced Frémont through "evil counsels."[11]

The testimony it heard during its two-week stay in Saint Louis in October 1861 convinced the Van Wyck committee that Blair was right about Frémont and McKinstry. "In the whole range of their investigations," wrote the congressmen in their first report, "your committee have found nothing so alarming as the defiance of law and the disregard of superior authority" that Frémont demonstrated when he directly gave contracts for the construction of fortifications to an old friend from California. McKinstry, the committee reported, had issued orders for horses, uniforms, and other goods without making any effort to encourage competitive bidding. Between February and October 1861, a single Saint Louis hardware wholesaling firm, Child, Pratt, & Fox, had sold McKinstry's depot over $825,000 worth of supplies of all kinds.

According to the proprietors of several Saint Louis firms, they had been unable to sell directly to McKinstry's depot; some of them had subcontracted with Child, Pratt, & Fox. That firm, the committee found, had enjoyed an informal arrangement with the quartermaster and his clerks, under which few prices were set and the company was to be allowed a "fair mercantile profit" to be determined after the fact. Such evidence, the Van Wyck committee members concluded in their first report, convinced them "that the quartermaster himself was in collusion with corrupt and unprincipled men, who combined together to swindle the government."[12]

The Van Wyck committee's visit to Saint Louis confirmed the worst fears of officials in Washington and other Northerners across the country. On 2 November, Frémont received a note from President Lincoln that removed him from command. Four days later, the *Chicago Tribune,* celebrating Frémont's removal, informed its readers that, during the committee's investigations in Saint Louis, "an amount of waste, extravagance, corruption, rascality, profligacy and downright and unblushing scoundrelism [has] been uncovered that is without parallel in the history of this country." Such cries echoed across the North. "Let Congress strike the plunderers of the public in high places," wrote the *Cincinnati Daily Commercial* in January 1862, "and the country will rejoice. There is nothing that has had a tendency to discourage the people, as the general conviction that much of the enormous war expenditure is not applied to its legitimate object, but goes to fatten the corruptionists." Days later, J. W. Jones of Schenectady, New York, wrote an encouraging letter to committee member Henry L. Dawes: "All our misfortunes since this war began," wrote Jones, "may be traced to incompetency [and] the rage for plunder. I do most sincerely hope that you will follow the work of exposing all the rascals engaged in these . . . contract robberies. You will do more good than any General in the field and will deserve a more lasting renown." At the beginning of March, the committee was portrayed positively by a cartoon in *Harper's Weekly* magazine, in which Van Wyck's bespectacled head appeared on the body of Uncle Sam's "favorite bull-terrier," using his jaws to dispatch large rats labeled with the names of the committee's major targets, including Morgan, Cameron, and Quartermaster McKinstry.[13]

But not everyone was so enthusiastic about the Van Wyck committee. Northerners of various political persuasions complained that it had failed to allow the men it accused to defend themselves. And many Radical Republicans objected to its treatment of Frémont, whose hard line on slavery the Radicals supported. By December 1861, newspapers across the country were beginning to question the committee's fairness. "The investigation itself was a disgrace to the members of Congress engaged in it—it was one-sided, unfair, and oppressive," wrote the editors of the *Missouri Republican* in Saint Louis. The fiercest attacks on the work of the committee came on the floor of the U.S. House in April 1862, as many Radical allies of Frémont claimed that the investigators had been biased and irresponsible. Illinois Representative Owen Lovejoy, among others, compared the Van Wyck committee to the Spanish Inquisition. William D. Kelley, the Philadelphia congressman, admitted that some guilty parties required punishment. "But do not imitate the cuttle-fish," Kelley advised the committee members. "Do not fling your vile odors around everybody, and make the names of contractor and agent so odious that honest men will shrink from dealing with the government." The most vivid of all the criticisms came from Thaddeus Stevens, another Pennsylvania representative and a leading Radical. "I can hardly believe that they should rejoice or take any pleasure in feeding upon the mangled carcasses of their fellow-citizens," Stevens said of the Van Wyck committee members, "for that would be ranking them with the furies of the French Revolution, who attended the scaffolds, who dipped their handkerchiefs in human gore, and gobbled up the garbage which lay around the foot of the guillotine."[14]

Such spirited opposition in Congress and in the press, combined with growing news coverage of the war's first large-scale battles, helped to bring the Van Wyck's committee's activities almost to a standstill after the summer of 1862, nearly a year before it was officially dissolved. Northerners looking for signs of reform in the military economy could take comfort in the fact that men like Cameron, Morgan, and McKinstry had been stripped of their procurement authority. In addition to its prosecution of individual men, however, the Van Wyck committee had developed a broader critique of the North's military economy, which would resonate through the rest of the war. Echoing the anti-middleman rhetoric that the *New York Tribune* and other Northern voices had expressed since the first weeks of the conflict, the committee lashed out repeatedly against profiteering intermediaries. For the congressmen, as for other Northerners, the context of the war amplified an existing discomfort with commercial profit taking in a political culture that valorized farmers and mechanics. But the war also transformed that tension by creating a giant purchaser—the national state—that provided an ideal customer for the country's virtuous producers.

Although some of the illegitimate intermediaries clearly owed their positions to their special treatment by public officials, middlemen also seemed to be part and parcel of the military contracting system. In October 1861, one Boston shoe manufacturer told the Van Wyck committee that "[t]he contracts that the government has given out have been more frequently given to middle men, who were not manufacturers, and had not a manufacturer's reputation to sustain; whose only interest was a large profit. . . . The result has been that they have given the government a low-priced article, and the government has been shockingly cheated." Soon after, a Saint Louis horse dealer testified that the army had paid too much for horses and mules there because quartermasters "do not deal with first parties, but through contractors and sub-contractors."[15]

The Van Wyck committee itself, as it denounced the activities of profiteering middlemen in the North's military economy, began to suggest that the problem was not a matter of misbehavior by individuals but rather a more general disease that had infected the entire North. In the introduction to its second report, the committee condemned "intermediate parties, or middle men, that bane and almost universal curse well nigh inseparable from all transactions in the supply of government during the war."[16]

When Van Wyck himself took the House floor in early February 1862 to discuss the work of the committee, he suggested that the problems described in the committee's report went beyond the immorality of a few individuals. "The mania for stealing," he declared, "seems to have run through all the relations of Government. . . . Nearly every man who deals with the Government seems to feel or desire that it would not long survive, and each had a common right to plunder it while it lived." The nation at war, according to Van Wyck, was being attacked by "harpies," "vultures," "fiends in human shape," "a bevy of cormorants," and "a crew of plundering leeches" on the home front. Two months later, when they were being criticized heavily by their Radical colleagues, Van Wyck's fellow committee member Henry Dawes addressed the House in more moderate language to make a similar point. There was "a system of commissions to middle men growing up . . . all over the country," according to the Massachusetts congressman, "stepping in between the producer and furnisher on the one side and the Government on the other" and costing the public "millions and millions of dollars."[17]

Even as the Van Wyck committee began to wrap up its work in 1862, the cries against the middleman echoed in the findings of two special executive commissions charged with auditing early army contracts. The first of these boards, which consisted of David Davis (the Illinois lawyer who had been Lincoln's campaign manager), Joseph Holt (the Kentuckian who had served as secretary of war during the last days of the Buchanan administration), and Hugh Campbell (a Saint Louis merchant), was charged with auditing the army contracts made in Saint Louis under the administration of Frémont and

McKinstry. Arriving in Saint Louis in November 1861, only days after the Van Wyck committee departed, this board quickly sustained the findings of the congressmen.

Just after he arrived, Davis wrote to his wife to express his outrage: "It is shocking, when we think, that when a Govt. is in throes of dissolution, that every scoundrel is trying to cheat [and] swindle it." In its final report, which was sent to Secretary of War Stanton in April 1862, the Davis-Holt commission followed the lead of the Van Wyck committee by blaming Frémont and McKinstry for wielding an "orientally despotic power" and turning Saint Louis into the "El Dorado of Army contractors." General Frémont, by directing contracts personally, had "virtually ignored the existence of the quartermaster's and the commissary's departments, and of the Ordnance bureau." Quartermaster McKinstry, however, had done no better: demonstrating a "complete indifference to the public interests," the commissioners charged, McKinstry had allowed "personal favoritism" to enrich a small "clique" of suppliers.[18]

Like the Van Wyck committee, the Davis-Holt commission summed up the problems in Saint Louis in 1861 by invoking the figure of the parasitical middleman who came between the wartime state and its legitimate suppliers. According to the commissioners, the "country merchants, farmers, and mechanics" in and around the city would have "dealt honestly" with the U.S. army "had not the inexorable 'middle-man' policy of the quartermaster's department excluded them from the honor of a direct commerce with the government of their country." In adjusting the nearly sixty-five hundred claims for over $9 million on army purchases in Saint Louis before mid-October, the Davis-Holt commission reported, "we have held that 'middle men' shall receive no profit on the supplies passing through their hands, and that the government shall pay for such supplies only their fair market value." In the end, the commission ended up adjusting or dismissing more than one thousand claims, reducing approved payments to a total of $7.6 million. Child, Pratt & Fox, whose books were seized by the commission, was allowed profits of 10 percent or more for hardware but none at all for the many other sorts of goods it had sold to the quartermaster's department. The Detroit banker S. P. Brady, a friend of McKinstry who had served as a contractor, was allowed "no profit whatever."[19]

The ideal of a military economy without middlemen, promoted across the North during the first year of the war, received more support in 1862 by another auditing commission. From March to July of that year, the accounts of the Union's early arms suppliers were reviewed by Joseph Holt (who had just completed his service on the Saint Louis claims board), the Indiana reformer Robert Dale Owen, and army ordnance officer Peter V. Hagner. Finding a "lack of system" in small arms orders during the first year of the war, this board reduced payments to 19 of the 107 suppliers under review. Echoing the language of other government investigators, the Holt-Owen commission complained that, by purchasing some imported and domestic arms on the basis of "private proposals," instead of competitive bidding, the Ordnance Department had encouraged "middle men" who were "not dealers in arms nor skilled in their value." Indeed, the use of contracts based on "private proposals" instead of advertising and competitive bidding had created "a class of 'middle men,' most of whom are speculators and adventurers, to whom, instead of to the manufacturers themselves," the military had given many of its orders. "The class of men referred to," the Holt-Owen commission continued, "are generally rapacious and unscrupulous, and thrust themselves between those whose interest it is to deal, and who ought in every case to deal, directly with each other."[20]

By the summer of 1862, when the Holt-Owen commission concluded its work and the Van Wyck committee published the second of its two major reports, there was some reason to think that the Union had succeeded in solving its problems with bad behavior in the military economy. Certainly, the supply system, now more fully under the control of the U.S. military bureaus, had been regularized. And the Northern press, which had often turned to the procurement process on the home front during the first months of the war, had turned to news from the field, which was increasingly shocking. In the West, the single battle of Shiloh in April produced some twenty thousand dead and wounded; in the East, the Seven Days' campaign at the end of June caused thirty thousand more casualties. As the Van Wyck committee faded away with a whimper, it seemed as if the struggles over corruption and illegitimate profit in the Union's war economy might also disappear. In fact, however, the campaign against the middleman was just beginning.

Quartermasters on Trial

High-ranking military supply officers in the North, who each handled contracts and payrolls that amounted to millions of dollars a year, were natural targets for investigations into corruption. By taking kickbacks from contractors or directly embezzling public funds that amounted to just a tiny fraction of their total business, senior officers in the Quartermaster's Department and other supply bureaus could have amassed large fortunes during the war. Additionally or alternatively, they could have helped Republican party officials to create a giant patronage machine, in which some of the huge procurement expenditures could be used systematically to reward the party faithful. Certainly, a few procurement officers, along with their inspectors, did use their positions for personal advantage or to benefit their relatives and friends. But most top supply officers avoided scandals involving simple corruption. The most celebrated investigations and dismissals of senior procurement officers in the North, which included extended trials by court-martial of two depot quartermasters, consisted largely of vigorous debates over the meaning of supply officers' obligations to avoid dealing with middlemen. These trials suggested that, while simple mismanagement or criminality and the influence of party politics were always of concern, many Northern observers were troubled above all by concentration of the military supply business into the hands of a small number of contractors who seemed to be inserting themselves between true producers and the nation at war.

Although only a few Northern supply officers ended up being prosecuted in military courts, most of them faced accusations and resentment, some of it from their fellow soldiers. There was a long tradition of mutual suspicion and jealousy in the army between line and staff. If line officers believed that staff officers enjoyed many comforts without having to withstand the hardships of field operations, their counterparts in the staff bureaus worried that only the line officers who attained glory in battle would be rewarded with promotions.[21] This divide between line and staff, which united volunteer and regular quartermasters in a common cause, could only widen during the Civil War as military operations became much larger and more specialized. As they read newspaper headlines every day that referred to the latest clash of Northern and Southern forces in the field, many quartermasters—whatever their differences—suspected that their important work on the material production of war was being overlooked. Worse, supply officers were frequently accused of incompetence, corruption, and disloyalty.

Insecure about their distance from the dangers and glories of the battlefield, supply officers at the rear depots worried constantly that they would lose out to their colleagues on the line or politically favored volunteers when it came to respect and promotions. In the fall of 1862, soon after he left his position as chief quartermaster for the Army of the Potomac for the New York City depot, Major Stewart Van Vliet wrote directly to Secretary of War Stanton to ask that he not be overlooked when it came time to make promotions. "My duties," claimed Van Vliet, "have been and are now ten times more arduous and responsible than those of a Brigadier General in command of a Brigade." In the same letter, Van Vliet told Stanton, "I have made my own way from early boyhood, and can call on no powerful political friends to aid me and in truth I would rather stand on my own merits when advancement in my profession is concerned." While Van Vliet may have been less well connected with the Republican Party in power than he would have liked, in fact he worked to overcome the relative obscurity of his staff assignment by maintaining contacts with politicians and other men of influence in both parties.[22]

Other top supply officers made similar efforts to keep themselves from losing out to line officers in the battle for rank. Philadelphia depot chief George Crosman, for example, was recommended for promotion by his longtime acquaintance Jay Cooke, the North's chief war bond salesman. As "a faithful officer" who had "saved millions to the Govt. since the war began," Cooke told one senator in early 1864, Crosman should not suffer when it came to promotion simply because he worked behind the lines.[23] In Washington, meanwhile, Quartermaster General Meigs himself worked steadily to keep his officers from being ignored for promotions. "The Quartermaster's Department," Meigs reminded Secretary of War Stanton in June 1864, "spends nearly one half the money used to carry on the war. It is of the highest importance that its officers should be encouraged to feel their positions are positions of honor in proportion to their labor and responsibility."[24]

Meigs knew that many of the North's top supply officers did not feel appreciated, not only because they were passed over for promotions but also because they were subject to constant attacks. In early 1864, Lewis B. Parsons—the railroad official turned western transport chief—complained to Quartermaster General Meigs that their department was "at least as important as any in the army while it is by far the most onerous and laborious. It disburses annually hundreds of millions; and any failures in performance in its duties involve the failure or defeat of our armies." Despite its importance, however, Meigs's bureau and its officers received only disrespect: "It is not only not as reputable as other Departments," continued Parsons, "but with the present state of public sentiment, is it not prima facia disreputable to belong to it? You can scarcely take up a paper or be present at any gathering that you do not see or hear innuendos or open ridicule and abuse of the entire Department." Many Northern procurement officers agreed with Parsons. As Saint Louis and Louisville depot chief Robert Allen put it in a report in late 1864, a supply officer behind the lines "fights no battles and has no gallant acts to chronicle and although without him the battles would not have been fought or without him the victory won, he has no record of it unless it is in the shape of censure, to which the slightest fault is sure to render him amenable."[25]

In addition to steady complaints about logistical failures and digs at their comfortable situations behind the lines, supply officials on the home front were subjected to accusations about their political bias. Given that the Republicans dominated party politics in the wartime North and appointed dozens of volunteer supply officers, it might be assumed that the charges about partisanship always came from Democrats who believed that procurement officers were helping to steer purchases to Republican party favorites. More often, in fact, top supply officers were accused of being insufficiently friendly to the party in power.

On more than one occasion, Northern depot chiefs—many of them regulars whose careers predated the origins of the Republican Party—were accused of political "disloyalty," meaning insufficient friendliness to the Republican administration or even sympathy for the South. In March 1863, the thirty-six-year-old general James G. Blunt, a volunteer with an abolitionist political background, wrote Washington to request the removal of Fort Leavenworth depot chief Langdon C. Easton, a senior quartermaster officer, "for the reason that in my opinion he is disloyal." According to Blunt, Easton had dealt with "disloyal" men as contractors and employees and had criticized Blunt for the general's vigorous pro-Republican activities. When the complaint reached Quartermaster General Meigs, he explained coolly to general-in-chief Henry Halleck that "any one who does his duty will make enemies." Although Easton himself requested an immediate transfer, Meigs waited until the end of the year to move him to the Department of the Cumberland; during the last year of the war, Easton would serve as chief quartermaster for the forces of General Sherman.[26]

One depot quartermaster who was accused repeatedly of political disloyalty was John Dickerson, who oversaw clothing and equipage procurement in Cincinnati for much of the war. As early as the summer of 1861, when Dickerson was beginning to assert U.S. authority over supply, Ohio's Republican governor William Dennison tried to have him removed on political grounds. But Quartermaster General Meigs protected his officer, telling Dennison that recent charges "impeaching the loyalty of Captain Dickerson" were probably "produced by disappointed bidders for contracts." In February 1862, while Dickerson was away from Cincinnati on a visit to his dying mother, his wife, Julia, tried to prevent Burnett House hotel staff from burning candles in the windows of their room as part of a celebration of Washington's birthday. Some interpreted her behavior as anti-Union; she was reportedly snubbed by other hotel guests in the hotel dining room the next day. When Dickerson returned to Cincinnati, he assaulted two men he accused of having insulted Julia and was soon compelled to pay a fine of thirty-five dollars. Just days later, Meigs ordered Dickerson to respond to charges that he had been employing "disloyal" men at the Cincinnati depot. In his reply, Dickerson told the quartermaster general that the charges amounted to "a gross perversion of facts" created by partisan enemies. "If I were an abolitionist," the Cincinnati quartermaster explained, "I think fewer charges would reach you." Although Dickerson suggested that he be reassigned to California, Meigs again decided to retain him at the same post.²⁷

Dickerson's political problems in Cincinnati continued into the summer of 1862, when pro-administration newspapers in the city complained that he was placing advertisements for contracts in the *Enquirer*, an opposition newspaper. Then, in the spring of 1863, the attacks on Dickerson came to a head. In April, Secretary of War Stanton urged Meigs to remove the quartermaster because Augustus E. Tracy, a brother-in-law of Dickerson, had been working as an agent for several eastern firms that served as leading textile and clothing contractors with the Cincinnati depot. But General Ambrose Burnside, the new commander of the Department of the Ohio, advised Stanton that "the charges against him are of a partisan character. . . . [H]e is an honest, capable, and loyal officer, but suffers from the misfortune of having a rebel wife." A military court of inquiry soon confirmed Tracy's role as a contractors' agent but never established any wrongdoing by Dickerson, who remained at the depot. Then, in July 1863, military police arrested Frank Hurtt, an assistant quartermaster at Cincinnati who was also a co-owner of the *Ohio State Journal*, the Republican newspaper at the state capital at Columbus. In the course of Hurtt's subsequent trial by court-martial, it became clear that he had tried energetically to remove Dickerson as a way to promote the interests of the Republican Party. "I sought," Hurtt declared, "to have the legitimate and proper influence of the Government patronage at one of its largest depots in the hands of unquestionable friends of the administration." When Dickerson resigned in March 1864, it was not clear whether he expected more specific charges that might lead to a court-martial or was simply exhausted by the earlier troubles and ready to try to make more money by joining his brother-in-law as a contractors' agent.²⁸

Supply officers' political struggles involved not only their own positions but also control over their many employees, including the powerful civilian inspectors of incoming supplies. In general, depot quartermasters seem to have been satisfied that Stanton and Meigs allowed them considerable autonomy in this field. In early 1862, Philadelphia depot chief George Crosman told the Van Wyck committee that, although the office of the secretary of war had formerly used the appointment of inspectors at the Schuylkill Arsenal as "a political machine to reward political favorites," this was not the case under the current administration. But the light touch of Stanton and Meigs did not keep Crosman from being accused of employing disloyal people, just as Dickerson had been in Cincinnati. In March 1863, Crosman responded to the complaints by telling Meigs that he had done his best to avoid hiring "any one who is in the least tainted with opinions adverse to the present administration." At the same time, however, Crosman refused to allow party officials to name his inspectors. In May 1863, Charles O'Neill, a Republican congressman from Philadelphia, asked Stanton to approve the appointment of a new clothing superintendent and a new chief inspector in Philadelphia. "I know," O'Neill told the secretary of war, that "the active men of our party, would be highly pleased if you would consent to have the changes made." Quartermaster Crosman, however, refused to budge on this issue. Responding to pressure for personnel changes that had been applied by Congressman Leonard Myers just before O'Neill's efforts, the quartermaster reminded a sympathetic Meigs that "I alone am responsible to the Government" for huge sums of money. As for clerks and inspectors, therefore, Crosman continued, "I claim the legal right to select [them].... [I] am not willing that these agents should be selected by others, however distinguished for political influence and power."[29]

Their uncertain relationship to the political parties was only one of several areas in which quartermasters could get into trouble. As Dickerson's problems in Cincinnati suggested, top procurement officers were sometimes accused of simple nepotism, as well as disloyalty. These charges were sometimes well founded. Although Dickerson may not have been breaking any existing laws, the activities of his brother-in-law as a leading contractors' agent in Cincinnati raised legitimate questions about favoritism.[30]

The military supply activities of public officials' brothers-in-law became the sources of contention on more than one occasion during the war. George Morgan, the ship broker who became a target of the Van Wyck committee in 1861, was a brother-in-law of Navy Secretary Welles. Less well known to Northerners was M. K. Moody, a hat and fur dealer from New York City who signed $2 million worth of contracts for forage from May to December 1863, making him the top prime forage contractor in the entire country during that period. These contracts were all negotiated by Samuel L. Brown, the young volunteer quartermaster from Chicago who had recently become the North's top buyer of forage. At the end of the year, the secretary of war's office discovered that Moody was Brown's brother-in-law. "It is not known that there is anything wrong in Captain Brown's transactions with Moody," Stanton's office told the Quartermaster's Department, "but the relationship existing between them makes it improper for Captain Brown to either buy, inspect or pay for Moody's forage." After the discovery, Moody received no more prime contracts. Brown continued to buy millions of dollars worth of forage, and a potential scandal was effectively buried by the War Department.[31]

But contracting with family members and friends was the exception rather than the rule. The vast majority of the army's purchases were made between officers and firms that had no personal connection. Furthermore, as the North's many large public military enterprises suggest, some officers actually went out of their way to avoid contracting with private firms. In the Quartermaster's Department and other supply bureaus, many officers demonstrated an ethic of public entrepreneurialism or professional public service that led them away from crude corruption or nepotism. In June 1862, Thomas Swords, the senior quartermaster who was then supervising the Louisville depot, informed one potential supplier that he was not interested in bribes or kickbacks. "In regard to my retaining 'whatever per cent I wish to pay me for my trouble,'" Swords wrote, "I have to inform you that I am paid by the United States, for trouble and everything else incident to my office."[32]

Even some volunteer quartermasters, who often came straight from the business world and owed their appointments to the new Republican administration, adopted this ethic. Louis B. Parsons, the prickly former railroad executive who worked in Saint Louis for much of the war as transport chief for the West, was certainly not a believer in large-scale public enterprise. But as he negotiated dozens of short-term contracts with railroads and boat owners, Parsons seemed to go out of his way to avoid giving special favors to his former company, the Ohio & Mississippi Railroad. Indeed, his former colleagues concluded by December 1861 that Parsons was actually doing his best to squeeze them, if not the entire railroad industry. "Parsons is working against the interest of our road," one of them reported after a visit in Saint Louis. "[He] is doing all in his power, at this time, to convince the Govt. that it is paying the Roads of the country too much for its transportation." Far from benefiting from Parsons's position, Ohio & Mississippi leaders soon found themselves working (unsuccessfully) to have the quartermaster removed. The relationship between Parsons and his former company reached new lows during the negotiations over one transport contract in early March 1862. After Ohio & Mississippi agent Mendes Cohen failed to respond immediately to a request to restate the road's bid in per-mile terms, Cohen reported, the quartermaster angrily ended their conversation by exclaiming, "God damn you go to Hell."[33]

While most Northern procurement officers became inured to a constant flow of accusations and resentful attacks, a few found themselves being tried by formal courts-martial. Like all soldiers, supply officers were subject to institutions of military justice, which had operated since the Revolution. During the summer of 1862, new congressional legislation created an official corps of judge advocate general officers to prosecute cases in military courts and review their proceedings; it also created the new position of judge advocate general, which was soon filled by Joseph Holt, formerly of the contract auditing committees. For the rest of the war, army lawyers working for Holt's department prosecuted thousands of soldiers in military courts.[34]

The vast majority of soldiers tried by courts-martial were common privates accused of desertion, disobedience of orders, abuse of alcohol, swearing, and other such crimes. But a few of the cases heard by military courts concerned alleged corruption by supply officers. In Detroit, commissary officer Horace Turner received a reprimand from his court-martial in October 1863 after he pled guilty to having acted as a partner in a meat company during the previous winter. One quartermaster convicted by court-martial was Frank Hurtt, the Cincinnati assistant quartermaster who tried to engineer the removal of clothing and equipage depot chief John Dickerson for political reasons. Hurtt was given a dishonorable discharge in early 1864, after letters seized by investigators revealed that he had been using his insider knowledge about impending army purchases for the financial benefit of himself and his friends. In Kentucky, one of several border states in which accusations of disloyalty and fraud were especially widespread, two other volunteer quartermasters were court-martialed and cashiered during the third year of the war. Henry J. Latshaw, an assistant quartermaster at the small Lexington depot, was dismissed after a military court found him guilty of embezzling five thousand dollars from public accounts and helping a forage contractor to defraud the government. In Louisville, the site of a major depot, Assistant Quartermaster Samuel Black was accused of conspiring with his son and other contractors to boost their profits by delivering adulterated forage. At the end of the court-martial proceedings, Black was dismissed, fined the large sum of ten thousand dollars, and sentenced to two years in prison.[35]

Not all military trials of supply officers generated such clear-cut accusations of simple corruption. Of the several court-martial cases involving Northern procurement officers, the most significant were the two in 1862–63 that involved the Quartermaster's Department depot chiefs at Saint Louis and Baltimore.[36] The Saint Louis quartermaster, Justus McKinstry, was well known to many Northerners from the reports of the Van Wyck committee. Like most of his peers among army regulars, McKinstry had spent much of the antebellum period in the West; more than most, he had a penchant for getting into trouble. An 1838 graduate of West Point, McKinstry joined the Quartermaster's Department during the Mexican War. While in Mexico, McKinstry was accused of accepting money from one of his local suppliers of blankets and clothing, but he was acquitted by a court-martial in early 1848. The next year, he was ordered to San Diego, California. Within weeks, he found a bitter enemy there in the person of Cave J. Couts, a younger army officer who was equally hot-headed. After he and Couts squabbled over gambling debts and fought a duel over alleged insults to Ysidora Bandini (the daughter of a prominent local family and Couts's future wife), McKinstry was court-martialed again in 1850; this time, he received a reprimand and a three-month suspension. In 1853, McKinstry was investigated by a court of inquiry after Couts—who had since resigned from the army to oversee the large ranches around San Diego that he had gained from his marriage to Bandini—wrote to Washington accusing the quartermaster of corruption and interfering with Couts's own political aspirations. McKinstry was relieved temporarily after the court of inquiry criticized his accounting methods and found "an unnecessary mingling of public and private interests" at his own ranch, but he was soon reinstated and stayed two more years in San Diego. During the second half of the 1850s, McKinstry served at posts in Florida and spent a leave in New York.[37]

After the Civil War began, it took only a few months for the trouble-prone McKinstry to encounter serious new problems. Already one of the army's central supply posts when McKinstry arrived in early 1860, the Saint Louis depot became a hive of activity after the war broke out in April 1861. In August, the Saint Louis correspondent of the *New York Tribune* referred approvingly to McKinstry as "one of the gallant officers" in the Mexican War, who "will be remembered by early residents of California as one of the most popular members of his profession in that State," and was managing supply efforts in Saint Louis skillfully with his "quick and ready mind." But soon after this report, McKinstry's popularity slipped. This fall may have been caused in part by his growing powers: already the chief quartermaster for Frémont's department, he proceeded to become provost marshal for Saint Louis and later accepted a commission as brigadier general of volunteers that allowed him to lead troops in the field. In September, McKinstry told an Irish Legion meeting in Saint Louis that he had voted for Democrat Stephen Douglas in the 1860 election. "I have ever been a Democrat in sentiment," the quartermaster reportedly announced, even as he insisted that he was a strong supporter of the Union and its war effort.[38] For Saint Louis Congressman Frank Blair and other conservative Republicans, this may have been at least as offensive as Frémont's recent order that seemed to support radical abolitionism. A month later, the Van Wyck committee arrived and quickly identified McKinstry and Frémont as the perpetrators of the procurement mess in Saint Louis. In mid-November, soon after Lincoln removed Frémont from command, McKinstry was arrested and imprisoned in the city's arsenal. His books seized, the North's most infamous quartermaster was now positioned to serve as a symbol of wartime corruption, vanquished by vigilant congressional investigators.

But McKinstry refused to accept the role established for him by Blair and the Van Wyck committee. Over the next year, he accused the congressmen of ignorance and bias and tried to pull apart the more abstract case against the middleman that was so central to Northerners' understanding of problems with their military economy. Signs that McKinstry would not go down easily came early. In the autumn of 1861, a grand jury assembled by the U.S. District Court in Saint Louis adjourned without charging McKinstry. In early 1862, as it questioned the methods of the Van Wyck committee, the *New York Tribune* published a series of letters defending the Saint Louis quartermaster. One explained that the expensive silver set given to McKinstry was a gift designed to replace goods stolen from his house after he had left for the field and not the improper donation from aspiring contractors that the congressmen had described.[39]

Then, in June, the quartermaster himself published *Vindication of Brig. Gen. J. McKinstry*, an extended reply to his accusers. Describing the Van Wyck committee as a "modern *Inquisition*," this pamphlet also reproduced correspondence from 1861 to suggest that Frank Blair had broken with the quartermaster because McKinstry had refused to award contracts to the men recommended by the congressman. At the same time, McKinstry's pamphlet made a broader argument against the many charges that he had improperly allowed the business of war in Saint Louis to be dominated by illegitimate intermediaries. On the contrary, the quartermaster argued, "It was fortunate that the Government found 'middle men'—capitalists—who could wait for payment."[40]

This statement was remarkable, in that it openly rejected the foundation of most contemporary criticisms of the North's procurement system. Denying the charges that he had paid too much for supplies during the first months of the war, McKinstry's *Vindication* mocked the idea that it would have been possible to acquire the goods he needed without turning to so-called middlemen. Like other quartermasters across the North in 1861, McKinstry had frequently dispensed with public advertising to acquire supplies more quickly; like his peers, he was compelled by the cash shortage to buy on credit. This meant that most orders would be taken "only by those much-abused 'middle-men' who could raise the means wherewith to buy for cash and wait indefinitely for their pay." Whereas the findings of the Van Wyck committee and the Davis-

Holt commission "would restrict Quartermasters to dealing with producers and manufacturers only," this had not been possible in Saint Louis in 1861, where—according to McKinstry—it was in fact cheaper and easier to deal with willing merchants with deep pockets. "The oft-repeated story of the *'middlemen'* is mere *twaddle,*" the quartermaster declared, "and unworthy of serious refutation. Any one familiar at all with the course of business or the practice of the Government must see that the learned Commission, in their dissertation on *'middle men,'* have 'strained at a gnat and swallowed a camel.'"[41]

With these words, McKinstry suggested the extent to which his case would serve not merely as a trial for alleged corruption by one public official but also as a referendum on the general operations of the North's procurement project. Indeed, McKinstry's trial in a military court turned into a debate about procurement practices, rather than clear-cut acts of corruption. In September 1862, ten months after McKinstry was arrested and three months after the publication of his pro-middleman *Vindication,* his elaborate court-martial began in Saint Louis. Open to the press, the four-month trial involved dozens of witnesses from the city's business community. The court consisted of eight senior army officers; the prosecution was handled by Judge Advocate A. A. Hosmer, while McKinstry was assisted by a team of five lawyers.

Far from being a convincing demonstration that McKinstry was guilty of nakedly corrupt acts such as bribe taking or embezzlement, this trial became a contest over the middleman. Without ever accusing McKinstry directly of obvious crimes, Hosmer tried to convince the court that the quartermaster had violated army regulations by failing in his open-market purchases to "adopt the usual course of individuals, in their private transactions, as the standard of official vigilance and fidelity." McKinstry had erred, according to Hosmer, by failing to "purchase in the market"; instead, he had turned to a few middlemen (such as the hardware company Child, Pratt & Fox), who "went into the market with his orders in their hands, bought at the market value, [and] immediately transferred the articles to him at a large advance." McKinstry's rebuttal, which he supported by calling a series of witnesses to testify about business conditions in Saint Louis in 1861, echoed the arguments he had already put forth in his *Vindication:* namely, that the "men of Capital" who had supplied his depot, far from being profiteering parasites, had been the only ones willing to sell him large quantities of goods on indefinite credit. Comparing himself to Saint Paul, "who likewise suffered under false accusations," McKinstry used his final statement at the trial to rehabilitate the figure of the much-reviled middleman.[42]

McKinstry soon learned the consequences of defying the producerist logic that so many Northerners used to decry the evils of intermediaries in the war supply system. Finding the quartermaster guilty of twenty-six of the sixty-one specifications named in the official charges, the military court concluded that he had wasted public funds by using only a few select suppliers of horses, tents, uniforms, and other goods, who subcontracted for most of the supplies and charged the army excessive markups. Notably, the military court never argued that McKinstry had gained personally from his office, only that he had violated standing rules governing public purchasing and allowed too much profit to his suppliers. The court's ruling, approved by President Lincoln at the end of January 1863, cast McKinstry out of the army. According to the final report of the Van Wyck committee, his dismissal showed that "no guilty officer, however high his rank, or however skilled in fraudulent practices, can escape . . . proper punishment." By the end of the summer, McKinstry was spotted making an anti-Republican speech in Illinois, which confirmed for some of his critics that he had always been a "copperhead"; no less damning, perhaps, in the eyes of Northerners who heard them, were reports that he had landed in New York City in 1864 as a stock broker.[43]

But McKinstry's trial had always involved something more than outright fraud or partisan differences. If the Saint Louis quartermaster had been lazier than his peers about identifying qualified suppliers and promoting competition in contracting, he was hardly the only one to rely upon mercantile intermediaries who charged a premium for taking on the financial costs of selling in bulk to a cash-poor government. Few supply officers, however, were as willing as McKinstry to accept and defend such arrangements. In the end, McKinstry's dismissal was the result not only of his own poor management and political judgment but also the belief of his accusers and judges that an ideal military economy would allow the wartime state to bypass middlemen and deal directly with true producers.

Just a few months after McKinstry was sentenced, many of the same debates were replayed in the court-martial of another senior depot quartermaster, James Belger. Like McKinstry, Belger had joined the Quartermaster's Department during the Mexican War. Although he never attended West Point, Belger had spent three decades in the army before the Civil War, mostly on the frontier. Before he was ordered to Baltimore in 1861 to head up the depot there, Belger had recently served as a quartermaster at posts in Texas and Arkansas. Like many of his peers, Belger was accused during the first months of the war of showing insufficient loyalty to the Union (or the Republican administration) in his purchases and hires but was sustained by Meigs.[44] In July 1862, after examining the Baltimore quartermaster's reports, the quartermaster general warned Belger that so many of his recent ship charters had been made with a single broker that the business was beginning to look like a "monopoly." But when senior quartermaster Charles Thomas was dispatched by Meigs to investigate, he reported that the Baltimore quartermaster's "business seems to be performed with an eye to economy."[45]

Then Belger, as McKinstry had been, found himself headed toward a court-martial after being censured by a congressional investigation. Belger's accuser was a small Senate investigating committee that operated during the winter of 1862–63. Led by Senator James A. Grimes of Iowa, this committee focused on reports of mismanagement and profiteering in the chartering of military transport vessels. According to the Grimes committee report on ship charters, Belger had failed to advertise sufficiently or deal directly with "the owners of vessels, or with their immediate, legitimate, established agents." The senators explained that, instead, Belger had frequently referred such owners and their agents to one broker, Amasa C. Hall, who had collected as much as forty-five thousand dollars in commissions on the military charters.[46] As military lawyers prepared for a court-martial of Belger, they added charges that he had paid too much for coal by buying from two dealers who did not offer the lowest possible rates.

Held in Baltimore in June and July 1863, Belger's court-martial coincided with the campaigns of Vicksburg and Gettysburg and the New York City draft riots. For reasons of timing alone, then, it was bound to receive less public attention than McKinstry's trial had. But Belger, like McKinstry, mounted a vigorous defense in which he argued that entrepreneurial intermediaries in the military supply business were essential to the Union's war effort. As a parade of coal dealers and ship brokers took the stand to inform the court about local business standards and their dealings with the Baltimore bureau, Belger and his lawyers disputed the prosecution's contention that he could have saved public funds by doing more to promote competition and discover "market" rates. Several expert witnesses testified that it was normal practice for ship owners to use the services of brokers and that they doubted that the army could have done better than the rates negotiated by Hall. Others explained that Belger's office was compelled to pay a premium above cash "market" rates for coal because of the prevailing discount on the certificates of indebtedness in which the army actually settled its contracts.[47]

Without producing many witnesses to sustain him, Judge Advocate W. L. Marshall dismissed such claims. Under Belger at Baltimore, Marshall told the court, "the patronage of the Government has become a monopoly" dominated by one ship broker and two coal dealers. "It has been suggested," said the judge advocate, "that to canvas the vendors and invite free competition may not be the most certain mode of purchasing cheaply." This defense, concluded Marshall, was nonsense.[48]

Belger and his lawyers countered by explaining that Marshall and other critics simply did not understand the business of war supply. Like McKinstry, Belger used his defense to reject the popular view that army procurement officers needed to take advantage of market competition and promote a producerist military economy in which original owners and manufacturers served as the prime contractors. Calling the trial a "farce," Belger explained that he should not be held to the standards of "a mercantile agent or clerk or factor, whose only duty is to make the best bargain for his principal. I am a public and not a private, a military and not a mercantile agent." In other words, because he had to meet sudden, huge demands from generals who could not wait for lengthy negotiations, the Baltimore quartermaster's enterprise was unique. For the prosecution to suggest that he should have paid the lowest daily "market price" as reported in newspapers, Belger argued, was absurd. Such a price, he explained, was merely "the generalized result of that day's struggle of competing buyers and sellers, speculators and debtors verging on bankruptcy; it does not state any sale nor all sales . . . it is varied by the caprice, or the greediness, or the liberality, or the personal necessities, or the personal feelings of the individual dealers." Because he was paying suppliers in vouchers that would be exchanged at some unknown later date for certificates of indebtedness, Belger continued, it was especially unreasonable to expect that he could find someone to fill his orders at the lowest price being quoted at the time.[49]

Belger's extraordinary defense culminated in an ode to contractors, who closely resembled the middlemen who had been so thoroughly demonized by the Van Wyck committee and other Northerners during the previous two years. "There is a class of men under all Governments, in times of great emergency like the present," Belger explained,

> who have means and wish to increase them, who like the gambling uncertainties of public supply and demand, who want great gains on great transactions and are willing to encounter the great risk incident to such dealings. . . . Without such men this Republic could not carry on this war for a day. England and France would have dropped their arms before they had worried each other a year during the Revolutionary Wars. Napoleon could not have moved his armies without them. They are not public benefactors, nor disinterested patriots; but as long as [the] Government can't pay in cash, must borrow to live, will indulge the luxury or are driven to the necessity of war . . . so long will this genus of men exist, be a necessity, supply the

public needs when the prudent merchant shuns the risk; and be vilified by men who envy their gains but lack the spirit or genius to render the service, or the courage to meet the risks they encounter.

According to Belger, it was actually speculating middlemen willing to embrace the financial risks of the military supply business, and not small producers or conservative merchants who normally demanded cash, who enabled the Union's war effort. "With such men I have dealt; by their aid I have met the requisitions of the Government in the greatest emergency of our national life," the Baltimore quartermaster boasted. "Through them, I have moved and supplied expeditions greater than any history records since that to Moscow."[50] In Belger's telling, speculating intermediaries were not parasites on the body of the nation at war but rather the true producers of the Union's military capacity.

Belger's defense of himself and the North's middlemen was more successful than McKinstry's—but only in the short run. On 14 July 1863, when Americans on both sides of the war were still digesting the news of more than fifty thousand total casualties at Gettysburg, the officers assigned to Belger's court-martial acquitted him of all thirteen specifications and the overall charge of "neglect and violation of duty." But this verdict did not sit well with Secretary of War Stanton, whose office had complained to Meigs before the trial that Belger's advertisements for proposals were too vague about quantities required and allowed too little time to promote competition among prospective suppliers.[51] A few weeks after the verdict was announced, Stanton convinced President Lincoln to overturn it. In November, Stanton kicked Belger out of the army with a dishonorable discharge. Even taking into account the prevailing discounts on vouchers and certificates, Stanton claimed, Belger still paid too much for coal. Purchasing so far above the "cash market price," according to Stanton, was a practice to which the War Department "attaches a high degree of criminality."[52]

The McKinstry and Belger trials were more complicated and more significant than subsequent courts-martial of quartermasters, such as those of Black in Louisville and Hurtt in Cincinnati in late 1863 and early 1864, which concerned charges of out-and-out corruption. Unlike those more junior officers, McKinstry and Belger were senior depot chiefs who were never convicted of using their public offices for personal gain, but whose trials turned into battles over distinct visions of the military economy. Ultimately, Belger and McKinstry lost their jobs not simply because they failed always to manage procurement energetically enough to minimize prices but also because they lost their gamble against the political power of the producerist vision of military economy. By pointedly rejecting the idea that they should have promoted competition and virtue by seeking out original manufacturers and dealers and by going so far as to claim that so-called middlemen and speculators were actually the key to mobilization, both quartermasters violated the basic principles of the ideal procurement system that had been built up in the North over the first months of the war. By departing so openly and stubbornly from this ideal—which was embraced by some of their own peers, as well as much of the press, the Congress, and the Northern public—both officers invited the punishment they received in 1863. Meanwhile, the businessmen who had allegedly robbed the public by acting as middlemen in Saint Louis and Baltimore remained outside the reach of the military courts. But this would soon change.

Contractors on Trial

Long after the Van Wyck committee faded from public view in 1862, Northerners continued to attack military contractors as immoral profiteers who enriched themselves while Union soldiers died by the thousands. Indeed, after most legislative investigations into contracting ended, mounting casualties and other continuing stresses of war led many Northerners to denounce contractors even more heartily than they had during the opening months of the war. As in the cases of the embattled quartermasters, this campaign against bad contractors was not simply an effort to punish individuals for specific acts of fraud or a cry against political favoritism. Rather, Northerners continued to rail against illegitimate profit taking by parasitical intermediaries in the war economy. During the second half of the war, one of the most dramatic effects of this continuing discontent was the expansion of the jurisdiction of military courts to cover contractors in the private sector.

Long before contractors were subject to courts-martial, Northerners had compiled a variety of complaints against men in the military supply business. In some cases, wrongdoing by contractors amounted simply to the flip side of corruption by public officials. Bribery was hardly unknown in the wartime North. The Van Wyck committee discovered that, during the first months of the war, railroad companies had paid officers of Wisconsin regiments for the privilege of transporting them. The committee also heard one inspector at the New York City depot say that he had been offered a $300 bribe by Ware & Taylor, a footwear contractor from Boston then struggling to fill a large order. Indeed, the supply bureaus' civilian inspectors—who influenced the initial evaluation of proposals and had the power to accept or reject shipments—were the people most likely to be approached by unscrupulous contractors. According to the quartermaster general, they were also the most likely to succumb to improper influence. "While some Quartermasters have been found guilty of fraud," Meigs informed the office of the secretary of war in January 1864, "many more inspectors have yielded to temptation." Like supply officers, inspectors could be caught up in partisan storms that generated constant complaints about favoritism, corruption, and disloyalty. "I have a drawer full of anonymous notes attacking my inspectors," Philadelphia depot chief George Crosman testified late in the war, "from the highest to the lowest." In some cases, at least, such accusations were well founded. Upon taking over the Cincinnati depot in late 1864 from Charles Moulton, Quartermaster William McKim reported to Washington that the recent administration of procurement there had been "defective and corrupt." One chief inspector, McKim told Meigs, was living in one of Cincinnati's finest hotels, "his wife amply supplied with diamonds and expensive apparel; his expenses for board being more than his pay."[53]

If such reports suggested that at least a few contractors paid bribes and kickbacks throughout the war, Northerners also had good reason to believe that some suppliers owed their military business mainly to political connections. Members of Congress, of course, often recommended firms in their districts to the War Department, with mixed success. Early in the war, the efforts of Philadelphia Congressman William D. Kelley apparently helped wagonmakers Wilson, Childs & Company to convince the Quartermaster's Department to give them the first of what would become a steady stream of large orders. Similar efforts by Congressman Isaac N. Arnold of Chicago, however, never succeeded in securing large tent contracts for Gilbert Hubbard & Company, which had been a major supplier to Illinois during the first year of the war. Congressmen also visited the supply bureaus and the Treasury to try to speed along the settlement and final payment of their constituents' contracts, again with mixed results.[54] For his part, Quartermaster General Meigs in Washington routinely deflected requests from elected officials by explaining that any prospective supplier must compete in terms of price and quality by submitting bids at the depot level.[55]

Although Meigs had considerable success in keeping contracting decentralized and depoliticized, well-informed Northerners would not have had to look far for signs that some military contracts seemed to be rewarding Republicans and their friends. William Calder, the leading horse and mule contractor from Pennsylvania, was a business partner of the son of Simon Cameron, Lincoln's first secretary of war. As Calder's own testimony before the Van Wyck committee in 1861 and 1862 revealed, the younger Cameron had arranged loans of $800,000 from their bank to allow Calder to pay his subcontractors during the first week of the war. Even after Cameron stepped down, his friends in Pennsylvania seemed to benefit from their Republican credentials. William Anspach and his son-in-law M. Hall Stanton, merchant-bankers from Philadelphia who were both active Republicans and acquaintances of Cameron, filled over $2 million worth of prime contracts for clothing during the war.[56]

Another leading clothing contractor, Thomas Carhart of New York City, could also claim solid Republican credentials. Carhart's silent partner was George Opdyke, who at the end of 1861 had been elected as New York's first Republican mayor. In fact, Carhart and Opdyke—both experienced clothiers who had worked before the war in New Orleans and New York—had entered the military contracting business during the first months of the war, before Opdyke became mayor. But nearly all of Carhart's over $4 million worth of contracts with the Quartermaster's Department were signed in 1862, when Opdyke was in office. While Carhart ran the business and made all contracts in his name, Opdyke supplied most of the capital and received a third of their $500,000 in profits.[57]

Although criticisms by Democrats of alleged Republican manipulation of military contracts constituted a surprisingly small part of the totality of anti-administration rhetoric, they did escalate during the 1864 presidential campaign. Over a year earlier, the *Cincinnati Enquirer,* a pro-Democratic newspaper, had claimed that the Holenshades—the North's leading wagon contractors—had told their workers to vote Republican and dismissed some who refused to do so. In September 1864, as the election approached, the *Enquirer* informed its readers "that the Administration intends to use its vast money patronage to perpetuate its power—and that it means to make the shoddyites, the contractors who have grown rich out of the war, and the office-holders, contribute liberally for that purpose." Across the North, other Democratic newspapers made similar charges. The Lincoln administration, their editors claimed increasingly as the election approached, was in cahoots with the "shoddy contractors" who used the war as a money-making opportunity and were therefore interested in seeing it continue. This theme was evident in an 1864 print entitled "Running the Machine" published by Currier & Ives, one of the nation's leading producers of popular images. By portraying Lincoln and his cabinet as incompetent and heartless administrators of the war, this print promoted the main theme of the campaign of Democratic presidential candidate George McClellan. But by picturing two anonymous contractors around a table with the Republican officials, demanding "more greenbacks" from the Treasury, "Running the Machine" demonstrated that one part of the Democratic campaign strategy was to complain that the contracting system was being used as a patronage tool.[58]

Fig. 5.1. "**Running the 'Machine.'**" Published by Currier & Ives, a leading American producer of popular prints, this visual attack on the Republican administration's handling of the war effort was created for the 1864 presidential campaign. The two figures in the foreground, sitting at the table with President Lincoln and several of his cabinet members, are identified only as "Contractor." Addressing Treasury Secretary William P. Fessenden, who is cranking "Chase's Patent Greenback Mill" (named for Fessenden's predecessor), the contractors demand, "Give us more Greenbacks," and "Give us more Greenbacks, compound interest." Arguing that incompetent Republican leadership was allowing a few well-connected contractors to pillage the public purse, the cartoon suggests that, well beyond the war's first chaotic months, Northerners remained deeply concerned about the problems of corruption and profiteering in the military economy. *Source*: Library of Congress, Prints and Photographs Division, reproduction number LC-USZ62-9407.

Even from the fragmentary evidence on this subject that survives, it is clear that there were cases in which Republicans indeed tried to milk the contracting system, especially during the 1864 campaign. Anspach & Stanton, the Philadelphia clothing contractors with close ties to Cameron, told the former secretary of war that they paid nearly $5,700 in "election expenses" for the Republican cause.[59] Perhaps the most remarkable case of electoral-year patronage politics in the North's military economy involved another supplier of uniforms from Philadelphia, E. Tracy & Company. Known to the business community as a manufacturer of watch cases, Tracy & Company had filled a few small orders for haversacks early in the war. But in October and November 1864, it suddenly signed three giant contracts for overcoats, worth a total of over $3 million. The Philadelphia depot officer who negotiated these contracts, a young army regular named Herman Biggs, admitted to Quartermaster General Meigs at the time that the extraordinary Tracy contracts were a result of pressure from the office of Secretary of War Stanton.[60]

Meanwhile, at the large depot in Nashville, volunteer quartermaster and U.S. Military Railroad administrator John C. Crane wrote a note on government stationery calling on local contractors to contribute to the Lincoln campaign, promising that "to those who respond cheerfully to this call . . . the patronage heretofore extended to them [in the form of jobs and contracts] will without doubt be continued." At the end of October, days after circulating this memo, Crane boasted to the private secretary of vice presidential candidate Andrew Johnson that he would raise as much as twenty thousand dollars for the campaign.[61]

However spectacular, these instances of partisan manipulation of the procurement system do not mean that Republicans used the North's war economy systematically as a patronage machine. Across the war economy as a whole, the relationships between leading contractors and the Republican Party were mixed. Some major contractors certainly identified themselves as Republicans. George Bullock, a partner in the Philadelphia firm that served as the North's leading supplier of woolen textiles, ran for Congress (unsuccessfully) in 1864 on the Republican ticket. Henry McComb, the leading Delaware contractor, wrote letters to Lincoln and Johnson during that year to express his support. But even these cases were more complicated than they appear at first sight. Bullock & Company was no wartime upstart but a wealthy and well-established firm with experience as a supplier of wool to antebellum contractors. While many contractors enclosed recommendations from politicians with their bids, the Bullocks' proposals were recommended by George Crosman, the army regular who headed the Philadelphia depot for most of the war.[62] And although

McComb supported the Republicans during the war, he had also won army contracts during Democratic administrations in the 1850s, when he sold leather to the Philadelphia depot.

Indeed, several of the top contractors in the North had been selling to the army long before the Republican ascendance. This group included Fox & Polhemus, the nation's top supplier of tents and duck; the Saxonville Mills, a leading blanket contractor; and most of the wartime North's top suppliers of wagons and heavy ordnance. In short, many of the Union's leading contractors were exactly those firms that one might have expected to fill the role under a Democratic administration. Other leading Civil War suppliers, including clothier John T. Martin (easily the king of all Northern contractors in dollar terms), seem to have avoided party politics altogether.[63] When the entire procurement system is considered, in other words, the influence of Republican patronage priorities in army contracting appears to be far from total.

Although partisanship was never entirely absent, the procurement process and Northerners' debates over its ethics always involved more than party politics. By the time of the 1864 election campaign, when Democratic editors attempted to tie the Republican administration to contracting abuses, Northerners had been exposed to three years' worth of intense public discussion of corruption and profiteering in the war economy. Even as the headlines became dominated by the activities of the armies in the field and legislative investigations of contracting became less frequent, many Northerners continued to denounce what they regarded as illegitimate profit taking during a wartime emergency. As the court-martial trials of quartermasters McKinstry and Belger in 1862 and 1863 showed, the middleman remained an important target, well after the heyday of the Van Wyck committee.

"Shoddy," another target defined during the first weeks of the war, also remained a keyword. In 1863, the Northern reading public was presented with Henry Morford's *The Days of Shoddy: A Novel of the Great Rebellion in 1861*. This novel tells the story of Charles Holt, a conniving New York City textile merchant who combines profiteering with the attempted seduction of the wife of one of his employees while that man is doing his military service. Although Holt receives his due in the end (he is killed by a sharpshooter hired by his wife as he stands at the window of a military prison in Richmond), Morford also used a series of asides in the novel to demonize profiteering contractors. While "shoddy" first appeared in 1861 "as a designation for swindling and humbug of every character," it had hardly been banished. Rather, Morford suggested, the Union was continuing to do battle with "an evil spawn of men" on the home front. "The leech has fastened upon the blood of the nation, and it will not let go until the victim has the last drop sucked away, or finds strength, in recovered health, to dash the reptiles from its bleeding sides." By seeking to take advantage of the crisis even as thousands of soldiers sacrificed their lives, Morford concluded, "[e]very shoddy contractor . . . has been a national murderer."⁶⁴

In some cases, the attacks on "shoddy" merchants functioned as anti-Semitic slurs. Members of a small ethnic minority and prominent in the clothing trade in many American cities, Jews in the wartime North experienced some frightening instances of persecution. At the end of December 1862, General Grant had issued a general order stating that "Jews, as a class, violating every regulation of trade established by the Treasury Department . . . are hereby expelled" from his military department—an area that then covered much of western Kentucky and Tennessee. Although President Lincoln revoked this order only days after it was issued, Grant had encouraged Northerners to see Jews as the chief practitioners of wartime commercial immorality. Meanwhile, as the historians Gary Bunker and John Appel have shown, several political cartoons in Northern periodicals used Jewish stereotypes in visual illustrations of unscrupulous "shoddy" contractors.⁶⁵

The continuing attacks on 'shoddy" and parasitical contractors were directed not only against Jews, however. During the New York City draft riots of July 1863, protesters targeted at least two establishments associated with the military supply business. One of these was a store of Brooks Brothers, the clothing firm that had been investigated after problems with the quality of its deliveries to the state of New York in 1861. Another was a small arms factory owned by Mayor Opdyke and his son-in-law.⁶⁶ Whether the rioters knew that Opdyke had been a silent partner in Thomas Carhart's huge 1862 clothing contracts is unclear. But all New Yorkers had the opportunity to learn of this in the winter of 1864–65, when Opdyke sued Albany newspaper editor Thurlow Weed (also a Republican) for libel. This suit came about after Weed wrote that "George Opdyke has made more money upon army contracts than any fifty Jew sharpers in New York" and accused him of presenting an inflated claim for riot damages to the arms factory. Seeking fifty thousand dollars in damages, Opdyke and his lawyers went to court.

They probably regretted it. Testimony during the trial elicited many details about the Carhart contracts, including the fact that Opdyke had personally cleared nearly sixty thousand dollars from them. Opdyke's lawyer, the prominent legal reformer David Dudley Field, tried to convince the jury that such gains were justified. "The furnishing of supplies to the army and navy on contracts fairly obtained and fairly executed," Field argued "is an honest and patriotic act." Honest contractors, he insisted, "are helpers of their country as truly as he who leads a battalion, or digs in the trench, or mounts the parapet in a storm of fire and leaden hail."[67] The jury, which hung, was not entirely convinced; Opdyke received nothing, and New Yorkers now knew a great deal more about his business operations while he was serving as a wartime mayor. Many could only conclude that Opdyke deserved the unfriendly words of Weed, if not worse. For the *New York Sun*, the trial served as a demonstration of the terrible inequalities that prevailed in the contracting system. The garments that generated huge profits for the former mayor, the newspaper noted, were "doubtless manufactured at low prices by the poor sewing women, who have recently appealed against the rich contractors, who realize enormous fortunes and refuse to pay living wages to those whom they employ."[68]

As the reaction of the *New York Sun* to the Opdyke-Weed trial suggests, many Northerners were disturbed by the spectacle of a few entrepreneurs reaping giant wartime profits while most of their fellow citizens suffered. For some highbrow critics, the rise of a "shoddy aristocracy" served as a new chapter in the old story of the rise of an uncultured new rich that did not know how to spend its money. In *The Days of Shoddy,* Morford complained that "[t]he wife of the shoddy millionaire will buy diamonds that she can neither appreciate nor value. . . . His daughters will struggle for incongruous marriages . . . and his sons will disgrace the country abroad." In 1864, a year after Morford's novel was published, similar complaints appeared in a *Harper's Monthly* magazine article by the etiquette writer and war journalist Robert Tomes. "The ostentatious nouveau riche, the fraudulent contractor who makes a display of his ill-gotten gains, and vulgar pretenders of all kinds," wrote Tomes, "will forever, in the popular eye, bear upon their emblazoned coaches, the fronts of their palatial residences, the liveries of their coachmen, and on their own backs of superfine cloths and glistening silks, the broad mark of SHODDY."[69]

As casualties in the field mounted and inflation cut sharply into real wages, many Northerners attacked contractors in stronger terms than this. The nation at war, many observers claimed, was being weakened by a few evil parasites who placed personal gain above the welfare of the Union. Even the *North American and U.S. Gazette,* a conservative Republican newspaper in Philadelphia that sometimes defended contractors as men who "have enabled us to carry on a great war," sensed this grave problem. Behind the tragic and glorious events in the field, its editors wrote in a March 1863 editorial, was the war's "dark background, where sordid and greedy plunderers are secreting the spoils wrung from their country's coffers at the hour of her utmost need." Even Tomes, concerned largely with putting the nouveau riche in their place, dropped at times into less subtle criticisms: those who regarded the war as a profitable time, he wrote in 1864, represented "the chuckling of gain over its pockets filling with the treasure of the country, while our brave soldiers are pouring out their blood in its defense. . . . They are at a banquet of abundance and delight . . . though the ghosts of the hundreds of thousands of their slaughtered countrymen shake their gory locks at them."⁷⁰

Although some of the most egregious acts of selfishness in the context of national sacrifice were not technically illegal, many Northerners believed that they must still be punished. In 1861, before the war had started in earnest, several enthusiastic critics of the early procurement effort had pointed to examples from the Napoleonic wars, in which contractors had been subjected to military discipline. Two or three years later, having experienced the trauma of mass death and injury, a more grim Northern public recommended similar policies. In June 1864, the *Newark Daily Advertiser,* a pro-Republican newspaper, concluded that corruption was rampant in the military economy. "Scarcely a day passes that we do not hear of some fresh fraud upon the Government," wrote its editors, "some attempt on the part of contractors or subordinate officials to add to their personal gains at the expense of the people." In response to these outrages, the *Advertiser* insisted, "no mercy" should be shown to any guilty contractor. "The people everywhere," it concluded, "demand that the punishment of all offenders in this direction shall be summary and severe."⁷¹

In fact, by the time this new wave of cries for punishment appeared, contractors were already starting to be subjected to military laws and tried by courts-martial. This remarkable development served as the final chapter in the North's continuing struggle over the ethics of war economy. It was possible because Congress redefined contractors as part of the "military establishment." The statutory redefinition of the boundaries of private and public in the war economy started as early as 1862, in the wake of the most celebrated operations of the Van Wyck committee. In June, Representative Frank Blair of Saint Louis—the same congressman who had urged the committee to go after Frémont and McKinstry—introduced an amendment into a large military pay and benefits bill that defined contractors as "part of the land or naval forces of the United States" and therefore subject to military justice. Among the supporters of the amendment, which was inserted without strong objections and passed along with the whole pay act in July, was Henry Dawes of the Van Wyck committee. Subjecting suppliers to military justice, Dawes noted, would allow them to be "punished promptly, severely, and immediately."⁷² Over the year that followed, however, it seems that no contractor was actually subjected to a court-martial. By that time, hundreds of other civilians were being arrested and tried for alleged disloyalty or other crimes by military commissions and courts, often in border states where the rule of normal legal institutions was most disturbed by the war.[73]

The application of military justice to contractors began only after a new statute, enacted in March 1863, reinforced and elaborated upon the July 1862 act. This law, an "Act to prevent and punish Frauds upon the Government of the United States," contained provisions for handling such cases in the existing federal courts; several of its sections, including a clause allowing whistle-blowers to collect up to half of the fines collected by the government, continued to define federal law in this field for decades afterward. But the same act also made it clear that contractors were part of the military establishment and could be tried by courts-martial, which could apply any fine or prison term other than capital punishment.[74]

Unlike the 1862 statute, the 1863 act was met with objections from several congressmen, who argued that contractors should be treated as civilians and tried in the regular courts. Senator Edgar Cowan of Pennsylvania, for instance, described the courts-martial provision as a "monstrous proposition," worthy of Caligula, that might force hundreds of individual artisans into army courts. But the majority of Cowan's colleagues supported the measure. "Contractors for furnishing supplies to the Army," Jacob Howard of Michigan reasoned, "are just as indispensable as soldiers and sailors in the prosecution of a war. Without supplies the Army and Navy could not exist." The Union needed to treat contractors as part of the military, Howard concluded, in order to prevent the many "frauds, corruptions, and peculations" in the war economy, just as Napoleon had done a half-century before.[75]

Soon after the "Frauds" bill was signed into law by Lincoln in March 1863, effectively providing new public authority over what previously had been regarded as purely private enterprise, military lawyers began to prepare court-martial cases against contractors. The defendants in these cases, now defined by Congress as part of the "military establishment," were often charged with "willful neglect of duty," in addition to fraud. During the second half of the Civil War, as many as several dozen Union contractors were tried by military courts.[76] Although these trials were formal affairs that often stretched out over many weeks as prosecution and defense called witnesses, they often featured passionate rhetoric. Judge advocates, taking up a rhetorical case that had been building in the North since the beginning of the war, often denounced the accused as evil profiteers and parasites. Although convicted contractors were not subject to capital punishment, several were sentenced to serve prison terms and pay large fines. Whereas during the first year of the war the Van Wyck committee and other legislative investigations had simply censured the public officials and contractors it accused of wrongdoing, now the military justice system was capable of handing out more substantive penalties.

The first court-martial trials of contractors began in the autumn of 1863. From the beginning, it was evident that the some of the accused individuals would become symbolic targets toward whom Northerners could focus their deep general discontent about inequality and immorality in the war economy. The first court-martial of a contractor to be widely reported in the press was that of John K. Stetler, who was charged with "willful neglect of duty" in the case of a May 1863 contract with the Subsistence Department depot at Baltimore for about $38,000 worth of coffee. Stetler's first misdeed was missing his delivery deadline; more importantly, when he offered to fill the order a few weeks late, the coffee was subjected to an unusually rigorous inspection (including analysis by chemists in Washington) and found to be adulterated. In early November, after hearing witnesses from both sides for over a week, the military court convicted Stetler and sentenced him to five years at the penitentiary in Albany. Several observers agreed that the Stetler case was being used as a general lesson to the contracting community. Following the trial from several hundreds of miles away one Cincinnati newspaper noted approvingly that "such exemplary punishment will have a most salutary influence and tend to check the terrible fraud and profligacy" in military contracting, "which . . . are a national scandal and reproach." Across town, however, another Cincinnati paper argued that Stetler was being used unjustly as a "scape-goat." In its internal review of the case, the office of the judge advocate general in Washington reached similar conclusions. "It was doubtless in view of the atrocious and wide spread frauds committed by government contractors since the beginning of the war," the review suggested, "that the court was led to make a marked example of the accused."[77]

Stetler was soon joined by other contractors, who also tried—often in vain—to claim that they were being scapegoated for minor offenses that did not deserve the attention of military courts. Correctly, they pointed out that, for late deliveries and quality problems, procurement officers were already using existing formal and informal mechanisms to fine negligent suppliers. In practice, supply officers and their inspectors regularly renegotiated deadlines, rejected certain lots of goods without canceling whole contracts, or accepted subpar goods with warnings or price reductions. Furthermore, contractors submitted formal bonds for all sizable orders, and most written contracts contained clauses stating that the prime contractor would have to cover any losses to the government resulting from nondelivery.[78]

With all of these mechanisms in place, some accused suppliers claimed, trials in military courts were overkill. William H. White, who was court-martialed for "willful neglect of duty" for failing to fill an 1863 haversack contract and delivering inferior goods, argued in his defense that "a rigorous application of the law to every case of supposed failure to fulfil a simple contract would defeat the whole object of all laws on the subject" by "extinguishing the whole race of contractors." But the army officers who served as the court in White's case rejected this argument, fining him three thousand dollars and sending him to join Stetler in the Albany penitentiary until the fine was paid. Again, according to the office of the judge advocate general in a review that approved the sentence, the court had imposed "exemplary punishment" on a relatively minor supplier.[79] Although some of White's peers in the clothing and equipage supply trade were acquitted in similar cases, it was clear that army lawyers and the Northern public as a whole were eager to identify individual contractors who could be punished for the sins of their peers.[80]

Ironically, in the context of the campaign against middlemen in the war economy, White and other small-scale suppliers of manufactured goods could seem even more culpable than leading contractors who reaped much larger profits. The stiff sentence imposed in White's case, the judge advocate general's office suggested in its review, was "doubtless in view of the great injury the service has suffered . . . through a class of speculators, who, without capital or ability to carry out extensive contracts become the willing 'men of straw' for parties possessed of more means, but not more integrity."[81] The paradox of small-scale contracting attracting more censure than large operations appeared even more clearly in the court-martial case of Benjamin C. Evans, a clothing supplier from Philadelphia. Evans had landed in a military court after he failed to deliver on an order for $60,000 worth of trousers. The judge advocate in his case, H. B. Burnham, asked the court to consider the plight of "the half-clad regiments who were fighting and marching" in the field while Evans carried out "reckless and profligate speculation" on the home front, "*trifling* with the government, and *periling* the great interests of the people in the *greed* of his *speculation*." Remarkably, Burnham drew a sharp contrast in the morality of the small and unsuccessful operations of Evans and the much larger dealings of George Bullock, the top textile contractor from Philadelphia who had testified briefly during the trial. The irresponsible acts of men such as Evans, the army lawyer claimed, meant that "the high-minded and enterprising manufacturers" such as Bullock, "men whose operations are based upon real capital and whose ability is commensurate with the wants of the government, are driven in disgust from dealing with the government!"[82] Employing a conservative strain of producerist logic, Burnham's argument implied that well-established manufacturers—unlike mere speculators who bid without means and relied wholly on subcontractors—could earn legitimate profits from the business of war.

Although the cases of White and Evans suggested that it was often minor contractors who became ensnared by the new push to punish the middleman, there was one part of the procurement system that was scarred deeply by the application of military justice. Starting in 1863, several of the Union's foremost suppliers of horses and mules were arrested, tried by military courts, and sentenced to heavy fines and prison terms. Even as many quartermasters were arguing that more open-market purchasing of animals would allow them to bypass big prime contractors to deal more directly with farmers, many of those major suppliers were being denounced and punished by courts-martial. For many Northerners, it is clear, middlemen in the horse and mule trade were particularly objectionable. Although large-scale mercantile intermediaries had been well established in many parts of the antebellum American economy, the traders in giant numbers of horses were called into existence only by the unprecedented military mobilization starting in 1861. By the second half of the war, these upstart horse and mule contractors had become the targets of some of the most strident attacks against profiteering ever heard in the North.

The first wave of trials of top horse suppliers began in the autumn of 1863, making them among the very first courts-martial of contractors in the North. In October and November, the same military court in Cincinnati heard the cases of two business partners, Charles W. Hall and Edward A. Smith. At the time, Smith ranked as one of the North's leading military contractors: during the first half of the war alone, he signed over $1 million worth of formal contracts and may have supplied close to $5 million worth of animals altogether as a contractor, subcontractor, and open-market supplier. Accused of offering bribes to inspectors at the Louisville depot, Hall and Smith denied the charges and blamed jealous competitors for inventing them. In both cases, a series of witnesses produced contradictory testimony about the alleged bribes; teams of lawyers for the contractors constructed extended defenses. But Hall and Smith encountered a particularly aggressive prosecutor, who encouraged the court to use their cases as an opportunity to strike back against immorality across the war economy as a whole.[83]

The Hall and Smith cases were notable not only because of the identity of the defendants but also for the rhetoric employed by the judge advocate who prosecuted them. Henry L. Burnett, the top army lawyer in the Department of the Ohio, peppered his closing arguments with anti-profiteering language as vivid as any that was ever voiced in the North. Only days before the Hall and Smith trials started, Burnett had stood before the same court to prosecute a minor forage supplier from Kentucky named J. C. Wilmore. In the Wilmore case, Burnett urged the court to treat the defendant harshly. "Let us," he demanded, "wherever we can, put our hands upon the leeches [and] vampires that are sapping the life blood of our army [and] the government."[84] Over the next few weeks, in the Hall and Smith cases, Burnett continued to denounce both the individuals on trial and military suppliers as a class. "The mass of these contractors," he told the court in the Hall case, "seem to look upon the Government as fair game, to be plucked at every convenient opportunity.... Their cold-blooded, covetous hearts know no pity, no remorse." Behind Hall's good grooming, Burnett warned the court, was an immoral man. "With impudent effrontery," he continued, "these contractors, made rich in a day, sport and flout their ill-gotten riches, their diamond pins, and gold chains, in the very face of the gallant but suffering army, whom they have robbed.... It is these respectable rascals that make rascality respectable."[85]

In the Smith case, Burnett used the trial of one of the Union's top contractors to continue an attack that echoed some of the most intense language voiced by Northern critics of the military economy over the previous two years. Dishonest contractors like Smith, Burnett claimed, were "assassins" of the nation who deserved far worse punishment than the military justice system could provide. "When the great Commander in Chief of the universe, shall sound the last revellie that shall call from their long restless sleep the thousands of slaughtered heroes to their final just Judgment," Burnett told Smith's court, "many . . . will point to these heartless contractors [and] corrupt officials as their murderers, and the Great Judge of all will assuredly mete out to them the punishment their iniquity deserves."[86] Although it could not begin to match the wrath of God, Burnett implied, the court could at least give the contractors a small taste of their ultimate fate by convicting them under the recent "Frauds" statute. In two of the very first military trials of contractors, the judge advocate won. The court first sentenced Hall to six months in prison and a $10,000 fine; then Smith, the prime contractor, received a sentence of one year in prison and a $20,000 fine. For the first time, one of the North's top contractors was punished by a military court.

By the time of Hall's trial in October 1863, the army had arrested Moses Brown and J. M. Bryant of Louisville. Like Hall and Smith, Brown and Bryant were top-ranking horse and mule suppliers who had filled over $1 million worth of prime contracts over the previous two years. Now they were accused of supplying underage animals under contracts with Louisville depot quartermaster

Walworth Jenkins. Reporting back to Washington, War Department inspector H. S. Olcott indicated that he was no less eager than Burnett to rid the war economy of these men. "One or two such severe lessons as this," Olcott reported of the arrest and likely upcoming trial of Brown and Bryant, "will teach contractors in that section of the country that the United States cannot always be swindled with impunity." When Brown and Bryant managed to leave prison thanks to the intervention of James and Joshua Speed, friends of President Lincoln from Kentucky, Olcott was dismayed. But the arrests had disrupted the business of the Louisville contractors. "The delay" in operations created by the arrests, Brown complained to the War Department only days after Olcott's communication, "has already seriously damaged our standing and credit," creating problems that "threaten us with bankruptcy."[87] Although it seems that they were never court-martialed, Brown and Bryant would receive no more significant prime contracts.

Having scored several major successes in 1863, the campaign against the North's top horse contractors continued into the next year. In February and March 1864, both Cincinnati depot chief Thomas Swords and General J. H. Wilson of the Cavalry Bureau made it clear to Quartermaster General Meigs that much of the new demand for horses and mules would be filled by open-market purchases. Meanwhile, Wilson issued a circular that established stricter inspection standards for animals presented to depots across the country. By ordering inspectors to hold and observe horses and mules for a day before accepting them and to brand rejected horses with an *R*, Wilson explained, he sought to "make it to the advantage of the contractor to identify his interests with those of the Government."[88]

Contractors soon complained that these shifts in procurement practice made it difficult for them to find subcontractors. This state of affairs led directly to a new series of courts-martial of leading contractors. These cases involved three top suppliers of horses and mules: John Spicer, Samuel Smoot, and Daniel Wormer. Each man had signed contracts with various depots in February 1864; all together, these agreements involved 9,100 animals worth over $1 million. These contractors found that, after the Wilson circular and the rise in open-market purchasing by the Quartermaster's Department, their suppliers demanded prices much higher than the prime contract rates of $124 to $135 per animal. Rather than filling their contracts at enormous losses, Spicer, Smoot, and Wormer all defaulted.

Within weeks, all three of these leading horse contractors found themselves being tried in military courts. A contracting officer, by taking advantage of standard contract clauses, always had the option to procure any missing goods through open-market purchases or by re-letting contracts, charging the original contractors for any extra costs. But thanks to the congressional legislation of 1862 and 1863, the military was also able to go beyond this by court-martialing the defaulting contractors for "willful neglect of duty." The War Department opted to take this path in the cases of Spicer, Smoot, and Wormer, who were arrested and jailed in Washington from April to June 1864 as they awaited trial. Their courts-martial featured testimony from army officials and suppliers about recent developments in the market for horses and mules. According to the contractors and their lawyers, shifts in procurement practices at the beginning of the year had so altered the state of the market that they were effectively released from their agreements. In the eyes of the prosecution, however, the defaults had shown the contractors for what they truly were: speculating intermediaries who enjoyed the fruits of dozens of profitable deals but fled when a single contract threatened them with losses. Testifying in the Spicer trial, Wilson himself said that the more stringent inspection standards created by his circular "looked to breaking down the system of middle-men, or the subletting of these contracts." Over two years after the Van Wyck committee had warned Northerners of the damage done by the rise of profiteering middlemen at the beginning of the war, defaulting horse contractors such as Spicer could be seen as perfect examples of these unpatriotic intermediaries. The military court that heard their cases agreed with the prosecution. Spicer, Smoot, and Wormer were each found guilty, ordered to pay substantial fines (of $2,000 to $10,000 each), and sentenced to remain in prison until they paid.[89]

THUS, BY THE SUMMER OF 1864, less than a year after the first courts-martial of contractors had commenced, several of the North's leading military contractors had not merely been forced out of the business but sat in prison. By this time, the Quartermaster's Department had issued a general order informing officers at depots across the country that contractors who failed to deliver, in addition to those suspected of outright fraud, could be arrested and tried in military courts.[90] The courts-martial of suppliers lasted through the end of the war in April 1865. Outside the field of horse and mule supply, it seems, these trials never seriously disrupted the business of military contracting. Certainly, the private, profit-seeking side of the North's mixed military economy was alive and well at the end of the war. Nevertheless, the courts-martial of contractors from 1863 to 1865 were an important sign of the virulence of the struggle in the North to investigate and punish alleged corruption and profiteering. Although the first year of the war saw the Van Wyck committee and others achieve important revelations and reforms, the problem of profit taking in the context of a national military struggle was too deep to be solved by the dismissals of a few individuals. By referring frequently to abstract targets such as shoddy, middlemen, speculators, and parasites, critics of the Northern war economy suggested that its worst problems were systemic. Under these circumstances, the defendants in military courts and other individuals singled out for punishment functioned as symbolic targets, who stood for a range of injustices that went well beyond their own particular alleged crimes.

While the war in the field still raged, the profound discontent many Northerners felt about their military economy could not be assuaged. When the Union was able to claim victory in April 1865, however, the situation changed. Suddenly, many of the tensions that accompanied the everyday operation of the supply system could be released. As the North digested the news of its victory and carried out an unprecedented demobilization, many contractors found that their status, if not their incomes, took a decided turn for the better. Indeed, in many ways, political and economic developments in the postwar era suggested that Americans quickly discarded many of the organizations and ideas that had only recently been so important in the North's war economy. There were some signs, however, that, even if they were not widely acknowledged, wartime developments did have important legacies.

CHAPTER 6

The Unacknowledged Militarization of America

To understand the legacies of the Union's giant military supply project, one must accept two distinct conclusions. On the one hand, no one could doubt that peace was radically different from war. When the Union declared victory in April 1865, the North's procurement project was called suddenly to a halt. In a matter of days and weeks, the previously voracious wartime state turned to a starvation diet. The demobilization involved not only severe slowdowns and stoppages but also some outright reversals. Buyers became sellers; thousands of public employees returned to the private sector; the U.S. military machine moved back to the West; defendants became plaintiffs; parasites might even become producers. The economic demobilization was remarkably swift and comprehensive—so much so that the effects of the North's military supply effort, which had been so unprecedented and far-reaching during the four years of war, soon seemed to have been erased completely by the return of peace.

But there was also a different ending to this story, one that has been much more strongly resisted in American memory and history.[1] Although the sweeping demobilization process starting in April 1865 did reverse many wartime developments, the war continued to affect the ways in which many Americans approached economic and political questions, in a wide variety of fields. The North's war procurement project, in particular, had been associated with two developments with potentially important long-run consequences. First, the war had exposed Americans for the first time to a truly massive national state, which featured a robust administrative bureaucracy and large-scale public enterprises. Second, by giving rise to a giant public procurement machine capable of bypassing some mercantile intermediaries and by highlighting economic inequalities, the war had evidently amplified the producerist critiques of commercial capitalism that had long been part of popular political economy in America, while suggesting that the national state might play a larger role in administering and regulating the economy than had previously been imagined.

Both of these developments associated with the business of Civil War in the North helped to shape postwar business and government. To be sure, it would be foolish to conclude that the Civil War was the prime mover for all significant developments in the late nineteenth century. But it is no less misguided to focus exclusively on other potentially powerful sources of influence, such as the rise of large-scale industrial capitalism and the growing ranks of reform-minded professionals. To some degree, at least, as a few historians have suggested, all of these things were interrelated.[2] For the most part, however, the Civil War and American military institutions have been seen as having little to do with the rise of industrial corporations, bureaucracy, and popular political movements during the last decades of the nineteenth century. Such a view overlooks evidence suggesting that memories of the war mobilization continued in the decades after 1865 to influence many Americans' practical and imaginative efforts to build and manage larger, more complex, and more powerful organizational forms in government, business, and society. Despite the thoroughness of the postwar demobilization and the importance of private corporations and reform-minded professionals, there were significant military foundations of American modernity—foundations that have remained mostly unacknowledged, from the time of the Civil War to the present day.

Running in Reverse

Through the spring of 1865, the Quartermaster's Department and other Northern military supply bureaus continued to buy millions of dollars worth of goods and services every month. But by the summer of that year, when it was evident that the formal peace would last, they found themselves with a new kind of work. To be sure, during the several months that it would take for most Union soldiers to muster out, supply officers continued to undertake substantial logistical feats of supply and transport. In many fields, however, the direction of transactions was now reversed; at the same time, tasks that had been going on throughout the war now attained a higher priority. Millions of dollars worth of goods that were not auctioned off needed to be warehoused; thousands of war claims demanded settlement. Among those pressing claims were several of the contractors who had been openly vilified or formally prosecuted during the war years. Remarkably, several of these claimants won their cases and recovered considerable sums. The reversals in the North's procurement machine, therefore, proved to be political and ideological as well as material. In the wake of victory, as previously disgraced contractors turned into victorious plaintiffs, the business of war became more respectable.

The North's demobilization effort started only hours after the surrender of General Lee at Appomattox, which took place from 9 to 12 April 1865. By 13 April, one day before President Lincoln was shot at Ford's Theatre, the main Quartermaster's Department depots were contacting dozens of their suppliers to suspend recent orders. On 28 April, the War Department ordered the supply bureaus to stop buying most goods and to begin dismissing employees.[3] As Quartermaster General Meigs recalled in 1866 in describing one important class of goods handled by his bureau, "[t]he manufacture and purchase of clothing ceased with the termination of the war, and all contracts which the United States could retire from without violation of public faith were discontinued."[4]

Within weeks, the effects were noticeable. By June 1865, one journalist reported that "the sight of soldiers in the streets is becoming rare. . . . The foundries have done their work; shot and shell are no longer turned out by the ton, but, figuratively speaking, the iron for them is cast into plowshares, and the sword has been beaten into the pruning hook." For Northerners who had just experienced four years of war, the visible effects of economic demobilization were signs of part of a larger, dramatic change. "The shipyards are busy on their contracts for merchants," the journalist continued, "the armories are disbanding their forces, the makers of ordnance are unemployed, and the whole tenor and tone of our daily lives is suddenly transformed from one of eager and vigilant activity for our national existence as if we had dropped from one sphere to another."[5]

As this report suggested, the summer of 1865 was a season of dramatic changes in the national labor market. Thousands of men and women laid off from their military employments looked for jobs in the civilian economy. The largest layoffs occurred in the Union military forces. By mid-November, 80 percent of the million-man force had been discharged. The reduction continued in the months to come, as Congress demanded a smaller army. In July 1866, the authorized strength of the army was cut to 54,000 men, about three times the size of the antebellum force. Three years later, this figure was reduced to 37,000; in the mid-1870s, it bottomed out at 27,000. Once a world-class military force, the U.S. Army in the late nineteenth century was only half the size of that of Belgium, a nation with a population only one-tenth of that of the United States.[6] In the military supply business, the layoffs were faster and equally dramatic. Two of the largest private small arms factories in the North, the Sharps plant in Hartford and that of the Norwich Arms Company in Norwich, Connecticut, closed entirely only months after the end of the war. Colt, another leading arms maker based in Connecticut, cut its work force from a wartime high of roughly 2,000 to about 600 immediately after the war.⁷

The military-industrial layoffs of 1865 were equally large on the public side of the North's mixed military economy. Many Quartermaster's Department manufacturing operations quickly ground to a halt. In Louisville, all 255 women employed by the depot on clothing production were dismissed by the end of May 1865. At Nashville, which as a U.S. Military Railroad hub had a large concentration of civilian employees, the number of men on Quartermaster's Department payrolls dropped from 13,000 in April to 2,000 by December 1865. At the Washington, D.C., depot, the bureau's civilian work force declined from 9,200 in June 1865 to under 1,400 a year later. Across the country and across the various bureaus, the downsizing followed similar patterns. By January 1866, the Springfield Armory of the Ordnance Department had a work force of just over 700, down from more than three times that a year before.⁸ After President Grant ordered further cuts soon after taking office, the Quartermaster's Department—which had employed some 100,000 civilians by the second half of the Civil War—by the end of 1869 counted just 4,000 civilians on its payrolls; this figure would drop to 2,700 by 1871.⁹

The army's demobilization efforts involved not only men and women but also animals and material goods. Auctions of war surplus, which had occurred on a small scale during the war, now became the arenas for transfers of large stocks of property from public to private hands. In the six years following the end of the war, army supply departments—led as usual by the Quartermaster's Department—received over $66 million through their sales.¹⁰

Horses and mules, with their large appetites for forage, had to be disposed of quickly. From May 1865 to August 1866, the Quartermaster's Department sold nearly 235,000 horses and mules for a total of close to $15.2 million. By the summer of 1866, the army was left with just 10,000 horses and 16,000 mules. Meanwhile, the Quartermaster's Department used other auctions to dispose of its own large fleet. By the summer of 1866, it had sold nearly all of the 170 ocean-going vessels and the 600 steamboats and river barges it had owned at the end of the war. A year after the end of the war, the sales of army vessels had brought in $3 million; the navy, for its part, garnered $10 million by 1868 by selling off some 440 vessels. The military telegraph and military railroads, which had operated as semi-autonomous divisions of the Quartermaster's Department during the war, were also put on the auction block. By 1866, some 15,000 miles of military telegraph had been sold. In the South, the army sold $11 million worth of railroad equipment to private parties, largely on credit terms.¹¹

The disposal of Ordnance Department surplus received special attention from Congress and the American public. The sales of public ordnance peaked during the 1871 fiscal year, which coincided with the Franco-Prussian War in Europe. In that year, the Ordnance Department sold nearly $10 million worth of surplus; during the previous five years, it had sold only $7 million worth. For Americans interested in the Franco-Prussian conflict, these sales of weapons (most of which went to France) created political tensions. They also concerned several congressmen and other Washington officials, who were wary of violating neutrality laws. By February, both branches of Congress had launched investigations into the ordnance sales. They found that the Turkish government had purchased $2 million worth of the public arms stocks; Austin Baldwin, & Company, a New York commercial firm, had bought $3 million worth. In addition, three leading Civil War arms contractors—Remington, the rifle manufacturer, and Schuyler, Hartley & Graham and Herman Boker & Company, two more New York import-export businesses—had together purchased $4 million worth. (Remington, Austin Baldwin, and Schuyler, Hartley, & Graham had all sold arms to France, while Boker & Company apparently tried and failed to supply the Germans.)[12] In a dramatic example of the kind of reversals occurring in the postwar military economy, leading Northern war contractors were now buying back from the United States nearly as many arms as they had sold it in 1861–65.

Clothing and equipage, which had accounted for such huge Quartermaster's Department outlays, were also shed. By 1871, the Quartermaster's Department had sold off nearly 1.3 million coats and jackets, over 800,000 blankets, over 400,000 shirts, over 350,000 pairs of trousers, nearly 630,000 hats and caps, and nearly 220,000 pairs of stockings. All in all, by early October 1873, the bureau had received $9 million from sales of clothing and equipage. Such numbers attracted the attention of the National Association of Wool Manufacturers, an important postwar industrial lobby, which used a sympathetic report from the U.S. Commissioner of Agriculture to complain that the large sales of surplus had damaged their industry. The manufacturers had also lost an important customer in the military, which had bought large quantities of textiles for its own manufactories during the war and had awarded at least a few modest contracts every few years throughout the antebellum era. From 1865 to 1877, however, the army's outlays for textiles were close to zero.[13]

As the absence of textile purchases during Reconstruction suggests, the army did not sell off everything. In September 1866, Washington depot chief Daniel Rucker noted that the energies of the Quartermaster's Department were now devoted to storing goods as well as getting rid of them. "Formerly," observed Rucker, "our greatest desire and effort was put forth to supply the material of war in the shortest practicable time, and of the very best that could be obtained; now, that effort is directed to protect and dispose of the immense amount of Government material of every conceivable kind, that has accumulated here."[14]

Protecting the goods deemed worth keeping turned out to be no easy task. This was demonstrated in dramatic fashion on 9 June 1865, when the army lost goods worth an estimated $4 million in a fire at the Quartermaster's Department depot in Nashville.[15] The threat of fire would be of great concern at the central postwar depots, which included a newer facility in Jeffersonville, Indiana, with a storage capacity of 1.6 million cubic feet. In the East, the major postwar storage facilities stood at the army's traditional supply center of Philadelphia, where the Quartermaster's Department now constructed an entirely new fire-resistant warehouse with a capacity of 1.5 million cubic feet. These two facilities held the bulk of the estimated $42 million worth of surplus clothing and equipage the bureau still held at the end of the 1860s. Along with fire, moisture and insects now stood among quartermasters' worst enemies. In 1873, Senator John Logan of Illinois informed his colleagues that much of the war surplus was now "rotting" in army warehouses. During the next fiscal year, the Quartermaster's Department spent $350,000 on antimoth and antimildew chemical treatments.[16] Once the lifeline for some of the largest armies in world history, supply officers now stood guard over a decaying heap.

While the flow of new contracts slowed to a trickle, War Department and Treasury officers and their clerks in Washington continued to drown in claims. During the first full fiscal year after the war, one observer noted in 1867, the second auditor of the Treasury—which handled many War Department contracts and accounts—settled more than ninety-one thousand separate demands for payment involving $178.5 million. "Thus immense," this author wrote, "are the proportions of what is known as the war-claim business."[17] This business, which gave rise to a new class of professional claims agents, comprised several distinct branches. One important group of postwar claims were those pressed by the various Northern state governments, which Congress had promised to reimburse for their early procurement expenditures in 1861. As the historian Kyle Sinisi has shown, the states eventually received roughly $44 million in payments, not only for their 1861 outlays but also for subsequent, emergency militia mobilizations. The states' efforts to secure this money, which could involve considerable clerical and lobbying work, lasted well into the 1870s and beyond.[18]

Even more complex was the settlement of claims by thousands of civilians from the border states and the South, who now demanded reimbursement for property seized or damaged by the Union armies. Until 1871, these claims were sent directly to the Quartermaster's and Subsistence bureaus. By 1874, the Quartermaster's Department alone had audited some 20,000 of these claims, two-thirds of which were rejected entirely; for those that were fully or partially accepted, it approved payments of $2.8 million. At this late date, however, there were another 12,000 claims still awaiting audits at the quartermaster general's office. Because these claims needed to show evidence of the claimant's wartime loyalty to the Union and were often poorly documented, the work of auditing them was both difficult and highly subjective. In 1874, Meigs explained that, after looking at the relevant paperwork, "I then make up my mind as to whether the claim is just, whether the claimant was loyal, and how much ought to be allowed in the claim; and I endorse my opinion on the back of it, and send it to the Third Auditor [of the U.S. Treasury], recommending what I believe to be the legal and just allowance." After Congress established a new board to audit common claims from Southern residents, Meigs and his colleagues were relieved of some of the burden of dealing with them. Between 1871 and 1880, the Southern Claims Commission considered over 22,000 petitions representing a total of over $60 million in claims, which were supported by the testimony and affidavits of some 220,000 witnesses. It eventually allowed a total of close to $5 million in payments.[19]

While the state claims and Southern claims were important, the dynamic of reversal in the military economy field was perhaps best illustrated by the handling of disputed claims by Northern contractors. The supply bureaus and the Treasury settled many contractors' claims on a routine basis, but some dissatisfied contractors sued for payment. This led them to the U.S. Court of Claims, which had been established as a kind of advisory board in 1855 and had become a real court in 1863, when Congress gave it authority to render binding judgments. During the postwar years, the Court of Claims emerged as a powerful institution. By 1880, it had considered more than thirteen thousand cases and had ordered awards of close to $20 million.[20] Although they brought only a small fraction of these cases, the dozens of military contractors whose suits reached the court during this period tended to be disputing unusually large sums of money. Moreover, the contractors' cases raised important questions about the proper relations between the wartime state and its suppliers in the private sector. Throughout the Reconstruction period, the Court of Claims tended to rule in ways that worked to bolster the reputations of contractors and the legitimacy of wartime profit taking.

Soon after it was under way, the Court of Claims' postwar rehabilitation of contractors was noted and defended by an anonymous writer in the *American Law Review*. "Much has been said to bring government contractors into odium," this author noted in 1869; now it was time to defend them. "We question the wisdom of denouncing, by the wholesale, any class of men whose services are essential to the public," he continued. In the afterglow of the Union's victory, Americans could now consider the contractors' wartime sacrifices. "While the grievances of Government, through the wrongful acts of contractors, is a thrice-told tale," this writer explained, "that of the grievances of contractors, through the wrongful acts of Government, is an unwritten chapter." Although Northerners might have assumed that all contractors made vast fortunes from the business of war, the truth was that many had suffered deeply: "Contractors have been subject to martial law. Their stores have been forcibly closed, their business broken up, and their credit ruined—all without the shadow of justice." In the continuing decisions of the Court of Claims, this author concluded, "the public may fairly expect to see old abuses receive a vigorous check."[21]

Although many Northerners might have rejected this author's sympathetic portrait of contractors, his prediction about their fate in the Court of Claims proved accurate. The court's friendly treatment of contractors was demonstrated in one group of cases involving suppliers who claimed that they had been wronged when army quartermasters refused to accept their deliveries at the end of the war. One important early case of this kind involved Brandeis & Crawford, the Louisville grain merchants who served during the war as major suppliers of forage. (Adolph Brandeis, one of the senior partners in this firm, was the father of Louis Brandeis, who would become one of the nation's most important lawyers and jurists.) During the second week of April 1865, immediately after Lee's surrender at Appomattox, quartermasters at depots around the Midwest had refused to accept further deliveries of forage. In the case of Brandeis & Crawford, the last 39,000 bushels of corn to be delivered on its 300,000-bushel contract of 31 December 1864 were rejected. Claiming that this rejection had cost it $31,000 (on a contract with a total value of close to $470,000), the company sued. In 1867, the Court of Claims ruled that the army had improperly broken its agreement and awarded Brandeis & Crawford $29,000—nearly the full amount it had requested. Over the next few years, the court continued to award large damages to contractors in similar cases. An associated group of forage contractors led by Oliver P. Cobb of Aurora, Indiana—perhaps the largest of all grain suppliers in the Union—received over $154,000 to compensate for rejected forage in an 1872 case; two years later, Cobb & Company received an additional award of nearly $100,000 more in a similar case involving different contracts. That same year, a group of mule dealers led by John A. Thompson was awarded nearly $109,000 to compensate for losses associated with the army's April 1865 rejection of animals.[22]

This friendly treatment of contractors by the Court of Claims was all the more remarkable when it involved suppliers who had been targeted by wartime investigators for alleged misdeeds. In one important group of cases, the court ruled that the contract audits and adjustments made during the first part of the war by the Holt-Davis and Holt-Owen boards had been illegal. According to the Court of Claims, the suppliers to the Saint Louis depot that had been denounced so forcefully in 1861–62 had in fact been the victims of great injustices. In a series of cases, the Court of Claims awarded damages to several early suppliers of the Saint Louis depot whose contracts had been audited and payments reduced by the Holt-Davis board.[23] It also approved payment of large claims of arms dealers whose payments had been reduced a few months later by the Holt-Owen commission.[24]

In these cases, the court not only awarded large sums to contractors but also went out of its way to criticize the wartime auditors. In awarding seventy-three hundred dollars to the Saint Louis clothing firm of Livingston, Bell & Company, Judge Charles Nott used his majority opinion to declare that this case was "one of that class which illustrate how a few public officers may drive from the governmental market all responsible contractors." Even more jarring, to those who recalled the discussions of the Saint Louis depot by the Van Wyck committee and the Holt-Davis board, was the Court of Claims' decision in the case of Child, Pratt & Fox. That hardware firm, identified by investigators in 1861–62 as the main beneficiary of improper procurement arrangements in Saint Louis, was now awarded the staggering sum of $163,000. Explaining its decision in the Child case, the court held that the "arbitrary and ex parte deductions made by the [Holt-Davis] commission . . . cannot be upheld by any principle of law or justice with which we are acquainted."[25]

The rehabilitation of suspect contractors by the Court of Claims continued when it considered the cases of the court-martialed horse contractors. In the spring of 1864, John Spicer, Samuel Smoot, and Daniel Wormer had each been fined and imprisoned by military courts after being convicted of "willful neglect of duty." In practice, however, these contractors did not suffer much punishment. With the approval of President Lincoln, Spicer was allowed to post a bond in June 1864 (just a month after his verdict) to avoid further incarceration. In September, after being encouraged to do so by Senator Alexander Ramsey of Minnesota, Lincoln pardoned Smoot.[26]

Having avoided the sentences imposed by their courts-martial, these contractors proceeded after the war was over to sue the United States in the Court of Claims. At first, it seemed that they would not be able to reverse their status from evil profiteer to aggrieved public servant. In an 1866 decision, the Court of Claims ruled against Spicer.[27] Soon, however, the horse contractors did better, perhaps because the Supreme Court's 1866 decision in *Ex Parte Milligan,* which declared that civilians should be tried in the civil courts whenever possible, called into question the legitimacy of the wartime courts-martial of contractors.[28] During its 1868–69 session, as it considered Wormer's suit, the Court of Claims decided that it had erred in the Spicer case: concluding that the army had effectively broken its contracts with these suppliers by imposing new inspection standards before delivery, the judges awarded Wormer nine thousand dollars. Building on this judgment, the court proceeded in its next term to award twenty thousand dollars to Smoot.[29]

Ultimately, the Court of Claims' generosity to previously disgraced contractors did not go unchallenged. Demonstrating an attitude more consistent with the wartime critique of profiteering middlemen, the Supreme Court reversed several key decisions. Given that one of the high court's justices was David Davis, a member of the Saint Louis auditing board that the Court of Claims had called illegitimate, this outcome was not entirely surprising. Starting with an 1868–69 case involving gunboat contractor Theodore Adams, the Supreme Court ruled that the Holt-Davis board had served as a noncoercive arbitrator with which contractors had voluntarily agreed to settle their Saint Louis depot accounts. Following the Adams decision, the high court reversed many of the decisions and awards made recently in the Court of Claims. Child, Pratt, & Fox lost their large award, and arms contractor Philip Justice was informed that, because the Holt-Owen board's wartime actions had in fact been legal, he would lose the eleven thousand dollars he had been promised by the lower court.[30] The horse contractors also lost their awards, as the Supreme Court revived some of the anticommercial rhetoric that had been so powerful in the wartime North. In a ruling against Smoot in 1873, Justice Samuel Miller's majority opinion held that because the supplier had never actually owned or delivered horses under the contract in question, the monies he sought to recover were merely "speculative profits."[31]

Despite the Supreme Court's unwillingness to endorse fully the generosity of the Court of Claims, Northerners who had followed the wartime debates over bad behavior in the military economy could not help but sense an important shift in the tone of public discourse. For those who sympathized with even the most controversial contractors, such as the writer for the *American Law Review,* this was a welcome development. Others, including former Van Wyck committee member Elihu Washburne, criticized the evident postwar ascendancy of private over public interests, of which the Court of Claims awards were one notable manifestation.³² Even after the Supreme Court overturned many of the lower court's awards, several of the controversial contractors managed to secure large monies. When the high court deadlocked in their case, one group of horse suppliers to the Saint Louis depot kept their Court of Claims award. Theodore Adams, the gunboat contractor, ended up recovering $112,000 through special congressional legislation.³³

Even suppliers who had been convicted by courts-martial could end up winning damages. This was true of William Cozens, a leading contractor from Philadelphia accused of delivering undersized and shoddy tents, whose trial had lasted past war's end into the summer of 1865. Although his court-martial sentenced Cozens to six months in prison and a $73,900 fine, he was released immediately after his trial and never paid a dollar. He and his friends, who had always held that Cozens was the victim of machinations by disappointed rivals, continued to appeal his case. In September 1867, President Johnson overturned the court-martial verdict. Then Cozens filed a claim at the War Department to demand reimbursement for losses he had suffered because of his arrest and the cancellation of his 1864 contracts. In 1870, the War Department paid him over $16,000. Cozens, who already held over $120,000 in government bonds, used this award to buy more bonds and railroad securities. For a man who had been among a handful of leading Northern contractors to receive harsh sentences in the military courts during the Civil War, life in the postwar era proved to be surprisingly agreeable.³⁴

Finally, in yet another reversal, the spirit of forgiveness toward court-martialed contractors was matched in the postwar treatment of disgraced quartermasters. In late 1863, Samuel Black, a captain and assistant quartermaster in Louisville, had been arrested after being accused of conspiring with his son and other forage contractors to sell adulterated grain to the army. In a January 1864 trial, Black had been found guilty by a court-martial, dismissed from the army, fined ten thousand dollars, and imprisoned. In May 1865, the judge advocate general's office, having concluded that the key witness against Black had held a grudge against the quartermaster, convinced President Johnson to overturn Black's sentence entirely and give him an honorable discharge.³⁵

Redemption took a little longer for James Belger, the wartime quartermaster at Baltimore who had been tossed out of the army by order of Secretary of War Stanton even after the military court in his case had acquitted him. After many friends and congressmen weighed in on his behalf, Belger's commission was restored in 1867; he returned to work as a depot quartermaster throughout the 1870s. Even Justus McKinstry, the disgraced Saint Louis depot chief, tasted from the pot of postwar forgiveness. Although his friends failed to get him reinstated in the army, McKinstry did succeed in securing a Mexican War pension. After the ex-quartermaster died in 1897, his widow Adelaide received a check from the U.S. government until she passed away twelve years later. Once among the symbols of bad behavior in the Northern war economy, McKinstry enjoyed at least some compensation through the vast postwar pension system.³⁶

Fading Westward and Away

After 1865, along with the reversals in the directions of military-industrial transactions and the status of contractors, the army also reversed field in a geographical sense, from East to West. This return to the West, which occurred in tandem with severe cuts in the size of military budgets, served to dissolve in short order the previously powerful presence of the military throughout much of America. This disappearing act involved key individuals as well as institutions. Many of the North's leading procurement officers and contractors, so prominent in wartime as the leaders of the Union's economic mobilization, faded steadily into obscurity.

The dismantling of the wartime state after Appomattox, which included the rapid mustering out of the North's military and military-industrial work forces, was the result of deliberate policy by fiscally conservative leaders in Washington. President Johnson, addressing Congress in December 1865, noted approvingly at that early date that the army would soon have only fifty thousand men, one-twentieth its wartime size. "The measures of retrenchment in each bureau and branch of the service," Johnson declared as he cited a similar decrease in the scale of annual military expenditures, "exhibit a diligent economy worth of commendation."³⁷ Congress, which established a Joint Select Committee on Retrenchment that helped it to cut taxes and military budgets repeatedly during the postwar years, shared Johnson's enthusiasm for military economy.³⁸

For many Democrats, the cuts were never steep enough. In the past, declared Democratic Representative Lewis Ross of Illinois, "we had a Government resting so lightly on the shoulders of the people that they hardly knew they were taxed." Now, he argued, "the old doctrine is being entirely reversed; we have been placed under a military despotism; a permanent military establishment has been created."³⁹ Given the extent of the reductions that were then being approved by Republican majorities and would continue into the future, Ross exaggerated. Indeed, many disinterested observers might have concluded that the United States was quite successful in preventing the giant

Civil War mobilization from ratcheting up the state's military outlays and activities in a significant way. By 1870, military spending accounted for less than 1 percent of national economic output, not much more than the fraction it had taken up in 1860.

The retrenchment project was supported not only by President Johnson and many congressmen but also by top military administrators. In early May 1865, Quartermaster General Meigs encouraged Saint Louis depot chief William Myers to cut spending drastically. "All reductions in your power," Meigs ordered, "should be effected by yourself as rapidly as possible." Three months later, in a letter to William T. Sherman, Meigs thanked his fellow general in advance "for any aid you can render the Government in the task of cutting down useless establishments and reducing expenditures to the scale of absolute necessity for the present state of military affairs." Later in the same year, Meigs told another Union war hero, General Philip Sheridan, that "[t]he present danger and difficulty of the country is financial." A decade after the war, an article in the magazine *The Galaxy* would claim that the U.S. army had never sought to maintain a large force after 1865. "Our military service," it stated while observing that the actual strength of the army had dropped to twenty-six thousand by 1875, "has always been remarkable for the entire absence of anything like a desire to perpetuate and extend its power. Army men have always been in advance of civilians in retrenching the military force, and West Point, instead of breeding a military autocracy or aristocracy, has shown that the most unselfish patriotism is taught within its walls."[40] If Meigs's behavior in 1865 is any guide, the gushing praise in *The Galaxy* was not far from the mark.

Supported by military and civilian authorities alike, the dismantling of the Union war machine seems even more remarkable when regional variations are taken into account. Predictably, the U.S. Army's numbers became especially tiny in the North. More surprisingly, the military presence in the South was also quite weak, despite the formal military occupation authorized by Congress during Reconstruction. The number of U.S. troops in Southern states (not including Texas), which had stood at 140,000 in September 1865, declined rapidly to fewer than 3,500 by 1872. By the middle of the 1870s, before Reconstruction was formally ended, the South absorbed roughly 10 percent of U.S. army troops and expenditures; the trans-Mississippi West, by contrast, accounted for 80 percent. As some historians have suggested, the speed of the Union demobilization and the army's surprisingly light presence in the postwar South probably doomed much of the congressional Reconstruction project from the outset.[41]

Although there were certainly significant military operations and violence in the West during the Civil War years, the vast majority of soldiers in America during 1861–65 worked east of the Mississippi.⁴² After Appomattox, however, the West quickly regained its status as the home region of the army. This was true not only in terms of troop deployments but also for procurement and logistics. By the end of the 1860s, virtually all army procurement efforts had left the South, and the main eastern supply depots at Cincinnati, New York, and Philadelphia were operating on a small scale. During the postwar era, the main centers of activity for the Quartermaster's Department included San Francisco, San Antonio, Fort Leavenworth, Omaha, and Saint Paul.⁴³ With the exception of significant new levels of activity in the northern plains, which were connected to U.S. Army conflict with the Sioux, this military geography was quite similar to the one that had existed in the 1850s.

The pattern of postwar contracting resembled its antebellum predecessor, in which overland transport in the West had been central. By the 1870s, when there were still large surplus stocks of clothing and equipage in their warehouses, army quartermasters spent well over half of their funds on transportation and animal feed in the West. As they had on a much larger scale during the Civil War, horses and mules continued to account for a large fraction of the army budget. By 1879, there were still about 11,000 horses and 10,000 mules in the army, nearly all of them in the West, serving along with some 25,000 men.⁴⁴ The hay and grain required to feed these animals, which cost some $3 million a year, absorbed about a quarter of all Quartermaster's Department funds during this period. Many of the nation's leading military contractors in the postwar era were merchants and landowners who resided near military installations in the West. These men included Nathanial Adams, a Manhattan, Kansas, merchant and mill owner who served as a leading forage contractor; Percival G. Lowe, a former army teamster who held local offices in Leavenworth, Kansas, while supplying quartermasters with mules and forage; and William A. Carter, the official army sutler at Fort Bridger (in southern Wyoming), who filled contracts for forage, firewood, and lumber.⁴⁵ Meanwhile, wagon freighting firms operating out of Leavenworth, Omaha, and San Antonio continued to serve as leading army contractors, just as they had in the 1850s.⁴⁶

For all the parallels between military logistics in the West of the 1850s and 1870s, however, there was one major innovation: railroad transport. Before the Civil War, when railroad networks in the East were already extensive, the absence of track in the trans-Mississippi West had prevented the army from taking advantage of this technology. In the postwar years, as rail networks spread across the West, this situation changed. To be sure, the limits of these networks meant that transport remained expensive: as Senator Logan of Illinois noted in 1873, the continuing difficulty of moving men and supplies across the West was one of the main reasons the United States spent over one thousand dollars per soldier per year, more than twice as much as European nations.⁴⁷

Even as Logan spoke, however, the steady substitution of cheaper rail transport for muscle-powered overland freighting was transforming the shape of the U.S. military economy. During the second half of the 1850s, the Quartermaster's Department had spent $4 million a year on transport, most of it for wagon trains in the West. During the 1870s, when the army was 50 percent larger, the bureau's transport outlays remained $4 million a year. By the end of that decade, however, rail transport accounted for over half of that total.⁴⁸ According to Quartermaster's Department calculations, the availability of railroads in the West had saved the country some $6 million by 1873. By the end of the decade, Justice Stephen Field of the Supreme Court suggested, in a generous estimate, this "immense saving of expense" amounted to as much as $5 million a year.⁴⁹

It was the Union Pacific Railroad, the first transcontinental line, which was most important to the postwar army. During the Civil War, a Republican Congress had promoted the Union Pacific by authorizing public land grants and bond guarantees. One of the perceived benefits of this enterprise, certainly, was national economic development.⁵⁰ But supporters of the transcontinental lines had also stressed their military benefits. This continued to be the case during the Reconstruction years, when both Congress and the Supreme Court referred to them as a "military necessity."⁵¹ Indeed, not long after the end of the Civil War, the Union Pacific became the nation's leading military contractor. Although the link to the Pacific coast was completed in 1869 and a direct connection to the eastern lines was not available until 1872, the Union Pacific took on military value no later than mid-1867, when its tracks stretched most of the way across Nebraska.⁵² In the year that followed, the army did over $1.4 million in business with the Union Pacific and its Eastern Division. By the end of the 1870s, the army had done a total of $6.6 million in transactions with the Union Pacific, along with $2.3 million with the Kansas Pacific and nearly $1 million with the Central Pacific.⁵³

As the army as a whole returned to an increasingly rail-accessible West after the end of the war, so did many of the officers who had overseen the North's vast wartime procurement project. In fact, many quartermasters returned to posts that they had known before the Civil War. Robert Allen, the senior supply chief who had spent the war years in Saint Louis and Louisville, went back to San Francisco, his home during much of the 1850s. Asher Eddy, the

U.S. quartermaster who had fought with Illinois state officials in 1861 and later oversaw depots in Tennessee, found himself back in Oregon. Even if they did not return to a previous post, many top officers were pulled in a westerly direction. William Myers, a Saint Louis depot chief, went to Omaha. James Ekin, the volunteer quartermaster who worked during the war in Indianapolis and Washington, became a fixture at the bureau's new warehousing depot at Jeffersonville, Indiana.[54]

Even for senior supply officers, the postwar transfers to western posts were not necessarily desirable. This was true for Joseph Potter, the volunteer quartermaster who served as Chicago depot chief for most of the war. After overseeing the Fort Leavenworth depot for two years, Potter (by now a colonel) was ordered in early 1867 to report to Galveston, Texas, the stormy and humid port town that held one of the main supply depots for the army's Department of the Gulf. Potter protested this assignment, complaining that his moving expenses alone would be "ruinous," given that he would have to take "over *one ton* of vouchers and public papers" accumulated at his previous posts. Given no choice, Potter proceeded to Texas. Soon after he arrived, he and his wife and child contracted yellow fever; only he survived. In despair, Potter told Quartermaster General Meigs that he might have been able to save his family if he had been granted the transfer he wanted. Instead, "I remained at my post and lost all that rendered life desirable." On the verge of resigning from the army, Potter decided to stay, perhaps because he found no better alternative. Over the next decade, he moved through posts in Texas, New Mexico, Indiana, Detroit, and New Orleans. Although Potter remarried and was able to retire to Ohio with a pension, at age sixty-two in 1878, the postwar years had not been easy.[55]

In moving to the West during the postwar period, top army procurement officers lost their status as leading economic actors based in the nation's largest cities. As the years went by, these officers gradually disappeared from public view altogether. Several of them, including the regulars, retired immediately after the war. David Vinton, George Crosman, and Thomas Swords—the top officers for most of the war at the North's three central wartime depots of New York, Philadelphia, and Cincinnati—all retired between 1866 and 1869 under a law providing for retirements (with pension) for officers at least sixty-two years old. Younger than these senior wartime depot chiefs, Quartermaster General Meigs held his post until 1882. His successor, wartime Washington depot chief Daniel Rucker, retired immediately after being named quartermaster general, also in 1882.[56]

Although it was clear during the Reconstruction years that industrial corporations were taking the place of the wartime military as the nation's largest economic enterprises, few top logistics officers joined the corporate world. Many of the regulars continued to serve in the army; occasionally, they tried their hands at other occupations, including writing, and toured the world.[57] In 1867–68, Meigs, then fifty-one, took an extended leave of absence to travel to Europe with his wife and two of their children; he would return to Europe in 1875–76 when President Grant sent him to study European armed forces. Meigs supplemented his postwar duties as quartermaster general by working on urban improvements in Washington and designing public buildings.[58] James Donaldson, a classmate of Meigs at West Point in the 1830s who became a wartime depot chief at Baltimore, continued to work as a quartermaster in Saint Louis and Boston during Reconstruction. In 1871, three years before he left the army, Donaldson published a historical novel of the second Seminole War (1835–42), called *Sergeant Atkins: A Tale of Adventure, Founded on Fact*. Before he died in 1885, he visited Paris. Many other top Union supply officers traveled the world after the Civil War. After a visit to Egypt with his family, Asher Eddy (the U.S. quartermaster who had clashed with Illinois officials in late 1861) died of dysentery at age fifty-five at the Imperial Hotel in Malta. Robert Allen (the senior wartime quartermaster at Saint Louis and Louisville), who had visited Japan in 1875, passed away twelve years later in Switzerland.[59]

Although a few stayed in the army, most leading volunteer quartermasters returned to their antebellum activities in business and party politics. Charles Moulton, who had been an important procurement officer at the Cincinnati depot, returned to his law practice and continued to promote the Republican cause. In 1875 Moulton also took the time to write an extended rebuttal of a critical review of General Sherman's memoirs. Another lawyer turned quartermaster, Roeliff Brinkerhoff, used his wartime experiences to become a successful war claims attorney. Brinkerhoff later moved into banking and eventually served as a president of the Ohio State Archaeological and Historical Society. Lewis Parsons, the wartime transport chief in Saint Louis and Washington, eventually resumed his antebellum activities as a top officer of the Ohio & Mississippi Railroad. Before this, in 1867–69, Parsons spent a full two years in Europe with his daughter, in a successful effort to recover from an illness. Meanwhile, he supervised his large farm in Illinois and joined his brothers in founding Parsons College, a Presbyterian institution located in Fairfield, Iowa. Parsons's rich postwar career also included an unsuccessful run as a candidate for Illinois lieutenant governor on the Democratic ticket in 1880. By 1898, when Parsons published a pamphlet on Civil War transport that highlighted his own achievements, most of his wartime peers were dead; in the collective memory of the American public, which was then turning its attention to the war with Spain, Civil War supply officers took up very little space.⁶⁰

In 1891, ten years after he retired, the seventy-six-year-old Stewart Van Vliet observed that, although his own health remained good, his generation of army officers was quickly disappearing. "The time is rapidly approaching," the former New York City depot officer wrote, "when our photos will be our sole representation." This statement was telling in more ways than one. Certainly, by the time Van Vliet died in 1901, there were few living members of the group of army officers who had graduated from West Point during the age of Jackson and went on to lead during the Civil War. Van Vliet—who had served in the West throughout much of the antebellum era and, as a midcareer quartermaster in 1857, had served as the U.S. emissary to Brigham Young on the eve of the Utah Expedition—had personal knowledge of the nation's history that few Americans alive at the turn of the century could share.⁶¹ But Van Vliet's statement about representation also applied more specifically to the fate of the managers of the North's procurement project. Working behind the lines during the Civil War, they were doomed to obscurity by conventional understandings of heroism and military history. Largely absent, personally, from the leadership of the most dramatic political and economic developments of the postwar era, they were even more easily forgotten.

The Military Foundations of Modern America

In virtually any general history of the United States, the late nineteenth century appears as the period in which modern organizational forms and attitudes took hold across broad areas of economy and society. Starting around 1880, bureaucratic structures and mentalities became increasingly influential. In the United States, it is often held, the most important promoters of bureaucratic administration were the managers of industrial corporations and reform-minded professionals. American modernity, therefore, was distinctly the creation of actors in the private sector. Whereas the pioneering analysis of bureaucracy by the German sociologist Max Weber had highlighted the formative role of governmental institutions as well as business enterprises, in the American case public institutions seemed irrelevant. As the historian Olivier Zunz put it, paraphrasing the leading business historian Alfred D. Chandler Jr., "Big business . . . not government, invented American bureaucracy."⁶²

Although many historians have challenged Chandler's influential general narrative of the rise of big business in the late nineteenth century, few have directly contested his argument about the roots of American bureaucracy.⁶³ By accepting the notion that modern bureaucratic administration in the United States was basically a corporate production, historians have consistently overlooked the influence of the Civil War and the military institutions of the American state. By using chronological frames that begin in the 1870s, many studies have ensured that the war years figure only as part of a shadowy background. Even those studies that reach back further in time have tended to search for the roots of American bureaucracy in virtually every field except the military.[64] In particular, the North's procurement project seems to have contributed little to the rise of big business and the more gradual development of modern governmental organizations. Given that the demobilization after 1865 was so rapid and comprehensive and that the postwar careers of most of the public managers of the North's war economy were so quiet, this interpretation seems at first glance to be entirely sensible.

Nevertheless, when they are seen together, wartime and postwar developments in the United States demonstrate that American modernity was not quite so free from military influences. More than is commonly acknowledged, economic and political developments in the United States in the decades before World War I can be described as a process of militarization. If this militarization was more indirect and took forms less formally militarist than it did in other settings, it still reflected the influence of the Civil War mobilization and the institutions that had swelled during that conflict.[65] Although the managers of industrial corporations and reform-minded professionals may have led the dramatic economic and political transformations in the United States during the Gilded Age and Progressive Era, they were influenced by military models of administration that had touched the lives of millions of Americans during the Civil War.

In the sphere of politics and government, the influence of military institutions and the Civil War mobilization was particularly evident in the civil service reform movement. Traditionally, American political history has treated the Civil War era as part of the heyday of the spoils system, in which virtually all government officials were party men. After the Civil War, the spoils system came increasingly under attack by civil service reformers, who realized some of their aims with the Pendleton Act of 1883. In practice, meaningful reform took many years and was never comprehensive. Most of the leaders of the civil service reform movement were Northern elites, including businessmen and politicians as well as intellectuals and professionals. For these reformers, there were several important models for the proposed revolution in American government. One of these was suggested by European nations such as England or Prussia, which used competitive examinations and higher education to select and train a corps of semipermanent public officials. Another model was American business enterprise, which seemed to value efficiency and common sense much more than did the existing American civil service. Yet another model, this one harkening back to the pre-Jacksonian era, was the virtuous republic, in which an enlightened citizenry stood vigilant against corruption.[66]

But it is rarely noted that an important model for American civil service reformers was the military. Particularly during the first decade of the movement, which started at the end of the Civil War, civil service reformers frequently cited the professional military establishment as an example of what the rest of American government might ideally become. Because the Pendleton Act was passed in 1883 and meaningful reforms came slowly after that, it may be tempting to dismiss these early efforts. (From 1865 to 1872, reform-minded Republicans sponsored eight civil service reform bills that made it out of congressional committees; none passed.) But civil service reform was an important political issue throughout the postwar era. President Grant, in an 1870 address, asked Congress to enact it. During the campaign of 1872, the call for civil service reform appeared not only in the platform of the reform-minded Liberal Republicans but also in the platforms of virtually every other political group—including the Democrats, the Republicans, and the Labor Reform and Prohibition parties.[67]

Above all, the movement's first decade is significant because it shows the intellectual origins of civil service reform in America. And it is clear that, from the beginning, when reformers tried to imagine the kind of government America should have, they looked to military institutions and the Civil War experience. In 1864, even before the war had ended, Senator Charles Sumner of Massachusetts and others had begun to raise the reform issue. The first serious national debates over civil service reform came in 1867, after a House committee led by Congressman Thomas A. Jenckes of Rhode Island submitted a report recommending it. In its first report, the Joint Select Committee on Retrenchment suggested that the American civil service should imitate not only the practices of European nations but also those of the U.S. military. Even before this report appeared, similar ideas had been discussed in the Senate. In June 1866, Senator B. Gratz Brown of Missouri told his colleagues that a resolution then circulating in the Senate was "an attempt to conform the civil service . . . to the regulations and conditions which govern the military service of the country." One of the admirable aspects of the military in comparison to the civil service, he suggested, was that army and navy officers had comparatively few "political appliances." When Jenckes himself first explained the case for civil service reform on the House floor, in January and February 1867, he extended this reasoning. "The civil service," said Jenckes "like the military and naval, should be conducted by the highest talent that can be procured." The Civil War, Jenckes stated openly, had made him understand for the first time the "great difference between the military and naval administrations and that of the civil departments." Now that the war was over, it was time for the United States to cultivate in its civil service the same competence and "esprit du corps" already found in its military. "A well-trained and thoroughly disciplined body of civil servants," Jenckes explained, "is as necessary to the preservation of the State as are an Army and Navy thoroughly trained and disciplined in arms."[68]

References to the military as a model for civil administration were hardly confined to Jenckes's unsuccessful efforts in Congress during the late 1860s. Most champions of civil service reform outside of government, including journalists and intellectuals, suggested that the rest of the American state should become more like the military. "Ours is probably the only country in the world," the prolific Retrenchment Committee staffer Julius Bing wrote in the *North American Review* in 1867, "where it does not exist in the civil service, though it exists in our military and naval service, the stringent discipline and efficiency of which are well known to all Americans."[69]

To be sure, many reformers mentioned European civil administrations and the American private sector as important models.[70] But those references were often supplemented by appeals to the military model that had been proven in the Civil War. American civil servants, argued leading reformer George William Curtis in 1869 and the editors of *Harper's Weekly* magazine throughout the Reconstruction era, needed to become more like soldiers: that is, more honest, more efficient, and more devoted to the public good.[71] E. L. Godkin, the influential pro-reform editor of the *Nation,* also held up the military as an example. One of the most "striking" aspects of the "vice and immorality" that prevailed in the civil service, Godkin wrote in 1869, "is the contrast it offers to the military and naval services." Military officers, he continued, "though they are drawn from the same race and the same rank of society as the officers of the civil service . . . are as much a credit and protection to the nation as the others are, as a class, a disgrace and a canker."[72]

The military model was promoted by not only politicians and journalists but also reform-minded civil servants themselves. During the Gilded Age, the Post Office ran an innovative railway mail service, which counted many Civil War veterans among its personnel. As the studies of Daniel Carpenter have shown, these men tended to imagine their operational achievements as similar to those of a successful army.[73] Earlier, just after the end of the war, many of the civil servants who supported Jenckes's ideas suggested that their offices would benefit from taking lessons from the Civil War experience. "I believe the civil service can be made more efficient and economical," wrote Alonzo Alden, postmaster of Troy, New York, to Jenckes's committee, "if established on the same basis with the military service with respect to appointments, tenure of office, and discipline."[74] According to John van Lear, an internal revenue assessor in Hagerstown, Maryland, the current objections to civil service reform were comparable to an irrational bias against army regulars that many Northerners had held early in the Civil War. Similar letters came to the Jenckes committee in 1867–68 from public officials across the country, from Assistant Secretary of the Treasury J. F. Hartley in Washington to Thomas Moonlight and J. C. Geer, internal revenue collectors in Kansas and Idaho.[75]

Although many civil service reformers were willing to hold up the regular military—the professionals—as an appropriate model for government, this idea met with fierce resistance. When leading civil service reformer Dorman B. Eaton published an essay in 1881 that referred to the army as "the pride of our statesmen" and a fine model for peacetime government, he was responding to an article published earlier in the same year by Albion W. Tourgée, the remarkable Union veteran from Ohio turned North Carolina carpetbagger judge, politician, and novelist. In his essay, Tourgée insisted that "[t]he regular army is, at best, an excrescence in a republican government."[76] Long before this, in 1867, Republican Congressman Frederick E. Woodbridge of Vermont had rejected the initial Jenckes bill as an "anti-democratic" measure, better suited to Europe than "free America."[77] Republican Congressman John Logan of Illinois, a former brigadier general and a leading advocate for Civil War veterans, argued that the Civil War experience showed that the spoils system was essential. Especially in wartime, Logan argued, elected officials needed to be able to dismiss the appointees of previous administrations and replace them with fully accountable and loyal ones.[78] The economic mobilization in particular, according to Republican Senator Matthew Hale Carpenter of Wisconsin, should not be remembered as a model of administration but rather as regrettable, excessive expansion of executive powers that had been accepted by the American people only in the panicky atmosphere of a war emergency.[79]

Even among those who saw the military as a model for the civil service, it was not common to hold up the management of the North's procurement project as an example of successful governance. Given the widespread complaints about corruption and profiteering in the war economy and the resentment many soldiers and others felt toward even the most competent procurement officers, this was not surprising. Nevertheless, many reformers understood that it was the operations of the military staff on the home front, rather than those of soldiers in the field, that most closely resembled civil administration. When they cited the wartime achievements of the regular (professional) military, reformers suggested implicitly that it was the management of the war economy, not only the management of men with guns by West Point graduates such as Grant and Sherman, that held key lessons for the postwar nation.

Although open references to the wartime supply bureaus as models for American government were rare, there is evidence that some civil service reformers were indeed thinking of the bureaucratic administration of the procurement project. In 1871, Senator Henry Wilson of Massachusetts, soon to be the vice president, joined many reformers in suggesting that the civil service movement was an effort to bring military discipline to peacetime government. As he did so, Wilson claimed that "more than one thousand million dollars were paid out during the war, much of it in the field, some of it under fire, and the Government lost less than two hundred fifty thousand dollars," which was only a fraction of what it had lost in the much smaller War of 1812. Wilson then asked, "How did it happen?" In front of his fellow senators, he answered his own question: "By reason of the good rules and regulations, the checks, the efforts made to hold men responsible, to make them honest; just what is now proposed here in this civil service reform."[80]

Some Northerners went even further. In one "Imaginary Conversation," a whimsical piece published by the New York City–based *Putnam's Magazine* in 1869, an anonymous author conjured up a discussion of postwar government and politics by a preacher, a soldier, and "The Chief of Men." At one point in the conversation, the soldier says, "I'll bet that an equal number of able quartermasters under you, Chief, would keep the business of the government not only clearer of arrears and cleaner of rascality, than any equal number of 'statesmen' that can be found, but simply clean of arrears and rascality." The preacher agrees, saying, "An army officer is trained to honor and to business. He is not trained to politics or to money-making. That is, he escapes the two worst itches of America, and he is educated in just that sentiment which American educations most lack."[81] Here, stated in no uncertain terms, was one of the more remarkable implications of using the Civil War mobilization as a positive example of governance: it implied that the military might serve as the basis for an important new force in America, which would balance or challenge the authority of the democratic parties and capitalist business enterprise.

Clearly, such a vision was far from universal. As one historian has concluded, despite the power of the Civil War experience to encourage a more active and professionalized government, "party politics, localism, and the prevalence of laissez-faire beliefs worked in the 1870s against the development of a strong and self-contained national state."⁸² By and large, the influence of the Civil War on the development of modern American government was indirect and incomplete. The military was not the only model for the civil service reforms that Congress authorized in the 1880s; in any case, the changes created by those reforms were hardly revolutionary. Nevertheless, the public bureaucracies that did emerge during the Progressive Era did have military roots. As the language of early civil service reformers suggests, the military that they knew from the Civil War served during the postwar decades as one important conceptual alternative to the spoils system. Over time, other American state institutions would become more like those that had developed in the military during the first two-thirds of the nineteenth century.

IT WAS NOT ONLY IN GOVERNMENT, but also in business, that the military model had important—if far from totalizing—effects. In many fields of business enterprise, the years following the Civil War were a time of radical transformation. One of the most important developments of the Gilded Age was the rise of big business. By the late nineteenth century, America was home to giant industrial corporations that employed tens of thousands of people, far more than any companies of the antebellum era. As early as 1873, a congressional committee declared that "[t]his country is fast becoming filled with gigantic corporations, wielding and controlling immense aggregations of money, and thereby commanding great influence and power."⁸³ Over the years that followed, as the corporations became larger and more numerous, many Americans became increasingly disturbed by the economic trends. Although their effects were limited, the acts passed by Congress to establish the Interstate Commerce Commission (1887) and a national antitrust law (1890) suggested that the rise of big business led Americans to create a more powerful national government.⁸⁴ But to what extent had the Civil War, and especially the Northern procurement project administered by military organizations in a greatly expanded national state, promoted the rise of big business in the first place?

One thing that is clear is that leading Civil War contractors, generally speaking, did not become the leaders of the giant industrial corporations of the Gilded Age. Certainly, many corporate leaders had made money during the war. But this did not mean that the war gave birth to big business. Although some historian-critics of business in the early twentieth century pointed to Civil War profiteering as an important source of capital for future robber barons, in fact very few of America's industrial leaders during the Gilded Age were major war contractors.⁸⁵ There were some exceptions. Henry McComb, the Wilmington, Delaware, leather man and general military contractor for the Union, was one of the directors of the Union Pacific, the first transcontinental railroad. William Deering, a clothing merchant from Maine who filled substantial contracts for uniforms, moved to Illinois after the war to establish an important mechanical harvester company that rivaled McCormick (the industry leader) and eventually became part of International Harvester, one of the world's leading industrial corporations in the twentieth century. Other contractors that became large industrial corporations included Proctor & Gamble, the Cincinnati firm that sold the Union armies some soap and candles, and Du Pont, the North's leading supplier of gunpowder.⁸⁶

Despite these exceptions, the North's leading contractors as a group—like its top military supply officers—played only a small role in the creation and management of the new industrial corporations. Some top contractors continued to be leaders in older industries, just as they had been before and during the Civil War. This was true of ready-made clothier William C. Browning and footwear wholesaler Benedict, Hall & Company, two New York City firms that turned their attention squarely to civilian markets after 1865. Other leading wartime suppliers had less success. Some contractors that had sold several millions of dollars worth of goods to the Union, including the formidable Philadelphia woolens concern of Benjamin Bullock's Sons and Boston wholesale clothier James H. Freeland, were forced to declare bankruptcy after the Panic of 1873. In the late 1860s, evidence from credit reports on Philadelphia businesses suggests, former war contractors were more liable to go bankrupt than were other firms.⁸⁷

Rather than supervising the increasingly vast corporate enterprises of the Gilded Age, many of the North's leading contractors reduced their active business operations after the war and turned to investing, politics, philanthropy, and leisure. William Calder, one of the Union's top suppliers of horses and mules, worked after the war for the Republican Party in Pennsylvania, as well as for hospitals and his Methodist Church. Cincinnati clothier Henry Mack, another Republican Party man, was a champion of Reform Judaism in his city. Another top Cincinnati contractor, the shoe dealer John Simpkinson, served after 1865 as the president of the Cincinnati Board of Trade, president of the local Society for the Prevention of Cruelty to Animals and Children, trustee of the Wesleyan Female College, and leading supporter of a national Arbor Day.⁸⁸

Perhaps most suggestive of all were the postwar activities of the king of all Northern contractors, the clothier John T. Martin. After the war, Martin invested in real estate, railroads, banks, and insurance companies. He served as a director of several financial institutions, including the Brooklyn Trust Company, the Nassau National Bank, and the Home Life Insurance Company. He was a co-founder of the Brooklyn Polytechnic Institute, a director of the Mercantile Library, and a member of the Long Island Historical Society. At the same time, Martin used the postwar years to amass a large art collection. On regular trips to Europe, he and his wife purchased paintings. In 1876, the Martins completed an addition to their Brooklyn Heights mansion—which Martin had purchased before the war after a successful antebellum career in the clothing business—that was designed as a dedicated art gallery. By the early 1890s, the gallery held close to one hundred paintings, including works by major artists such as Diaz and Corot. When Martin died in 1897, he was eighty-one and still wealthy.[85]

As Martin's postwar career suggests, in 1865 many of the North's leading contractors were ready for retirement. Thanks in part to the length of the war and the Union's military success, the risks of military contracting for many of those who entered the field early and on a large scale were outweighed by the eventual rewards. For many of these men, especially those who had already reached middle age by the time of the war, continuing to work long hours in active day-to-day management of firms in competitive industries was unnecessary and unattractive. Although several of these contractors invested in postwar railroads and banks and perhaps even in some manufacturing companies, the movers and shakers in the big businesses of the Gilded Age tended to be younger men working in new fields, such as John D. Rockefeller in petroleum and Andrew Carnegie in steel.

But if neither Union supply officers nor contractors were well represented in Gilded Age corporate board rooms and field offices, this did not mean that the Civil War experience and military organizations had no effects on postwar business thinking and practices. In 1871, in an essay entitled "An Erie Raid," Charles Francis Adams noted that the United States had succeeded after 1865 in returning to a peaceful condition that seemed at first to offer remarkably few signs of war-related transformation. "Yet while these superficial indications of change would be sought in vain," wrote this aspiring railroad expert who hailed from one of America's most eminent families, "other and far more suggestive phases of development cannot but force themselves on the attention of any thoughtful observer. The most noticeable of these is perhaps to be found in a greatly enlarged grasp of enterprise and increased facility of combination." The giant public enterprises associated with the Civil War, Adams suggested, had educated the private sector. "The great operation of war," he continued, "the handling of large masses of men, the influence of discipline, the lavish expenditures of money, the immense financial operations, the possibilities of effective cooperation were lessons not likely to be lost on men quick to receive and apply all new ideas." Five years after Appomattox, with the North's state-managed wartime projects a distant memory, it was now private business corporations who were acting like nations. "These modern potentates," Adams wrote of the postwar corporations, "have declared war, negotiated peace, reduced courts, legislatures, and sovereign States to an unqualified obedience to their will, disturbed trade, agitated the currency, imposed taxes, and, boldly setting both law and public opinion at defiance, have freely exercised many other attributes of sovereignty."[90] In a remarkably short time, Adams suggested, the locus of large-scale enterprise in America had passed from the military to business.

It was in the railroad industry, the focus of Adams's attentions, that postwar business leaders were most likely to imagine their enterprises as comparable to military organizations. On one level, the military metaphor served simply to describe the fierce competition among the Gilded Age railroads and their leaders.[91] But railroad leaders also thought of the military as a model for the increasingly complex organizations they managed. The work force that built the Union Pacific in the late 1860s, in the eyes of chief engineer Grenville Dodge, was "a well trained army."[92] In an 1875 memo, Charles Elliott Perkins of the Chicago, Burlington, and Quincy explained that "[t]he working organization of a Railroad, the *daily machine, so to speak,* may I think be modeled on that of the Army, the Co. being the unit!"[93]

Ten years later, a widely circulated industry directory referred to "the railway army in America," which "numbered about 600,000 men, commanded by some 5,000 general and division officers."[94] And at the end of the 1880s, Henry B. Ledyard, president of the Michigan Central Railroad, explained, "It would be impossible to manage a railroad and subversive of discipline, if every time a subordinate received an order which he did not think was exactly right, he should be constantly appealing over the head of his immediate superior, to those higher in authority."[95] Such language suggested that, for at least some of the architects and managers of modern American corporations, the hierarchical military institutions they remembered from the Civil War provided inspiration.

Although many of them referred to the army in a general way, some business leaders of the late nineteenth century seem to have been thinking about the lessons of military economy in particular. The greatest of all the industrial titans, John D. Rockefeller, hinted at this in an interview conducted in 1917, during the First World War. By that time, Rockefeller's Standard Oil empire had been known to Americans for a generation as the archetypal "trust." As he discussed longstanding criticisms that his company had unfairly used special discounts negotiated with railroads to destroy its competitors, Rockefeller suggested that his use of high-volume transactions resembled those of military procurement bureaus. "So much of the clamor against rebates and drawbacks," he complained, "came from people who knew nothing about business. Who can buy beef the cheapest—the housewife for her family, the steward for a club or hotel or the commissary for an army?"[96] With this rhetorical question, Rockefeller suggested that he and other corporate leaders imagined themselves as more closely related to wartime military procurement officers than to the small business proprietors of an earlier era.

As Rockefeller suggested, the national, integrated enterprises born in the late nineteenth century owed their huge size in part to the lower costs that increased scale and scope could sometimes bring. At the same time, the words of Rockefeller and other industrial leaders also indicate that big business was driven by an ideological project that sometimes conflated efficiency, size, and hierarchy. While Rockefeller and others could argue with some justification that the concentration in their industries was natural and efficient, it is also possible to see the modern American economy as having been shaped excessively by an enthusiasm for national territorial domination and bureaucratic organizational structures that reproduced many of the features of military institutions.[97]

Many American businessmen, however willing they were to use the military as a model for industrial organizations, assumed that most of those organizations would remain private, for-profit enterprises. Although some business leaders supported limited government regulation of industry and even public ownership of utilities, especially before the 1880s, few openly referred to state enterprise as an appropriate model for good business practices. This was true even of aspiring regulators like Adams, who regarded the prospect of national ownership of railroads as something that would go against "the fundamental principle of our political system."[98] Some postwar business leaders, influenced directly or indirectly by the work of Darwin, favored the notion that economy and society were driven by "natural laws"; for them, significant government intervention or regulation seemed unnatural and wasteful, and government was not likely to hold promising lessons for business. A few corporate leaders, including Perkins of the Chicago, Burlington & Quincy, went so far as to denounce Gilded Age proposals to nationalize the railroads or telegraph as "communistic."[99]

More commonly, those who defended the emergence of large industrial corporations stressed their economic origins and benefits, in the form of higher efficiencies and lower costs to the consumer.[100] In certain ways, it was possible to argue, the new enterprises helped to solve the problem of the middleman that had so disturbed Americans before and during the Civil War. By integrating manufacturing and marketing or by reducing the number of links in the chain between maker and final consumer, the bigger businesses of the late nineteenth century provided one way to answer the producerist call to eliminate parasitical intermediaries. Some businessmen made this claim openly. At the end of the nineteenth century, for instance, corporate leaders such as James J. Hill and Charles Schwab celebrated the success of big business in eliminating "unproductive middlemen" from the American economy, to the benefit of all.[101]

With its references to eliminating nonproducers from the economy, this corporate vision of industrial progress recalled the widespread attacks in the wartime North on illegitimate middlemen in the military supply sector. But in the blisteringly contentious political culture of the Gilded Age, many Americans had a different understanding of the middleman's postwar career. When Milton George of the Farmers' Alliance told a Senate committee in 1886 that the country was suffering from "a surplus of middle-men at the expense of the producer and consumer," for instance, he was not calling for more powerful integrated industrial corporations.[102] Instead, George and others imagined the corporations themselves as greedy, monopolistic intermediaries that stood between workers and consumers. From the 1870s to the 1890s, hundreds of thousands of Americans joined postwar organizations such as the Patrons of Husbandry (or Grange), the Knights of Labor, the Farmers' Alliance, and the People's Party (Populists). Advocating producers' cooperatives that could sell directly to consumers, these groups rejected the corporate solution to the middleman problem.

Especially after the 1870s, these groups broke with many business leaders over the question of public enterprise by endorsing state ownership of the telegraph, railroad, mines, crop warehousing systems, and other enterprises. To be sure, labor and populist support for state enterprise was often overshadowed by different programs and goals, many of them at the level of local communities; by some measures, they demonstrated little interest in using the national state to achieve their ends.[103] Nevertheless, one of the ways in which they departed from their antebellum forebears was in their greater willingness to endorse state regulation and public enterprise. Many Americans supported this stance. In the 1892 elections, the People's Party, calling openly for the nationalization of the railroads and telegraph, received over 1 million votes, or 8 percent of the national vote in a political system that did not favor third-party movements.[104]

Especially when developments on the Civil War home front are taken into account, such support for public enterprise seems less radical than many have suggested and more in line with American political tradition. Along with the Post Office, which was far larger than its private-sector competitors in the communications field throughout the century, the military served as an obvious example of large-scale, government-run enterprise.[105] The record of the North's war supply effort suggests that the mixed political economy supported by groups including the Knights and the Populists, although criticized by some as un-American, was no less traditional than the laissez-faire liberal economy championed by their opponents.

The growing ideological gap in the late nineteenth century between business leaders and labor and agrarian groups was hardly limited to differences over the question of national ownership of certain large-scale enterprises. Instead, there seemed to be more general conflict, often described in terms that recalled the Civil War. By the late 1860s, National Labor Union leader William Sylvis was already claiming that workers and their employers were locked in an "irrepressible conflict."[106] By the mid-1880s, when the country was reeling from disputes between workers and their employers that were unusually numerous and violent, even by international standards, many Americans may have been inclined to agree with the minister and social reformer Washington Gladden, who said simply that "the state of industrial society is a state of war."[107]

Indeed, the political economy of the Gilded Age struck many Americans as dangerously unbalanced—in part, perhaps, because of their memories of the Civil War. Although many workers could not help but conclude from postwar developments that military organizations seemed to work inevitably to the benefit of capital, it is also worth entertaining the possibility that it was the very weakness of the national military, rather than its strength, that exacerbated business-labor conflict in the late nineteenth century. Certainly, as workers and their supporters frequently pointed out, business was often assisted during many major strikes of this era—as it had been in some cases during the Civil War itself—by the U.S. Army, state national guards, and a variety of public and private police forces.[108]

It was also true that postwar business leaders, some of whom were increasingly determined to bring military discipline to their work forces, differed from their Civil War predecessors both in terms of the scope of the enterprises they managed and because they were not similarly balanced by the national state and the official political culture. In the economic mobilization for Civil War, robust public enterprises and official criticisms of profit taking at public expense had served to constrain the material and cultural authority of business. After

1865, even as business itself became more militarized, the wartime institutional constraints fell away. Although there was no lack of criticism of the rise in relative power of business over government and there was considerable popular support for public enterprise, such sentiments were no longer well supported by state institutions. Relatively unconstrained, the rampant industrial conflict of the late nineteenth century seemed to some Americans to be potentially even more destructive than the Civil War had been.

As many cultural historians have suggested, one measure of Americans' imaginative responses to the political and industrial developments of the late nineteenth century are the many utopian and dystopian novels published during that era.[109] The author of one of the most remarkable of these novels was Ignatius Donnelly, who during and after the Civil War had served as a Republican representative from Minnesota in Congress. During the Gilded Age, Donnelly moved into populist politics: in 1892, he would write much of the People's Party platform. Two years before that, when Donnelly was already an officer in the Farmers' Alliance, he published *Caesar's Column: A Story of the Twentieth Century*. Projecting one possible outcome of the ongoing conflict between labor and capital in Gilded Age America, Donnelly imagined in this novel a future world in which a cynical plutocracy ruled over a miserable proletariat. By 1988, the year in which the novel is set, this plutocracy is opposed by a populist terrorist organization known as the "Brotherhood of Destruction," which is planning a global uprising. In the end, a terrible war between the governing elites and the Brotherhood, followed immediately by the disastrous excesses of Caesar (the Brotherhood president) himself, kills three-quarters of the global population, destroying all of modern civilization in the process.[110]

Intended by its populist author as a warning against the inequalities and unresolved tensions of the Gilded Age, *Caesar's Column* also stood as evidence that the Civil War loomed large in the imaginations of many Americans as they grappled with the problems of the late nineteenth century. One lesson from the record of the Northern military economy was that the fantastic productivity of industrial economies contained terrible destructive potential. Mark Twain had pointed to this dark side of American enterprise in *A Connecticut Yankee in King Arthur's Court* (1889), in which the modern Yankee who finds himself in the distant past ends up using his knowledge in the service of mass killing. Donnelly, in conjuring up his future world, stressed a similar theme: in the 1980s of *Caesar's Column*, technological advances allow the aerial bombing of urban populations, killing millions. But Donnelly's novel also referred to the Civil War as an event that exposed and amplified economic inequalities and greed. To illustrate the kind of "cunning" that his imagined twentieth-century plutocracy used to take over the world, one of the novel's characters refers to alleged wartime profiteering by A. T. Stewart, the leading New York City dry goods wholesaler and retailer. A prime contractor to New York State early in the war, who probably acted later as a subcontractor, Stewart figured in Donnelly's novel as the model of a millionaire who used the wartime emergency to further his private interests instead of the public good. In the decades that followed the Civil War, *Caesar's Column* suggested, the continuing machinations of such selfish capitalists served to concentrate all wealth and power in the hands of a tiny cabal.

More significantly, the Civil War loomed in the background of the utopian novel that had inspired Donnelly to write *Caesar's Column* in the first place, Edward Bellamy's fantastically popular *Looking Backward*. Published in 1888, Bellamy's novel was the best-selling book of its era. Like Donnelly, Bellamy believed that the Gilded Age crisis in American industrial relations required radical solutions. But even more than did Donnelly's book, Bellamy's novel clearly looked back to the Civil War as a source of inspiration, even though its author was only fifteen when that conflict ended. As it did for the older Donnelly, the Civil War stood in Bellamy's eyes as a moment when America was introduced to scandalous new levels of inequality and greed—which would persist into the Gilded Age. "The speculative opportunities offered by the war," Bellamy explained in an 1892 magazine article, "had developed the millionaire and his shadow, the tramp. Contrasts of wealth, luxury, and arrogance with poverty, want, and abjectness, never before witnessed in America, now on every side mocked the democratic ideal and made the republic a laughing-stock."[111]

But the mobilization for Civil War, and military institutions more generally, also served as the model for Bellamy's utopia. As the scholars John L. Thomas and Wilfred M. McClay have shown, Bellamy consciously used the Union army as the inspiration for the "industrial army" that serves in *Looking Backward* as the backbone of his imagined socialist society of the year 2000. Bellamy originally planned to begin his novel by describing masses of his future soldiers of public industry marching through the streets, a scene in which he intended to recall the actual "Grand Review" of Union soldiers held in Washington at the end of the Civil War in 1865.[112] Although this scene was scrapped in the novel's final version, the industrial army remained. Staffed by a service program that drafted all adults for twenty-five years of their lives, this national work force created an egalitarian economy of abundance through large-scale public enterprises in distribution as well as manufacturing. Bellamy's narrator, whose unusually long sleep transports him from the Gilded Age to the end of the next century, is shocked to wake up into a world that has overcome war as well as want.

In the wake of his novel's spectacular success, Bellamy used his utopia as the basis for a "Nationalist" political program that joined the Knights of Labor, the Farmers' Alliance, and other major movements of the day in calling for public control of large economic enterprises, many of which seemed to be drifting steadily toward monopoly. Like many members of these other movements, Bellamy suggested that Americans begin by nationalizing telegraphy and the railroads, key industries that seemed especially well suited to government control. Citing the Post Office as one working model for large-scale public enterprise in a field of national concern, Bellamy and others called for a mixed economy for the modern age.[113] These ideas were echoed by Donnelly, both in fiction and in politics. At the end of *Caesar's Column,* the novel's heroes set up a utopia in Uganda (where they have escaped the world war by airship) that relies upon public ownership of mines, transport, and communications and in which the economic "producers" control the most powerful house of the legislature.[114] Two years after the novel appeared, Donnelly's 1892 platform for the People's Party called for a national political economy with a similar array of public enterprises.

Even more than most American proponents of public enterprise during the Gilded Age, however, the best-selling Bellamy saw the military as the key model for the economic administration of the modern world. When the prominent economist and former Union general Francis A. Walker, in a long critique of *Looking Backward* and the Nationalist movement, accused Bellamy of being overly enchanted with militarism, the novelist denied that his ideal economy would be any more authoritarian than the one that already existed in the Gilded Age. Like some civil service reformers before him, however, Bellamy admitted in an 1890 response to Walker that he admired the devotion of military men to "public service" rather than private gain, an attitude that was becoming increasingly rare. It was not a love of war or authoritarianism that attracted him to the military, Bellamy explained, but rather his interest in finding a usable model of national service in America that might be applied to the industrial sphere.[115]

Ultimately, it was not only military ethics and organization in a general sense that inspired Bellamy but also the business of military procurement in particular. Given that Bellamy (who lived in nearby Chicopee) worked for several years as an editor for a newspaper based in Springfield, Massachusetts, site of the Ordnance Department armory that had long been one of the most important public enterprises in the American military economy, his interest in

this area was predictable.[116] Curiously, Bellamy did not refer to the Springfield Armory in his arguments for Nationalism; in fact, in one article he even suggested that the Northern military economy had lacked significant government-run factories. "If our government had manufactured the soldiers' supplies in the Civil War," Bellamy wrote in 1894, having noted Britain's extensive use of public arsenals to make clothing and other goods, "it would have saved a vast sum of money." He immediately went on, however, to argue that the United States should use military arsenals as the seed for the nationalization of all major industry. These arsenals would begin by supplying goods of high quality and low cost to soldiers and other public workers; then, "as the number of [public] employees increased" and they were supplied by the expanding "product of publicly conducted industries," the country would arrive at the socialist political economy that Bellamy called Nationalism.[117]

For Bellamy, the business of war—which most of his readers associated with the events of 1861–65—provided the most important model for the task of large-scale public economic administration that would come with Nationalism. In an 1890 piece written for a British journal, as he responded to a French reviewer of *Looking Backward*, Bellamy stated clearly that military economy "involves the constant solution of problems of business administration on a far greater scale than they are presented by the affairs of the largest of industrial or commercial syndicates, and that, as a matter of fact, the work of the epauletted administrators is done with an exactitude and fidelity unequaled in business." Joining some civil service reformers, Bellamy saw the military—more than private enterprise—as the most appropriate model for government. "Upon this administrative and essentially business side of the great modern military organizations," Bellamy continued, "the advocate of the practicability of Nationalism may properly lay peculiar stress." The general staff of a European army, with its various supply bureaus and the public arsenals they oversaw, he argued, was a successful existing organization that showed that socialist political economy was quite possible.[118]

THE ULTIMATE FAILURE OF Bellamy's Nationalist movement to achieve revolutionary change, despite the considerable popular interest in his ideas, was only one indication that the militarization of America after the Civil War was far from total. The military was never the only model for reform in business, government, and society. Nevertheless, for better and for worse, the model of military organization and administration—mainly as it was known through the Civil War experience—did influence the modern American imagination to an extent that is rarely recognized. In several fields—including civil service reform, the management of large industrial corporations, populist movements, and popular culture—the military was seen by many as the most important past practitioner of large-scale administration.

Although Civil War history was dominated from the beginning by accounts of battles and generals, memories of the economic project that supported the Civil War armies remained important in the minds of postwar reformers. As time passed, however, memories of the wartime project grew increasingly dim—and conveniently so. During the Progressive Era, energetic new professionals and corporations tended to see the small contemporary army as backward and had little interest in acknowledging their own debts to the military model. In the twentieth century, many Americans became interested in constructing a unique national history, in which references to the past importance of military bureaucracy were unwelcome.[119] One thing that this exceptionalist history overlooked was that, although the United States had certainly maintained much smaller standing forces and used much less of its national wealth for military ends than did contemporary European states in the years before the First World War,[120] modern America, as a child of the Civil War, had nonetheless been shaped by a partial militarization of business and government. Although this process did not necessarily promote a militarist national culture, it surely enhanced American war-fighting capacities during the era of the world wars and beyond, when the military foundations of the modern United States would shape the triumphs and horrors of the "American century" and its aftermath.

APPENDIX A

Note on the Value of a Dollar in the Civil War Era

Because this book constantly mentions sums and prices measured in dollars, some readers may be interested in the value of a dollar during this period and how it has changed over time. Economic historians routinely use several standard methods, none of them perfect, to measure changes in the value of a currency. One of the most common methods is to use a long-run index of consumer prices, which attempts to trace the changing average costs over time of a similar bundle of consumer goods. A standard consumer price index for the U.S. dollar suggests that the purchasing power of $1 in 1861 is equivalent to that of about $20 today, in the early twenty-first century. But by other measures, such as those that compare average wage rates or per-capita incomes, $1 in 1861 can be regarded as the equivalent of $200 or more in our day. For a discussion of long-run price indices and the many difficulties associated with them, see John J. McCusker, *How Much Is That in Real Money? A Historical Price Index for Use as a Deflator of Money Values in the Economy of the United States,* 2nd ed. (Worcester, Mass.: American Antiquarian Society, 2001). As of the summer of 2005, there was a good electronic resource on this subject, maintained by EH.NET and located at the following URL: http://eh.net/hmit/.

One way to gauge the meaning of a dollar to the Americans who lived through the Civil War era is to consider contemporary wages and living costs. Just before the Civil War, in 1860, one dollar was a typical *daily* wage (for what was often a ten-hour workday) for a man doing what some economic historians call "unskilled" work or "common labor"—ditch digging, for example. Skilled male artisans, such as blacksmiths, could earn closer to $1.75 or $2.00 a day. Farm hands were typically paid about $13 a month plus board—exactly what privates in the Union army were paid for most of the Civil War (along with their clothing allowance). Female factory workers, who were among the highest-earning women in the country, often made around $0.50 a day. Many American households subsisted on less than $600 a year, nearly half of which went for groceries. For a small minority of Americans, earnings were significantly higher. Urban professionals, as well as senior military officers (such as army majors and lieutenant colonels), might earn $2,000 a year. President Lincoln, like his predecessors in office, enjoyed a salary of $25,000. Some leading manufacturers, merchants, and bankers enjoyed incomes close to that of the president; a tiny handful, including wildly successful businessmen such as New York City dry goods king A. T. Stewart, saw annual incomes of six or even seven figures. For further reading on wages and the cost of living in 1860, see Edgar W. Martin, *The Standard of Living in 1860* (Chicago: University of Chicago Press, 1942); Clarence D. Long, *Wages and Earnings in the United States, 1860–1890* (Princeton: Princeton University Press, 1960); and Robert A. Margo, *Wages and Labor Markets in the United States, 1820–1860* (Chicago: University of Chicago Press, 2000).

Because of wartime inflation, the value of a dollar changed significantly between 1861 and 1864. Average consumer prices in the North rose by approximately 75 percent during this period, meaning that it took $1.75 in 1864 to buy the same goods that had sold for $1.00 in 1861. (In the Confederacy, where prices rose by close to 5,000% over the same period, inflation was far more severe.) One reason for the inflation, along with the higher costs of transport and insurance that the war inevitably created, was an increase in the money supply. In December 1861, Northern banks had ceased specie (gold and silver) payments, effectively taking the United States off the gold standard. In early 1862, Congress authorized the printing of greenbacks, currency notes that were not backed by gold. The introduction of large amounts of new paper money, along with a variety of other Treasury notes and bonds, allowed the North to pay for the war, but it also promoted inflation. Historians have long known that wartime wages in the North did not rise as fast as prices. On average, real wages (adjusted for inflation) in the North declined by close to 25 percent between 1861 and 1864. This meant that for many workers, the war meant economic hardship. The classic studies of greenbacks and wartime inflation are two books by the pioneering economic historian Wesley C. Mitchell: *A History of the Greenbacks: With Special Reference to the Economic Consequences of Their Issue* (Chicago: University of Chicago Press, 1903) and *Gold, Prices, and Wages under the Greenback Standard* (Berkeley, Calif.: University Press, 1908). For more recent contributions, see Philip R. P. Coelho and James F. Shepherd, "Differences in Regional Prices: The United States, 1851–1880," *Journal of Economic History* 34 (1974): 551–91; Coelho and Shepherd, "Regional Differences in Real Wages: The United States, 1851–1880," *Explorations in Economic History* 13 (1976): 203–30; and Paul R. Auerbach and Michael J. Haupert, "Problems in Analyzing Inflation during the Civil War," *Essays in Economic and Business History* 20 (2002): 57–70.

Like all Northern buyers, military procurement officers observed significant rises in the prices they paid for goods. For example, contract registers show that the Quartermaster's Department was generally paying between $100 and $120 per horse at depots around the country in 1862 but rarely paid less than $150 in 1864. Under an August 1862 contract, the North's leading clothing contractor, John T. Martin, supplied infantry trousers to the New York City depot for $3.37 a pair; under an August 1864 contract signed by the same contractor, the price was $5.20 a pair. Prices sometimes rose fast:

between early 1864 and late summer of the same year, quartermasters in Philadelphia saw contract prices for six-mule wagons rise from about $150 to over $300 per vehicle. (For more on the effects of inflation on contracting, see chap. 4; for a discussion of the Quartermaster's Department contract registers, see app. C.)

APPENDIX B

Leading Northern Military Contractors in Selected Industries

Table B.1 Leading Small Arms Suppliers to the U.S. Army, 1861–1866

		Value of Purchases by Calendar Year (in thousands of dollars)[a]						
Supplier Name(s)	Location	1861	1862	1863	1864	1865	1866	Total, 1861–66
Colt's Patent Fire Arms Co.	Hartford, CT	480	1,502	1,805	900			4,687
Naylor & Co. (importer)	New York, NY	63	1,449	2,295	4			3,811
E. Remington & Sons	Ilion, NY	24	282	525	1,232	521	253	2,837
Herman Boker & Co. (importer)[b]	New York, NY		2,693	73		41	2	2,809
Sharps Rifle Manufacturing Co.	Hartford, CT	179	652	620	574	375		2,400
Marcellus Hartley (U.S. agent)	(Europe)		2,147					2,147
George L. Schuyler (U.S. agent)[b]	(Europe)		2,094					2,094
Burnside Arms Co.	Providence, RI	38	128	590	489	835		2,080
Spencer Repeating Rifle Co.	Boston, MA		20	525	805	703	25	2,078
Alfred Jenks & Son	Philadelphia, PA		126	800	800	234		1,960
Providence Tool Co.	Providence, RI		50	540	597	247		1,434
Starr Arms Co.	Yonkers, NY	25	213	268	634	100		1,240
Lamson, Goodnow, & Yale	Windsor, VT		97	474	419			990
Schuyler, Hartley, & Graham (importer)	New York, NY	412	421	83				916
Thomas Poultney / Poultney & Trimble	Baltimore, MD	141	195	218	176	173		903
Norwich Arms Co.	Norwich, CT			138	329	279		746
Total, 16 leading suppliers								33,132
Total, all army small arms purchases, 1861–66								47,060
Share of 16 leading suppliers								70%
Share of top 4 suppliers								30%
Top 4 suppliers, domestic private manufacture only								41%

Sources: "Ordnance Department," House Exec. Doc. 99, 40th Congr., 2nd Sess., ser. 1338. Total purchases by category from "Report of the Chief of Ordnance," in U.S. War Department, 1866 *Annual Report of the Secretary of War, 1866* (Washington, D.C., 1867).

[a]Ordnance Department purchases, in thousands of nominal dollars.

[b]For Boker and Schuyler, 1861–62 purchases were reported together; they are combined in the 1862 column.

Table B.2 Leading Heavy Ordnance (Artillery and Projectiles) Suppliers to the U.S. Army, 1861–1866

Supplier Name(s)	Location(s)	Value of Purchases by Calendar Year (in thousands of dollars)[a]						Totals, 1861–66
		1861	1862	1863	1864	1865	1866	
Robert P. Parrott / West Point Foundry	Cold Spring, NY	230	580	1,141	1,624	895	247	4,717
Charles Knap / Fort Pitt Foundry	Pittsburgh, PA	157	255	468	178	1,032	645	2,735
Cyrus Alger / South Boston Iron Works	Boston, MA	104	356	336	667	807	333	2,603
Hotchkiss & Sons	Sharon, CT / Bridgeport, CT	15	227	572	471	200		1,485
Seyfert, McManus & Co.	Reading, PA			3	270	228	150	651
James T. Ames / Ames Manuf. Co.	Chicopee, MA	292	51	162	77	1		583
Total, 6 leading suppliers								12,774
Total, all army small arms purchases, 1861–66								17,640
Share of 6 leading suppliers								72%
Share of top 4 suppliers								65%

Sources: Same as for table B.1. Ames & Co. supplied an additional $950,000 worth of swords and sabres, making it the Union's largest supplier of these items by far.

[a] Ordnance Department purchases, in thousands of nominal dollars.

Table B.3 Leading Wagon and Ambulance Suppliers to the U.S. Army, 1861–1864

Supplier Name(s)	Location	Value of Contracts by Calendar Year (in thousands of dollars)[a]				Totals, 1861–64
		1861	1862	1863	1864	
J. C. C. Holenshade / Holenshade, Morris & Co.	Cincinnati, OH	26	55	461	278	820
Henry Simons & Co.	Philadelphia, PA		263	164	191	618
Wilson, Childs, & Co.	Philadelphia, PA		150	108	134	392
Jacob Rech	Philadelphia, PA	1	104	29	59	193
Adonijah Peacock & Son	Cincinnati, OH			70	76	146
Louis Espensheid	St. Louis, MO		68	56		124
James L. Haven	Cincinnati, OH		10	65	23	98
Philip Dom	Cincinnati, OH		6	14	72	92
Total, 8 leading suppliers						2,483
Total, all large QMD contracts, 1861–64 (see appendix C)						3,053
Share of 8 leading suppliers						81%
Share of top 4 suppliers						66%

Sources: Manuscript and published Quartermaster's Department contract registers: vols. 13–15, e. 1238, RG 92, National Archives; "Contracts—War Department for the Year 1861," House Exec. Doc. 101, 37th Congr., 2nd Sess., ser. 1136; "Contracts Made by the Quartermaster's Department," House Exec. Doc. 84, 38th Congr., 2nd Sess., ser. 1230. See also appendix C. Because the Quartermaster's Department purchased some wagons in relatively small lots, further analysis of small contracts would likely provide a richer portrait of suppliers in this sector.

[a] Quartermaster's Department prime contracts, in thousands of nominal dollars.

LEADING NORTHERN MILITARY CONTRACTORS

Table B.4 Leading Boot and Shoe Suppliers to the U.S. Army, 1861–1864

		Value of Contracts by Calendar Year (in thousands of dollars)[a]				
Supplier Name(s)	Location(s)	1861	1862	1863	1864	Totals, 1861–64
John Mundell/ Mundell & Harman	Philadelphia, PA		109	383	1,165	1,657
Benedict, Hall & Co./ Hall, Southwick & Co.	New York, NY/ New Canaan, CT	130	223	678	433	1,464
R. M. Pomeroy & Co.	Cincinnati, OH		13	208	1,187	1,408
A. Simpkinson & Co.	Cincinnati, OH		48	227	957	1,232
Albert Jewett & Co./ George Chapin	New York, NY			316	557	873
James B. Harmer & Co.	Philadelphia, PA		133	437	273	843
H. S. Downs	Boston, MA		237	217	332	786
Howes, Hyatt & Co.	New York, NY	20	471	49		540
Kimball, Robinson & Co.	Boston, MA/South Brookfield, MA	10	86	295	109	500
Joseph Bickta	Wilmington, DE			32	460	492
Zenas Sears/Faxon & Elms	Boston, MA	388	31	10		429
Ware & Taylor	Boston, MA/ Jersey City, NJ	379				379
Seth Bryant	Joppa, MA		64	86	214	364
Daniel & William Temple	Trenton, NJ			198	143	341
Jenkins Lane & Sons	Boston, MA/ Abington, MA		147	4	184	335
E. P. Fenton & Co.	Syracuse, NY/ Auburn, NY			139	186	325
Total, 16 leading suppliers						11,968
Total, all large QMD contracts, 1861–64 (see appendix C)						20,272
Share of 16 leading suppliers						59%
Share of top 4 suppliers						28%

Sources: Same as for table B.3. Some contractor locations from *The Boot and Shoe Trade: Containing a List of the Boot and Shoe Manufacturers and Dealers in the United States* (New York: Wm. F. Bartlett, 1859).
See also appendix C.
[a]Quartermaster's Department prime contracts, in thousands of nominal dollars.

Table B.5 Leading Woolen Textile and Blanket Suppliers to the U.S. Army, 1861–1864

		Value of Contracts by Calendar Year (in thousands of dollars)[a]				
Supplier Name(s)	Location	1861	1862	1863	1864	Totals, 1861–64
Benjamin Bullock's Sons	Philadelphia, PA	612	1,566	3,718	1,739	7,635
W. C. Houston	Philadelphia, PA		600	1,035	2,970	4,605
Jones Bros./G. W. Jones	Cincinnati, OH	2	203	83	2,284	2,572
Saxonville Mills	Framingham, MA	718	244	377	505	1,844
Thomas Murphy	New York, NY			923	900	1,823
Lewis, Boardman, & Wharton	Philadelphia, PA			728	867	1,595
Edward T. Shaw	Philadelphia, PA		1,200	22		1,222
John Dobson	Philadelphia, PA	25			1,092	1,117
C. W. Freeland & Co./A. Rose	Boston, MA				953	953
Kunkel, Hall & Co.	Philadelphia, PA			76	867	943
E. S. Higgins & Co.	New York, NY	27	263	263	300	853
C. B. Mount	Philadelphia, PA		310	215	283	808
J. W. Dimick	New York, NY		116	206	405	727
Cronin, Hurxthal, & Sears	New York, NY	28	300	44	375	747
A. T. Lane	Philadelphia, PA		46	62	547	655
Sevill Schofield	Philadelphia, PA				638	638
Total, 16 leading suppliers						28,737
Total, all large QMD contracts, 1861–64 (see appendix C)						51,871
Share of 16 leading suppliers						55%
Share of top 4 suppliers						32%

Sources: Same as for table B.3. See also appendix C.
[a] Quartermaster's Department prime contracts, in thousands of nominal dollars.

Table B.6 Leading Clothing Suppliers to the U.S. Army, 1861–1864

Supplier Name(s)	Location	Value of Contracts by Calendar Year (in thousands of dollars)[a]				Totals, 1861–64
		1861	1862	1863	1864	
John T. Martin/Martin Bros./ John E. Hanford	New York, NY	270	2,263	799	9,895	13,227
Hanford & Browning/ Browning, Button & Co./ W. B. Button	New York, NY	2,265	1,563	1,263	1,623	6,714
C. W. & J. H. Freeland & Co./ Alvin Rose	Boston, MA		412	1,417	3,687	5,516
John Boylan/James B. Boylan	Newark, NJ		126	1,259	3,641	5,026
Thomas F. Carhart/ George Opdyke	New York, NY		3,064	105		3,169
E. Tracy & Co.	Philadelphia, PA				3,123	3,123
Joseph Lee	New York, NY		330	1,428	1,168	2,926
Slade, Smith & Co./Lewis, Boardman, & Wharton	Philadelphia, PA	65	1,265	719	640	2,690
Albert Jewett & Co./ George Chapin	New York, NY			593	2,031	2,624
Rockhill & Wilson	Philadelphia, PA	41	562	19	1,579	2,201
Anspach & Stanton	Philadelphia, PA	2	128	513	1,377	2,020
William Deering	Portland, ME	20	213	530	1,225	1,988
Joseph F. Page	Philadelphia, PA		75	710	1,167	1,952
Amos Clark	New York, NY	11	1,102	556	281	1,950
Mack, Stadler, & Glaser	Cincinnati, OH	352	1,273	163		1,788
John R. Evans	Philadelphia, PA			156	1,472	1,628
Total, 16 leading suppliers						58,542
Total, all large QMD contracts, 1861–64 (see appendix C)						102,997
Share of 16 leading suppliers						57%
Share of top 4 suppliers						30%

Sources: Same as for table B.3. See also appendix C.
[a] Quartermaster's Department prime contracts, in thousands of nominal dollars.

APPENDIX C

Note on Data Collection and Record Linkages

Published and unpublished contract registers of the Quartermaster's Department, which were used to generate the tables in appendix B, contain valuable information about military contracting during the Civil War era. Typically, each entry in the registers contains the following data for each contract: date of contract, name of contracting officer, place of contract, name of contractor, item to be supplied, quantity, and price. But readers should understand that the data contained in these contract registers has been processed selectively for the purposes of this book and that the registers themselves are imperfect sources of information about Northern procurement.

The Quartermaster's Department contract registers list very large numbers of transactions, some of which were very small in dollar value. Because of time and resource limitations, I did not compile every single registered contract. Instead, I used minimum cutoffs to capture the larger contracts. Registered contracts were excluded in the following way: for the 1856–60 period, only those with a minimum value of $500 were included; for 1861, the minimum used was $1,000; for 1862–63, $5,000; for 1864, $10,000; and for 1866–70, $1,000. Although the use of different minimums for different years may seem arbitrary, it was intended as a crude way to adjust for inflation and changes in the scale of military contracting over the course of the period. In the end, this method captured some 7,850 separate contracts for 1856–70 period, with a total value of approximately $338,500,000. [Readers interested in tables generated with the antebellum and postwar Quartermaster's Department contract data should consult my doctoral dissertation: Mark R. Wilson, "The Business of Civil War: Military Enterprise, the State, and Political Economy in the United States, 1850–1880" (Ph.D. diss., University of Chicago, 2002). The dissertation also includes tables of leading contractors with the various Northern states, generated from state auditors' reports and similar documents.] Because the wartime Quartermaster's Department tended to buy in bulk, only a small fraction of the total value of all registered contracts was excluded by these methods. Nevertheless, the cutoffs did exclude a significant fraction of registered contracts in some major military industries in which bulk purchasing was less common, such as those that supplied animals and wagons. Finally, such methods largely fail to capture the activities of contractors selling items of relatively low cost, such as thread, buttons, and hardware.

There are inherent problems with using the contract registers as a guide to U.S. procurement activity. Some of these registers list ship leases or charters, but they rarely provide information about the size of transactions. Payments to railroads are not indicated in the contract registers. This means that transactions with rail and water-going transport contractors, which constituted a very large part of the Northern military economy, cannot be reconstructed using this source. Equally important, readers must recognize that the contract registers do not indicate all purchases. Leaving aside the problem of contracts that were never reported or registered for some unknown reason, there is the larger problem of open-market purchases. As chapter 4 suggests, open-market purchases were used by the army to acquire large quantities of goods. But open-market purchases, unlike formal contracts, were not normally recorded in the registers. Finally, because the contract registers list initial agreements rather than final payments, even the transactions to which they refer may not always have been closed exactly according to the terms listed in the registers. All of this is to say that, although the contract registers (and any tables that might be generated from them) can be a rich source of information about army-supplier transactions in the Northern military economy, they certainly do not comprehend all such transactions.

Finally, readers should note that information about Ordnance Department suppliers (presented in tables B.1 and B.2) is contained not in Quartermaster's Department contract registers but in distinct Ordnance Department lists of purchases. These sources, which were eventually published for Congress, are cited in the tables.

Some of the maps and tables presented in this book use data compiled from annual U.S. Treasury "Receipts and Expenditures" reports, which were sent regularly to the U.S. Congress and published along with other government documents. The expenditures portion of the "Receipts and Expenditures" reports proved valuable for several reasons. Because they are organized according to particular U.S. government bureaus, they may be used as a source of information about departmental expenditures. In some cases, when departments' own reports lack clear annual expenditure figures, the "Receipts and Expenditures" figures were used as a substitute.

Because the "Receipts and Expenditures" reports break down disbursements to individual officers within the various bureaus, they may also be used to trace the flow of U.S. funds to particular persons, such as quartermasters. But readers should realize that the disbursements listed in these reports do not simply show how much spending was controlled directly by a given officer. In practice, parts of these disbursements were forwarded to subordinate officers or subdepots in other locations. For example, some of the funds disbursed to a supply officer in Louisville might be forwarded to lower-ranking officers and subdepots elsewhere in Kentucky or other nearby states. Table C.1 lists the "Receipts and Expenditures" reports used as data sources for the maps and tables in this book.

Table C.1 "Receipts and Expenditures Reports" Used as Data Sources

Fiscal Year	Document Number	Congress and Session	Serial Set Volume
1845	House Doc. 14	29 Cong. 1 Sess.	ser. 482
1846	House Doc. 10	29 Cong. 2 Sess.	ser. 498
1847	House Exec. Doc. 7	30 Cong. 1 Sess.	ser. 514
1848	House Exec. Doc. 11	30 Cong. 2 Sess.	ser. 540
1853	House Exec. Doc. 112	33 Cong. 1 Sess.	ser. 727
1854	House Exec. Doc. 10	33 Cong. 1 Sess.	ser. 782
1855	House Exec. Doc. 40	34 Cong. 1 Sess.	ser. 852
1856	House Exec. Doc. 86	34 Cong. 3 Sess.	ser. 908
1857	House Exec. Doc. 13	35 Cong. 1 Sess.	ser. 947
1858	House Exec. Doc. 20	35 Cong. 2 Sess.	ser. 1003
1859	House Exec. Doc. 7	36 Cong. 1 Sess.	ser. 1045
1860	House Exec. Doc. 12	36 Cong. 2 Sess.	ser. 1096
1861	House Exec. Doc. 36	37 Cong. 2 Sess.	ser. 1129
1862	House Exec. Doc. 8	37 Cong. 1 Sess.	ser. 1187
1863	House Exec. Doc. 84	38 Cong. 1 Sess.	ser. 1195
1864	House Exec. Doc. 73	38 Cong. 2 Sess.	ser. 1229
1865	House Exec. Doc. 12	39 Cong. 2 Sess.	ser. 1288
1866	House Exec. Doc. 315	40 Cong. 2 Sess.	ser. 1346

The "Receipts and Expenditures" reports were also used to estimate the flows of military money through particular places. Unfortunately, these reports and many of the other sources of quantitative data used in this study failed to identify individual officers and contractors by post or residence. The Quartermaster's Department contract registers, for instance, do not usually indicate the contractors' places of residence. To present geographical information in several of the tables, therefore, I linked quantitative and qualitative data from certain sources with qualitative data from separate sources. A variety of published and unpublished Quartermaster's Department records, for instance, make it possible to trace the movements of officers from post to post over time. When this information is linked with quarterly "Receipts and Expenditures" data, it becomes possible to arrive at a good approximation of the geographical distribution of Treasury disbursements to supply depots around the country, which are represented in the maps in this book.

The references in several tables to the geographical locations of contractors were derived from a wide range of sources. In a small number of cases, the contract registers themselves contain references to place. In other cases, contractors appear in contemporary city directories, state business directories, or biographical encyclopedias. Some contractors and their primary business addresses were located in the indices of the contemporary credit reports compiled by R. G. Dun & Company (which survive in the form of a large set of manuscript volumes held at Baker Library, Harvard University). In a few cases, information about contractors' residences emerged in legislative investigations or army supply department correspondence.

Notes

ABBREVIATIONS USED IN THE NOTES

ARSW	U.S. War Department, *Annual Report[s] of the Secretary of War* (Washington, D.C.: various years)
Ct. Cl.	U.S. Court of Claims Reports
Dun	R. G. Dun & Co. Collection, Baker Library, Harvard Business School
GovCon 1	"Government Contracts," Part 1, House Report 2, 37th Congress, 2nd Session, ser. 1142 (U.S. Serial Set)
GovCon 2	"Government Contracts," Part 2, House Report 2, 37th Congress, 2nd Session, ser. 1143 (U.S. Serial Set)
ISA-CPR	Indiana State Archives, Commission on Public Records
M421	National Archives Microfilm Publication M421: Letters Sent by the Secretary of War to the President and Executive Departments, 1863–70
M494	National Archives Microfilm Publication M494: Letters Received by the Secretary of War from the President, Executive Departments, and War Department Bureaus, 1862–70
M688	National Archives Microfilm Publication M688: U.S. Military Academy Cadet Application Papers, 1805–66
M745	National Archives Microfilm Publication M745: Letters Sent by the Office of the Quartermaster General, Main Series, 1818–1870
NA	National Archives and Records Administration, Washington, D.C.
LC	Library of Congress, Washington, D.C.
OR	U.S. War Department, *The War of the Rebellion: A Compilation of the Official Records of the Union and Confederate Armies* (Washington, D.C.: GPO, 1880–1901)
RG 92	National Archives Record Group 92: Records of the Office of the Quartermaster General
RG 94	National Archives Record Group 94: Records of the Adjutant General's Office, 1780s–1917
RG 153	National Archives Record Group 153: Records of the Office of the Judge Advocate General (Army)

RG 393	National Archives Record Group 393: Records of U.S. Army Continental Commands, 1821–1920
SHSW	State Historical Society of Wisconsin, Madison

INTRODUCTION

1. The classic analysis of the cost of the Civil War, which estimates indirect as well as direct costs on both sides, is Claud a D. Goldin and Frank D. Lewis, "The Economic Cost of the American Civil War: Estimates and Implications," *Journal of Economic History* 35 (1975): 299–326. Although I use 1860 dollars here to better compare Civil War expenditures with earlier ones, most of the dollar figures used in this book are in nominal dollars (not adjusted for inflation). For a brief note on the value of a dollar during the Civil War era, see app. A.

2. For U.S. expenditures, see U.S. Department of Commerce, Bureau of the Census, *Historical Statistics of the United States: Colonial Times to 1970*, pt. 2 (Washington, D.C.: GPO, 1975), 1114–15. Evidence of spending on the Crimean War is difficult to find. Based on state expenditure statistics, I estimate that Britain and France together spent a total of roughly 100 million British pounds on the Crimean War. The United States (the central government in the North) spent approximately the same amount during each year of the Civil War. See B. R. Mitchell, *British Historical Statistics* (Cambridge: Cambridge University Press, 1988), 587–88; Mitchell, *European Historical Statistics* (New York: Columbia University Press, 1976), 698. For an interpretive survey of wars around the globe during this era, see Michael Geyer and Charles Bright, "Global Violence and Nationalizing Wars in Eurasia and America: The Geopolitics of War in the Mid-Nineteenth Century," *Comparative Studies in Society and History* 38 (1996): 619–57.

3. Detailed accounts of wartime purchases appear in "Report of the Quartermaster General," 1865 *ARSW*; "Report of the Chief of Ordnance," 1866 *ARSW*; "Report of the Commissary General," 1867 *ARSW*.

4. Bruce Porter, *War and the Rise of the State: The Military Foundations of Modern Politics* (New York: Free Press, 1994), 258.

5. Many political historians who regard the Civil War era as part of the heyday of a "party period" have implied that Republicans, who controlled the White House and the Congress during the war years, used the surge in military spending to boost their patronage powers. Paul P. Van Riper, *History of the United States Civil Service* (Evanston, Ill.: Row, Peterson & Co., 1958), 43, 61; Stephen Skowronek, *Building a New American State: The Expansion of National Administrative Capacities, 1877–1920* (New York: Cambridge University Press, 1982), 30; Adam I. P. Smith, "Beyond Politics: Patriotism and Partisanship on the Northern Home Front," in *An Uncommon Time: The Civil War and the Northern Home Front*, ed. Paul A. Cimbala and Randall M. Miller (New York: Fordham University Press, 2002), 169. Much of this story rests (uneasily, I believe) on a small part of an older study: Carl Russell Fish, *The Civil Service and the Patronage* (New York: Longmans, Green & Co., 1905), 158–72.

6. By examining the organization as a whole rather than its wartime leader, this book adds significantly to the account of the bureau provided in Russell F. Weigley, *Quartermaster General of the Union Army: A Biography of M. C. Meigs* (New York: Columbia University Press, 1959). For a fine overview of the history of the Quartermaster's Department that contains a substantial section on the Civil War, see Erna Risch, *Quartermaster Support of the Army:*

A History of the Corps (Washington, D.C.: Office of the Quartermaster General, 1962). For a synthetic history of the Northern and Southern mobilizations, see Paul A. C. Koistinen, *Beating Plowshares into Swords: The Political Economy of American Warfare, 1606–1865* (Lawrence: University Press of Kansas, 1996).

7. Unfortunately, histories of important governmental administrative organizations and their officers in the early United States are rare. The pioneering work in this field by Leonard White skips the Civil War years. See Leonard D. White, *The Jacksonians: A Study in Administrative History, 1829–1861* (New York: Macmillan, 1954), and White, *The Republicans: A Study in Administrative History, 1869–1901* (New York: Macmillan, 1956). Other important exceptions include Malcolm J. Rohrbough, *The Land Office Business: The Settlement and Administration of American Public Lands, 1789–1837* (New York: Oxford University Press, 1968); Richard R. John, *Spreading the News: The American Postal System from Franklin to Morse* (Cambridge: Harvard University Press, 1995); Daniel P. Carpenter, *The Forging of Bureaucratic Autonomy: Reputations, Networks, and Policy Innovation in Executive Agencies, 1862–1928* (Princeton: Princeton University Press, 2001).

8. For the claim that the Northern political economy was based on an "unregulated capitalist market" but dominated by a "Republican party-state," see Richard Franklin Bensel, *Yankee Leviathan: The Origins of Central State Authority in America, 1859–1877* (New York: Cambridge University Press, 1990), 233–37. Nineteenth-century European states also used mixed military economies. For evidence of a turn toward private firms before World War I, see David Stevenson, *Armaments and the Coming of War: Europe, 1904–1914* (Oxford: Clarendon Press, 1996), 15–40.

9. For pioneering studies in the history of military enterprise in America, see the work of Merritt Roe Smith, including *Harpers Ferry Armory and the New Technology: The Challenge of Change* (Ithaca: Cornell University Press, 1977) and "Introduction" and "Army Ordnance and the 'American System' of Manufacturing, 1815–1861," in *Military Enterprise and Technological Change: Perspectives on the American Experience*, ed. Merritt Roe Smith (Cambridge: MIT Press, 1985), 1–86.

10. For two excellent studies of business and war in Philadelphia that emphasize the participation of relatively small-scale firms in prime contracting and subcontracting, see Philip Scranton, *Proprietary Capitalism: The Textile Manufacture at Philadelphia, 1800–1885* (New York: Cambridge University Press, 1985), 272–313, and J. Matthew Gallman, *Mastering Wartime: A Social History of Philadelphia during the Civil War* (New York: Cambridge University Press, 1990). See also Stanley J. Engerman and J. Matthew Gallman, "The Civil War Economy: A Modern View," in *On the Road to Total War: The American Civil War and the German Wars of Unification, 1861–1871*, ed. Stig Förster and Jörg Nagler (New York: Cambridge University Press, 1997), 217–47, and J. Matthew Gallman, "Entrepreneurial Experiences in the Civil War: Evidence from Philadelphia," in *American Development in Historical Perspective*, ed. Thomas Weiss and Donald Schaefer (Stanford: Stanford University Press, 1994), 205–22, 299–303.

11. Among the promoters of this view are historically minded social scientists who have recently revitalized the field of U.S. political history. See, e.g., Skowronek, *Building a New American State*, 19–35; Charles Bright, "The State in the United States during the Nineteenth Century," in *Statemaking and Social Movements: Essays in History and Theory*, ed. Charles Bright and Susan Harding (Ann Arbor: University of Michigan Press, 1984), 121–58; Theda Skocpol, *Protecting Soldiers and Mothers: The Political Origins of Social Policy in the United States*

(Cambridge: Harvard University Press, 1992), 86–87; Matthew G. Hannah, *Governmentality and the Mastery of Territory in Nineteenth-Century America* (Cambridge: Cambridge University Press, 2000), 34. For a critique that suggests that this view is becoming increasingly untenable in light of recent research, see two articles by Richard R. John: "Governmental Institutions as Agents of Change: Rethinking American Political Development in the Early Republic," *Studies in American Political Development* 11 (1997): 347–80, and "Farewell to the 'Party Period': Political Economy in Nineteenth-Century America," *Journal of Policy History* 16 (2004): 117–25.

12. As Gregory Hooks pointed out, early products of the movement that began in the 1980s to focus on the American state tended to overlook military institutions. See Hooks, *Forging the Military-Industrial Complex: World War II's Battle of the Potomac* (Urbana: University of Illinois Press, 1991), 4. For a more recent call urging political historians to pay more attention to the military in early America, see Ira Katznelson, "Flexible Capacity: The Military and Early American Statebuilding," in *Shaped by War and Trade: International Influences on American Political Development*, ed. Ira Katznelson and Martin Shefter (Princeton: Princeton University Press, 2002), 82–110.

13. The question of the Civil War's effects, no less than the question of its causes, has generated considerable debate for decades. Many historians have argued, convincingly, that the war did not do much to stimulate most industries or national economic growth overall. Others have insisted, also convincingly, that the war's effects in the field of banking and finance, at least, were profound. On the general question of how revolutionary was the war, see James M. McPherson, *Abraham Lincoln and the Second American Revolution* (New York: Oxford University Press, 1990), and Roger L. Ransom, "Fact and Counterfact: The 'Second American Revolution' Revisited," *Civil War History* 45 (1999): 28–60. On the question of the war's economic effects in the North in particular, see Engerman and Gallman, "The Civil War Economy," and Phillip Shaw Paludan, "What Did the Winners Win? The Social and Economic History of the North during the Civil War," in *Writing the Civil War: The Quest to Understand*, ed. James M. McPherson and William J. Cooper Jr. (Columbia: University of South Carolina Press, 1998), 174–200.

14. For treatments of the war that have pointed to a similar hypothesis, see Allan Nevins, "A Major Result of the Civil War," *Civil War History* 5 (1959): 237–50; Nevins, *The War for the Union*, 4 vols. (New York: Charles Scribner's Sons, 1959–71); George M. Fredrickson, *The Inner Civil War: Northern Intellectuals and the Crisis of the Union* (New York: Harper & Row, 1965); Fredrickson, "Nineteenth-Century American History," in *Imagined Histories: American Historians Interpret the Past*, ed. Anthony Molho and Gordon S. Wood (Princeton: Princeton University Press, 1998), 164–84; and Wilfred M. McClay, *The Masterless: Self and Society in Modern America* (Chapel Hill: University of North Carolina Press, 1994), 25–26.

CHAPTER 1. THE RISE AND FALL OF A FEDERAL SUPPLY SYSTEM

1. Harold S. Wilson, *Confederate Industry: Manufacturers and Quartermasters in the Civil War* (Jackson: University Press of Mississippi, 2002), 109–22.

2. This $2 billion in military spending is in nominal dollars (not adjusted for inflation). For a brief note on the value of a dollar during the Civil War era, see app. A.

3. This aspect of the history of the Northern procurement project complements traditional accounts of the more general rise in the powers of the national government that

came with the war. For an extreme view of President Lincoln's wartime "victory over the governors," see William B. Hesseltine, *Lincoln and the War Governors* (New York: Alfred A. Knopf, 1948), 391. For a more measured account of the emergence of a "new federalism, with the balance away from the states to the nation" by 1865, see James A. Rawley, *The Politics of Union: Northern Politics during the Civil War* (Hinsdale, Ill.: Dryden Press, 1973), 187. A more recent, original interpretation that emphasizes the important innovations in the financial sphere is Richard Franklin Bensel, *Yankee Leviathan: The Origins of Central State Authority in America* (New York: Cambridge University Press, 1990).

4. James M. McPherson, *Battle Cry of Freedom: The Civil War Era* (New York: Oxford University Press, 1988), 322–23.

5. James M. McCaffrey, *Army of Manifest Destiny: The American Soldier in the Mexican War, 1846–1848* (New York: New York University Press, 1992), 16–18; Paul Foos, *A Short, Offhand, Killing Affair: Soldiers and Social Conflict during the Mexican-American War* (Chapel Hill: University of North Carolina Press, 2002), 31–43.

6. See the fine discussion of precedents in Kyle Scott Sinisi, *Sacred Debts: State Civil War Claims and American Federalism, 1861–1880* (New York: Fordham University Press, 2003), 3–9. Although there is no adequate study of the Mexican War supply system, some details may be found in the following sources: Erna Risch, *Quartermaster Support of the Army: A History of the Corps* (Washington, D.C.: Office of the Quartermaster General, 1962), 253–54; James A. Huston, *The Sinews of War: Army Logistics, 1775–1953* (Washington, D.C.: Office of the Chief of Military History, U.S. Army, 1966), 127–32; C. Edward Skeen, *Citizen Soldiers in the War of 1812* (Lexington: University Press of Kentucky, 1999), 144–46; McCaffrey, *Army of Manifest Destiny*, 23–37; Richard Bruce Winders, *Mr. Polk's Army: The American Military Experience in the Mexican War* (College Station: Texas A&M University Press, 1997), 68–76, 108–12, 233.

7. Lewis Perrine to Simon Cameron, 16 Apr. 1861; Edwin D. Morgan to Cameron, 17 Apr. 1861, in *OR*, ser. 3, 1:73, 1:83; Alexander Randall to Cameron, 19 Apr. 1861, box 7, ser. 49, SHSW. On New York's governor, see James A. Rawley, *Edwin D. Morgan, 1811–1883: Merchant in Politics* (New York: Columbia University Press, 1955), 139–44.

8. Individuals and voluntary associations in the North raised as many as seventy-five regiments during the early months of the war. Rawley, *Edwin D. Morgan*, 150.

9. Cameron order, 23 Apr. 1861; Cameron to (Indiana governor) Oliver Morton, 26 Apr. 1861; Cameron to Andrew Curtin, 19 Apr. 1861, all in *OR*, ser. 3, 1:107, 1:116, 1:131–33; "Report of the Board of Officers Named in the Act of April 16, 1861," *Documents of the Assembly of the State of New York*, 85th sess. (Albany, 1862), 2:201.

10. Rawley, *Edwin D. Morgan*, 135; Ellis Paxon Oberholtzer, *Jay Cooke, Financier of the Civil War* (Philadelphia: George W. Jacobs & Co., 1907), 1:103, 1:110; George H. Porter, *Ohio Politics during the Civil War Period* (New York: Longmans, Green & Co., 1911), 75–77.

11. *U.S. Statutes at Large* 12 (1863): 276. For a discussion of how this law was interpreted by Secretary of the Treasury Salmon P. Chase, see Sinisi, *Sacred Debts*, 9–12.

12. "Report of the Quarter Master General," *Documents of the State of Indiana*, 40th sess., pt. 1 (Indianapolis, 1859), 263–73.

13. Harry N. Scheiber, *Ohio Canal Era: A Case Study of Government and the Economy, 1820–1861*, new ed. (Athens: Ohio University Press, 1987). 70–74, 303–4; Mark Wahlgren Summers, *The Plundering Generation: Corruption and the Crisis of the Union* (New York: Oxford University Press, 1987), 125–37.

14. Louis Hartz, *Economic Policy and Democratic Thought: Pennsylvania, 1776–1860* (Cambridge: Harvard University Press, 1948), 151–56; Ronald E. Shaw, *Erie Water West: A History of the Erie Canal, 1792–1854* (Lexington: University of Kentucky Press, 1966), 169–72.

15. George B. Wright, "War Experiences at Columbus," in Alfred E. Lee, *History of the City of Columbus, Capital of Ohio* (New York: Munsell & Co., 1892), 2:169; Noel Fisher, "Groping towards Victory: Ohio's Administration of the Civil War," *Ohio History* 105 (1996): 25–45; Robert E. Colebird Jr., "John Williams: A Merchant Banker in Springfield, Illinois," *Agricultural History* 42 (1968): 259–65.

16. *Philadelphia Daily Evening Bulletin*, 26 Apr. 1861, 2 May 1861, 8 May 1861; *North American and U.S. Gazette*, 23 Apr. 1861, 25 Apr. 1861, 8 May 1861, 13 May 1861; "Report of the Quartermaster General of the Commonwealth of Pennsylvania for the Year 1861," *Pennsylvania Legislative Documents* (Harrisburg, 1862), 424–30; J. Matthew Gallman, *Mastering Wartime: A Social History of Philadelphia during the Civil War* (Cambridge: Cambridge University Press, 1990), 287; Wisconsin accounts in folder 2, box 2, and folder 6, box 15, ser. 1159, SHSW.

17. "Payroll Accounts for Cartridges Backed," R.G. 301.082, Illinois State Archives; *Journal of the Constitutional Convention of the State of Illinois, Convened at Springfield, January 7, 1862* (Springfield, 1862), 368; "Report of the Quartermaster General [1861]," *Ohio Executive Documents*, pt. 1 (Columbus, 1862), 536; Morton to Cameron, 11 Oct. 1861, Morton telegram books, private dispatch book 1, ISA-CPR; *Report of the Adjutant General of the State of Indiana* (Indianapolis, 1869), 1:413–26.

18. "Report of the Quartermaster General of the State of Illinois," in *Reports Made to the General Assembly of Illinois*, 23rd sess. (Springfield, 1865), 3:14–15; "Indiana State Arsenal Report," undated manuscript, folder 107-17, drawer 107, ISA-CPR; "Payroll Accounts for Cartridges Backed," R.G. 301.082, Illinois State Archives.

19. Glen A. Gildemeister, *Prison Labor and Convict Competition with Free Workers in Industrializing America, 1840–1890* (New York: Garland, 1987).

20. *Annual Report of the Board of Inspectors of the Massachusetts State Prison, October 1, 1861* (Boston, 1861), 7–8, 27; Heg to H. W. Lawrence, 21 June 61, box 2, and Heg to Wisconsin quartermaster general, 27 Dec. 61, box 12, both in ser. 1159, SHSW. On sources for Massachusetts and Pennsylvania supply expenditures by recipient, see notes 27 and 28.

21. William Schouler, *A History of Massachusetts in the Civil War* (Boston, 1868), 219–20; Allan Nevins, *The Improvised War, 1861–1862* (New York: Charles Scribner's Sons, 1959), 342–69; Owen to Morton, 24 July 1861, Oliver P. Morton Papers, ISA-CPR; *Report of the Adjutant General of . . . Indiana*, 1:426–44.

22. "Report of the Quartermaster General," *Ohio Executive Documents*, pt. 1 (Columbus, 1862), 586–87; unpublished Wisconsin arms commissioners report, 31 Dec. 1861, box 12, ser. 49, SHSW.

23. William Levering to J. Vajen, 4 July 1861, 17 July 1861, box 1, Adjutant General Civil War General Correspondence, ISA-CPR; Levering to Vajen, 10 July 1861, 15 July 1861, Morton Papers, ISA-CPR; William W. Tredway to William Mears, 20 Aug. 1861, Mears to Tredway, 29 Aug. 1861, 30 Aug. 1861, box 5, ser. 1159, SHSW.

24. "Report of the Quartermaster General of the Commonwealth of Pennsylvania, for the Year 1861," in Pennsylvania *Legislative Documents* (Harrisburg, 1862), 427; *Journal of the Assembly of Wisconsin*, 14th sess. (Madison, 1862), 1409; Vajen testimony in GovCon 2, p. 1075.

25. L. J. & I. Phillips to Randall, 30 May 1861, box 8, ser. 49, SHSW.

26. Indiana firm-by-firm transaction data calculated from "Report of John H. Vajen, Quartermaster-General" and "Report of Ashael Stone, Commissary General," in *Documents of the General Assembly of Indiana*, 42nd sess., pt. 2 (Indianapolis, 1863), 1:661–75, 1:811–98; *Report of the Adjutant General of the State of Indiana*, 1:434. On Glaser Bros., see Indiana, vol. 67, p. 107, and Ohio, vol. 78, p. 731, Dun.

27. Massachusetts firm-by-firm transaction data calculated from "Report of the Auditor of Accounts of the Commonwealth of Massachusetts for the Year Ending December 31, 1861" and "Report of Master of Ordnance," Public Docs. 6–7, in *Public Documents of Massachusetts . . . for the Year 1861*, vols. 1–2 (Boston, 1862). On Benjamin Cheney, see Henry Hall, ed., *America's Successful Men of Affairs* (New York: New York Tribune, 1895–96), 2:159–61.

28. On Stewart, see Stephen N. Elias, *Alexander T. Stewart: The Forgotten Merchant Prince* (Westport, Conn.: Praeger, 1992). New York firm-by-firm transaction data calculated from "Report of the Majority of the Select Committee on So Much of the Governor's Message as Relates to the Transactions of the State Military Board," doc. 194, *Documents of the Assembly of the State of New York*, 85th sess. (Albany, 1862), 7:1–635; Pennsylvania firm-by-firm transaction data calculated from "Report of the Auditor General on the Finances of the Commonwealth of Pennsylvania for the Year Ending November 30, 1861," in Pennsylvania, *Reports of the Heads of Departments* (Harrisburg, 1861); "Report of the Auditor General . . . for the Year Ending November 30, 1862," in Pennsylvania, *Reports of the Heads of Departments* (Harrisburg, 1863); Michigan firm-by-firm transaction data calculated from payment lists in "Report of the Special Committee Appointed to Inquire into the Military Expenditures of the State during the Last Year," in *Documents Accompanying the Journal of the Senate of the State of Michigan* (Lansing, 1863), 47–99; fragmentary firm-by-firm transaction data for Wisconsin calculated from Quartermaster General's Ledger. vol. 3, ser. 1165, SHSW. Kohner testimony is in "Report of the Joint Select Committee," *Journal of the Assembly of Wisconsin*, 14th sess. (Madison, 1862), 1439–44. For a more detailed discussion of leading contractors to the various states, see Mark R. Wilson, "The Business of Civil War: Military Enterprise, the State, and Political Economy in the United States, 1850–1880" (Ph.D. diss., University of Chicago, 2002), chap. 2.

29. Edward Everett, "A Narrative of Military Experience in Several Capacities," *Transactions of the Illinois State Historical Society* 10 (1905): 230–36; "Report of the Joint Select Committee," 1441.

30. *Journal of the Constitutional Convention of the State of Illinois*, 368; *Report of the Quartermaster General of the State of Wisconsin* (Madison, 1862), 4; *Journal of the State Assembly of Wisconsin*, 14th sess. (Madison, 1862), 1407, 1437–38.

31. "Report of the Majority of the Select Committee," 243, 290–316.

32. Manuscript returns, U.S. Census, Products of Industry, 1860, New York State Archives [microfilm]; William E. Devlin, "Shrewd Irishmen: Irish Entrepreneurs and Artisans in New York's Clothing Industry, 1830–1880," in *The New York Irish*, ed. Ronald H. Bayor and Timothy J. Meagher (Baltimore: Johns Hopkins University Press, 1996), 169–92, 616–19.

33. "Report of the Majority of the Select Committee," 243, 290–316.

34. "Report of the Board of Officers Named . . . April 16, 1861"; "Report of the Majority of the Select Committee," 69–71; GovCon 1, 425–38.

35. *Journal of the House of Representatives of the State of Indiana during the Special Session*

of the General Assembly (Indianapolis, 1861), 213–18; *Report of the Adjutant General of the State of Indiana;* Emma Lou Thornbrough, *Indiana in the Civil War Era, 1850–1880* (Indianapolis: Indiana Historical Society, 1965), 168–69.

36. GovCon 2, 1071–72.

37. GovCon 2, xxviii–xxxix, 1000–1006, 1020–26, 1035–39, 1048–50, 1058–74. For more on the work of the congressional contracts committee, see chapter 5.

38. Richard Yates to John Wood, 5 Nov. 1861; Martin H. Cassell to Yates, 16 Nov. 1861 and 23 Nov. 1861, all in box 4, Richard Yates Papers, Illinois State Historical Library. Illinois firm-by-firm transaction data calculated from "Report of the Board of Army Auditors," in *Illinois Reports to the General Assembly* (Springfield, 1865), 2:593–808. For testimony on the connection between Cole & Hopkins and A. T. Stewart & Co. in 1861, see GovCon 2, 1016–19. For a very brief summary of the wartime business record of Sarmiento & McGrath, see J. Matthew Gallman, "Entrepreneurial Experiences in the Civil War: Evidence from Philadelphia," in *American Economic Development in Historical Perspective,* ed. Thomas Weiss and Donald Schaefer (Stanford: Stanford University Press, 1994), 217.

39. Martin H. Cassell to Randall, 21 Aug. 1861, box 5, and Napoleon Van Slyke to U.S. Quartermaster at Philadelphia, 12 Sept. 1861, box 21, both in ser. 1159, SHSW.

40. Owen to Vajen, 28 Aug. 1861, Morton Papers; Wisconsin arms commissioners report, 31 Dec. 1861, box 12, ser. 49, SHSW.

41. James M. McPherson, *Ordeal by Fire: The Civil War and Reconstruction* (New York: Alfred A. Knopf, 1982), 165.

42. Meigs to E. D. Morgan, 7 Aug. 1861, pp. 167–68, vol. 56, roll 36, M745.

43. "Report of the Board of Officers Named . . . April 16, 1861," 126, 160–96; "Annual Report of the Quartermaster General," *Documents of the Assembly of the State of New York,* 85th sess. (Albany, 1862), 13.

44. "The War Frauds in Pennsylvania," *New York Tribune,* 5 June 1861; "Army Contractors Indicted," *New York Tribune,* 18 June 1861; *Philadelphia Daily Evening Bulletin,* 30 Sept. 1861, 10 Oct. 1861; *North American and U.S. Gazette,* 15 Oct. 1861; Pennsylvania, *Journal of the House of Representatives* (Harrisburg, 1862), 755–59; "Testimony Taken before the Select Committee of the House of Representatives, Appointed to Inquire into Alleged Frauds in Army Contracts," doc. 80, *Miscellaneous Documents Read in the Legislature of the Commonwealth of Pennsylvania, during the Session Which Commenced at Harrisburg, on the Seventh Day of January 1862* (Harrisburg, 1862); "Report of the Special Committee Appointed to Inquire into the Military Expenditures of the State during the Last Year," *Documents Accompanying the Journal of the Senate of the State of Michigan* (Lansing, 1863).

45. "Report of the Joint Select Committee," 1345, 1362–87, 1493–94; *Cincinnati Daily Commercial,* 7 June 1861, 26 June 1861.

46. Traditional accounts, which tend to focus on scandals involving national officials, often overlook state-level problems. For classic works that emphasize the energy and efficiency of state governors such as Andrew and Morton, in contrast to early waste and confusion at the national level, see William B. Weeden, *War Government, Federal and State, in Massachusetts, New York, Pennsylvania, and Indiana, 1861–1865* (Boston: Houghton, Mifflin & Co., 1906), 74, 155–56, 165–66; Hesseltine, *Lincoln and the War Governors,* 148–52, 161–62.

47. Meigs to Sprague, 20 Sept. 1861, p. 434, vol. 56, roll 36, M745.

48. Meigs to F. Myers, 13 Nov. 1861, pp. 241–42, vol. 57, roll 36, M745; letters copied in

E. S. Sibley to J. Dickerson, 21 May 1862, pp. 121–28, vol. 60, roll 38, M745; Fisher, "Groping towards Victory," 41–42.

49. Meigs to Washburn, 12 Oct. 1861, pp. 80–81, vol. 57, roll 36, M745; see also Michael W. Whalon, "Israel Washburn and the War Department," *Social Science* 46 (1971): 79–85.

50. Morton to U.S. Commissary General J. P. Taylor, 9 Aug. 1861, Morton telegram books, private dispatch book 1; Morton to Seward, 1 Sept. 1861, Morton telegram books, general dispatch book 1, ISA-CPR.

51. Morton to Seward, undated telegram [probably Sept. 1861], Morton telegram books, private dispatch book 1, ISA-CPR.

52. Morton to Meigs, 25 Nov. 1861, Morton telegram books, general dispatch book 2, ISA-CPR. For a letter in which Quartermaster General Meigs urged Quartermaster Montgomery to bridge his differences with Governor Morton and work toward "a cordial cooperation," see Meigs to Montgomery, 7 Oct. 1861, p. 51, vol. 57, roll 36, M745.

53. Meigs to Morton, 19 Oct. 1861, Morton telegram books, general dispatch book 2, ISA-CPR; Meigs to Morton, 30 Oct. 1861, p. 166, vol. 57, roll 36, M745.

54. Meigs to Eddy, 27 Sept. 1861, and Eddy to Meigs, 16 Oct. 1861, reproduced in *Journal of the Constitutional Convention of Illinois*, 677–78.

55. Eddy to Yates, 9 Nov. 1861, and Yates to Eddy, 9 Dec. 1861, reproduced in *Journal of the Constitutional Convention of Illinois*, 157–59.

56. A report on early reimbursements appeared in *Bankers' Magazine and Statistical Register* (new ser.) 11, no. 5 (Nov. 1861): 398.

57. Eddy to Yates, 11 Dec. 1861, reproduced in *Journal of the Constitutional Convention of Illinois*, 157–59; *Illinois State Register* article reprinted as "A Jar With the Quartermaster at Springfield," *Missouri Republican*, 22 Dec. 1861; Yates to William Kellogg (and other members of the Illinois congressional delegation), 12 Dec. 1861, box 5, Yates Papers; Meigs to Eddy, 16 Dec. 1861, p. 426, vol. 47, roll 26, M745; Cameron to Yates, 16 Dec. 1861, in *OR*, ser. 3, 1:748.

58. Meigs to Edwin Stanton, 12 Feb. 1862, vol. 3, roll 2, Stanton Papers, LC.

59. *Journal of the Constitutional Convention of Illinois*, 32, 90, 157–64, 674–88; *Congressional Globe*, 37th Congr., 2nd Sess. (30 Jan. 1862), 565.

60. McPherson, *Battle Cry of Freedom*, 605. On the importance of bounties at the local level, see Thomas R. Kemp, "Community and War: The Civil War Experience of Two New Hampshire Towns," and Robin L. Einhorn, "The Civil War and Municipal Government in Chicago," in *Toward a Social History of the American Civil War: Exploratory Essays*, ed. Maris A. Vinovskis (New York: Cambridge University Press, 1990), 31–77, 117–38. Studies of state and local government assistance to soldiers' families include Joseph E. Holliday, "Relief for Soldiers' Families in Ohio during the Civil War," *Ohio History* 71 (1962): 97–112, and Russell L. Johnson, "'A Debt Justly Due': The Relief of Civil War Soldiers and Their Families in Dubuque," *Annals of Iowa* 55 (1996): 207–38.

61. Thanks to its large annual expenditures on horses and weapons, New Jersey spent roughly $1.5 million on military supplies from the beginning of 1862 to the end of 1864. See *Annual Report[s] of the Quartermaster General of the State of New Jersey*, 1862–65 (Trenton, 1863–66).

62. Sinisi, *Sacred Debts*, 41; Thornbrough, *Indiana in the Civil War Era*, 170–76.

63. "Report of the Quartermaster General" [1863], *Ohio Executive Documents*, pt. 2 (Columbus, 1864), 537; "Indiana Arsenal," in *Report of W. H. H. Terrell, Financial Secretary, to the*

Governor, May, 1864 (Indianapolis, 1864), 16–21. The history of the Indiana bakery, which made bread valued at $72,000 between September 1862 and December 1864, may be gleaned from the following sources: folders 53–62, drawer 106, Adjutant General Civil War Miscellany, ISA-CPR; "Report of Ashael Stone, Commissary General," in *Documents of the General Assembly of Indiana*, 42nd sess., pt. 2 (Indianapolis, 1863), 1:803–9; "Report of Ashael Stone, Quartermaster-General," in *Documents of the General Assembly of Indiana*, 43rd sess., pt. 2 (Indianapolis, 1865), 207–10; *Report of the Adjutant General of the State of Indiana*, 1:359, 1:450.

64. On the states' war claims, see Sinisi, *Sacred Debts*.

65. Stanton to Andrew, cited in Weeden, *War Government, Federal and State*, 213–14.

CHAPTER 2. THE FORMATION OF A NATIONAL BUREAUCRACY

1. Harry J. Carman and Rienhard H. Luthin, *Lincoln and the Patronage* (New York: Columbia University Press, 1943).

2. Eddy's circumstances before he attended West Point are described in file 60, roll 115, M688. For summaries of the careers of Eddy and other West Point graduates, see George W. Cullum, *Biographical Register of the Officers and Graduates of the U.S. Military Academy*, 3rd ed., vols. 1–2 (Boston: Houghton, Mifflin & Co., 1891).

3. Chris Cook and Brendan Keith, *British Historical Facts, 1830–1900* (London: Macmillan, 1975), 185; Paul Kennedy, *The Rise and Fall of the Great Powers: Economic Change and Military Conflict from 1500 to 2000* (New York: Random House, 1987), 154.

4. Marcus Cunliffe, *Soldiers and Civilians: The Martial Spirit in America, 1775–1865* (Boston: Little, Brown & Co., 1968).

5. U.S. Department of Commerce, Bureau of the Census, *Historical Statistics of the United States: Colonial Times to 1970* (Washington, D.C.: GPO, 1975), 1114–15; Ira Katznelson, "Flexible Capacity: The Military and Early American Statebuilding," in *Shaped by War and Trade: International Influences on American Political Development*, ed. Ira Katznelson and Martin Shefter (Princeton: Princeton University Press, 2002), 91–93.

6. See, esp., Francis Paul Prucha, *The Great Father: The United States Government and the American Indians* (Lincoln: University of Nebraska Press, 1984); Merritt Roe Smith, *Harpers Ferry Armory and the New Technology: The Challenge of Change* (Ithaca: Cornell University Press, 1977); and Robert G. Angevine, *The Railroad and the State: War, Politics, and Technology in Nineteenth-Century America* (Stanford: Stanford University Press, 2004).

7. Erna Risch, *Quartermaster Support of the Army: A History of the Corps* (Washington, D.C.: Office of the Quartermaster General, 1962), 1–240; James A. Huston, *The Sinews of War: Army Logistics, 1775–1953* (Washington, D.C.: Office of the Chief of Military History, U.S. Army, 1966), 112, 131–32; Cynthia Ann Miller, "The United States Army Logistics Complex, 1818–1845: A Case Study of the Northern Frontier" (Ph.D. diss., Syracuse University, 1991). For a broader view of army activities and reforms during this period, see Russell Weigley, *History of the United States Army*, enlarged ed. (Bloomington: Indiana University Press, 1984), 67–118.

8. Chester L. Kieffer, *Maligned General: The Biography of Thomas Sidney Jesup* (San Rafael, Calif.: Presidio Press, 1979).

9. Jesup to Calhoun, 5 June 1818, copy in box 3, Oscar F. Long Collection, Huntington Library, San Marino, Calif.

10. Kieffer, *Maligned General*, 67–118; Risch, *Quartermaster Support*, 186.

11. For one study that emphasizes Jesup's accomplishments while situating them in a broader context, see Samuel J. Watson, "Professionalism, Social Attitudes, and Civil-Military Accountability in the United States Army Officer Corps, 1815–1846" (Ph.D. diss., Rice University, 1996), 616–31. See also Robert G. Angevine, "Individuals, Organizations, and Engineering: U.S. Army Officers and the American Railroads, 1827–1838," *Technology and Culture* 42 (2001): 299.

12. Leonard D. White, *The Jacksonians: A Study in Administrative History, 1829–1861* (New York: Macmillan, 1954); Matthew A. Crenson, *The Federal Machine: Beginnings of Bureaucracy in Jacksonian America* (Baltimore: Johns Hopkins University Press, 1975); Richard R. John, *Spreading the News: The American Postal System from Franklin to Morse* (Cambridge: Harvard University Press, 1995).

13. White, *The Jacksonians*, 57–61, 354; Richard Bruce Winders, *Mr. Polk's Army: The American Military Experience in the Mexican War* (College Station: Texas A&M University Press, 1997), 17; 1857 ARSW.

14. Jesup to Calhoun, 5 June 1818, copy in box 3, Oscar F. Long Collection.

15. Risch, *Quartermaster Support*, 198, 240, 330; typescript summary in box 3, Oscar F. Long Collection; Durwood Ball, *Army Regulars on the Western Frontier, 1848–1861* (Norman: University of Oklahoma Press, 2001), xxii.

16. James L. Morrison Jr., *"The Best School in the World": West Point, the Pre–Civil War Years, 1833–1866* (Kent, Ohio: Kent State University Press, 1986); William B. Skelton, *An American Profession of Arms: The Army Officer Corps, 1784–1861* (Lawrence: University Press of Kansas, 1992), 154–74.

17. File 63 (Meigs), roll 81; file 351 (Swords), roll 34, both in M688; "American Publishers," *DeBow's Review* 1 (Jan. 1855): 140.

18. File 290 (McKinstry), roll 65; file 327 (Holabird), roll 150, both in M688.

19. File 93 (Donaldson), roll 82; file 49 (Allen), roll 67, both in M688.

20. Edward M. Coffman, *The Old Army: A Portrait of the American Army in Peacetime, 1784–1898* (New York: Oxford University Press, 1986); Skelton, *An American Profession of Arms*, 193–202; Ball, *Regulars on the Western Frontier*, 57–58.

21. "Report[s] of the Quartermaster General," in ARSW (1847 and 1848); K. Jack Bauer, *The Mexican War, 1846–1848* (New York: Macmillan, 1974); Risch, *Quartermaster Support*, 253–56, 292; Huston, *The Sinews of War*, 131–32; Emmett M. Essin, *Shavetails and Bell Sharps: The History of the U.S. Army Mule* (Lincoln: University of Nebraska Press, 1997), 28–29.

22. Typescript summary in box 3, Oscar F. Long Collection.

23. *National Cyclopaedia of American Biography* (New York: James T. White & Co., 1897), 4:542; Harry C. Myers, ed., "From 'The Crack Post of the Frontier': Letters of Thomas and Charlotte Swords," *Kansas History* 5 (1982): 184–213; George Walton, *Sentinel of the Plains: Fort Leavenworth and the American West* (Englewood Cliffs, N.J.: Prentice Hall, 1973), 47–71.

24. "Report of the Quartermaster General," in 1848 ARSW; Risch, *Quartermaster Support*, 278–80.

25. Neal Harlow, *California Conquered: War and Peace on the Pacific, 1846–1850* (Berkeley: University of California Press, 1982), 201, 274–75.

26. Ball, *Regulars on the Western Frontier*, xx–xxi.

27. Jesup to Vinton, 23 May 1855, p. 162, vol. 48, roll 31, M745; Lance E. Davis and John Legler, "Government in the American Economy, 1815–1902: A Quantitative Study," *Journal of*

Economic History 26 (1966): 514–55; Robert W. Frazer, *Forts and Supplies: The Role of the Army in the Economy of the Southwest, 1846–1861* (Albuquerque: University of New Mexico Press, 1983); William A. Dobak, *Fort Riley and Its Neighbors: Military Money and Economic Growth, 1853–1895* (Norman: University of Oklahoma Press, 1998); Thomas T. Smith, *The U.S. Army and the Texas Frontier Economy, 1825–1900* (College Station: Texas A&M University Press, 1999).

28. For a more detailed account of antebellum sources of supply, including public arsenals and contractors, see Mark R. Wilson, "The Business of Civil War: Military Enterprise, the State, and Political Economy in the United States, 1850–1880" (Ph.D. diss., University of Chicago, 2002), chapter 2.

29. Ball, *Regulars on the Western Frontier*, 28–29; "Receipts, Expenditures, and Appropriations from 1789 to 1857," House Exec. Doc. 60, 35th Congr., 1st Sess., ser. 955; "Report of the Quartermaster General." in 1856 ARSW.

30. 1848 and 1855–56 ARSW.

31. On the early history of the PMSC, see John Haskell Kemble, *The Panama Route, 1848–1869* (Berkeley and Los Angeles: University of California Press, 1943).

32. Forrest R. Blackburn, "Fort Leavenworth: Logistical Base for the West," *Military Review* 53 (1973): 3–12; Essin, *Shavetails and Bell Sharps*, 53–55; Smith, *Texas Frontier Economy*, 52; Risch, *Quartermaster Support*, 308–16; Henry Pickering Walker, *The Wagonmasters: High Plains Freighting from the Earliest Days of the Santa Fe Trail to 1880* (Norman: University of Oklahoma Press, 1966).

33. Howard Lackman, "George Thomas Howard, Texas Frontiersman" (Ph.D. diss, University of Texas at Austin, 1954), 280–331; Smith, *Texas Frontier Economy*, 53–54, 179.

34. Raymond W. Settle and Mary Lund Settle, *War Drums and Wagon Wheels: The Story of Russell, Majors, and Waddell* (Lincoln: University of Nebraska Press, 1966), 33–48.

35. Norman F. Furniss, *The Mormon Conflict, 1850–1859* (New Haven: Yale University Press, 1960), 1–63.

36. Jesup to Thomas, 21 June 1857, pp. 90–91, vol. 51, roll 33, M745.

37. Jesup to Babbitt, 28 May 1857, p. 569, vol. 50, roll 33, M745; Jesup to Babbitt, 1 June 1857 and 10 June 1857, vol. 15, e. 999, RG 92, NA.

38. Jesup to Crosman, 29 May 1857, p. 570, vol. 50, roll 33, M745; T. L. Brent manuscript report entitled "Abstract of Qr. Master's Stores Furnished to Cheyenne and Utah Expeditions, etc. at Fort Leavenworth, K.T., in 1857," and Charles Thomas report of 18 Dec. 1857, both in box 1174, e. 225, RG 92, NA; Samuel W. Ferguson, "With Albert Sidney Johnston's Expedition to Utah, 1857," *Kansas Historical Collections* 12 (1912): 303–12; Audrey M. Godfrey, "Housewives, Hussies, and Heroines. or the Women of Johnston's Army," *Utah Historical Quarterly* 54 (1986): 157–78; Furniss, *The Mormon Conflict*, 101–4, 169–71.

39. "Utah Expedition," House Exec. Doc. 71, 35th Congr., 1st Sess., ser. 956; Dickerson to Jesup, 20 Nov. 1857, and Dickerson, "Abstract of Quartermaster's Property in Possession of the Army of Utah," 31 Dec. 1857, both in box 1174, e. 225, RG 92, NA; Jesup to P. F. Smith, 17 Apr. 1858, pp. 72–93, vol. 52, roll 34, M745; Furniss, *The Mormon Conflict*, 112–18; W. Turrentine Jackson, *Wagon Roads West: A Study of Federal Road Surveys and Construction in the Trans-Mississippi West, 1846–1869* (Berkeley and Los Angeles: University of California Press, 1952), 132–34.

40. Furniss, *The Mormon Conflict*, 148–67; Jesup to Russell, Majors, & Waddell, 21 Jan.

1858, 30 Jan. 1858, 27 Feb. 1858, pp. 438–39, 449–50, 501, vol. 51, roll 33, M745; Settle and Settle, *War Drums and Wagon Wheels,* 80–85.

41. Jesup to Wayne, 15 Jan. 1858, 2 Mar. 1858, 15 Mar. 1858, vol. 16, e. 999, RG 92, NA; Jesup to P. F. Smith, 17 Apr. 1858, pp. 72–93, vol. 52, roll 34, M745; "Contracts—Utah Expedition," House Exec. Doc. 99, 35th Congr., 1st Sess., ser. 958; "David D. Mitchell," House Report C.C. 281, 37 Congr., 2nd Sess., ser. 1146.

42. *Congressional Globe,* 35th Congr., 1st Sess. (2 and 9 Apr. 1858), 1451–53, 1548; "Deficiencies in the Quartermaster's Department," House Misc. Doc. 22, 35th Congr., 1st Sess., ser. 961; Furniss, *The Mormon Conflict,* 148–73.

43. Settle and Settle, *War Drums and Wagon Wheels,* 119.

44. M. C. Meigs circular, 16 May 1862, pp. 49–54, vol. 60, roll 38, M745; Risch, *Quartermaster Support,* 334–35, 382–93.

45. James M. Ashley to Cameron, 29 May 1861, box 700, e. 225, RG 92; Jos. S. Fay to Meigs, 11 Jan. 1862, roll 12, Montgomery C. Meigs Papers, LC.

46. William G. Le Duc, *Recollections of a Civil War Quartermaster: The Autobiography of William G. Le Duc* (St. Paul: North Central Publishing Co., 1963); Ezra J. Warner, *Generals in Blue: Lives of the Union Commanders* (Baton Rouge: Louisiana State University Press, 1964), 360–61.

47. For one reference to the Moulton-Sherman relationship, see J. K. Butterfield to John Sherman, 12 Apr. 1864, roll 1, John Sherman Papers, LC.

48. *Biographical History of Eminent and Self-made Men of the State of Indiana* (Cincinnati: Western Publishing Co., 1880), 1:13–16; Lenette S. Taylor, *The Supply for Tomorrow Must Not Fail: The Civil War of Captain Simon Perkins, Jr., a Union Quartermaster* (Kent, Ohio: Kent State University Press, 2004); W. H. Brown to Lincoln, 19 Oct. 1861, 2215 ACP 1882, RG 94, NA; Roger D. Hunt and Jack R. Brown, *Brevet Brigadier Generals in Blue* (Gaithersburg, Md.: Olde Soldier Books, 1990), 83; Frederic Cople Jaher, *The Urban Establishment: Upper Strata in Boston, New York, Charleston, Chicago, and Los Angeles* (Urbana: University of Illinois Press, 1982), 461–62.

49. *Historical Encyclopedia of Illinois* (Chicago: Munsell Publishing Co., 1906), 1:31; *American Biographical History of Eminent and Self-Made Men . . . Michigan Volume* (Cincinnati: Western Biographical Publishing Co., 1878), 47–48; *Biographical Cyclopaedia and Portrait Gallery of . . . Ohio* (Cincinnati: John C. Yorston & Co., 1879), 458–59.

50. Ball, *Regulars on the Western Frontier,* 82.

51. William Skelton, "Officers and Politicians: The Origins of Army Politics in the United States before the Civil War," *Armed Forces and Society* 6 (1979): 22–48; Skelton, *An American Profession of Arms,* 139, 290; Coffman, *The Old Army,* 42–103; Ball, *Regulars on the Western Frontier,* 62–66, 76–82.

52. Samuel J. Watson, "Manifest Destiny and Military Professionalism: Junior U.S. Army Officers' Attitudes towards War with Mexico, 1844–1846," *Southwestern Historical Quarterly* 99 (1996): 466–98.

53. Allen to Polk, 22 Jan. 1846; Allen to Jesup, 6 Mar. and 13 Apr. 1846, all in box 27, e. 225, RG 92, NA.

54. Crosman to Eaton, 4 Nov. 1858, box 13, Eldridge Collection, Huntington Library; Crosman to Dreer, 10 Dec. 1866, Ferdinand J. Dreer Autograph Collection, Historical Society of Pennsylvania, Philadelphia.

55. David W. Miller, *Second Only to Grant: Quartermaster General Montgomery C. Meigs* (Shippensburg, Penn.: White Mane Publishing Co., 2001), 75.

56. M. C. Meigs to Charles D. Meigs, 4 May 1840, roll 4, Meigs Papers.

57. Russell F. Weigley, *Quartermaster General of the Union Army: A Biography of M. C. Meigs* (New York: Columbia University Press, 1959), 131, 159.

58. "The Quartermaster-General," *New York Tribune*, 8 June 1861.

59. Lincoln to Winfield Scott, 5 June 1861, copy in box 10, David Davis Papers, Chicago Historical Society; *The Diary of George Templeton Strong*, vol. 3, *The Civil War, 1860–1865*, ed. Allan Nevins and Milton Halsey Thomas (New York: Macmillan, 1952), 173; "Bureaucracy," *New York Tribune*, 25 Dec. 1861.

60. M. C. Meigs to Charles D. Meigs, 22 June 1861, roll 5, Meigs Papers; Meigs to Lincoln, 21 Dec. 1861, roll 30, Abraham Lincoln Papers, LC.

61. Meigs to E. H. Rollins, 16 July 1861, pp. 76–77, vol. 56, roll 36, M745; Meigs to Cameron, 27 Aug. 1861, roll 12, Meigs Papers; Meigs to Mary Lincoln, 4 Oct. 1861, p. 481, vol. 56, roll 36, M745.

62. *Revised United States Army Regulations of 1861* (Washington, D.C., 1863); Walworth Jenkins, *Q.M.D., or a Book of Reference for Quartermasters* (Louisville: John P. Morton & Co., 1865); Roeliff Brinkerhoff, *The Volunteer Quartermaster: Containing a Collection of the Laws, Regulations, Rules, and Practice Governing the Quartermaster's Department of the United States Army, and In Force May 9, 1865* (New York: D. Van Nostrand, 1865). See also R. F. Hunter, *Manual for Quartermaster and Commissaries* (New York: D. Van Nostrand, 1864), and George Patten, *Patten's Army Manual* (New York: J. W. Fortune, 1862).

CHAPTER 3. THE MAKING OF A MIXED MILITARY ECONOMY

1. Sarah Jacobs to Abraham Lincoln, 17 Oct. 1861, box 963, e. 225, RG 92, NA.

2. For a study that emphasizes the importance of the clothing industry and the political and economic activities of seamstresses in Philadelphia during the war, see Rachel Filene Seidman, "Beyond Sacrifice: Women and Politics on the Pennsylvania Homefront during the Civil War" (Ph.D. diss., Yale University, 1995). On the mostly unpaid work by women in voluntary organizations, which procured significant stocks of food and medical supplies, see Jeannie Attie, *Patriotic Toil: Northern Women and the American Civil War* (Ithaca: Cornell University Press, 1998); Judith Ann Giesberg, *Civil War Sisterhood: The U.S. Sanitary Commission and Women's Politics in Transition* (Boston: Northeastern University Press, 2000). For a recent survey, see Nina Silber, *Daughters of the Union: Northern Women Fight the Civil War* (Cambridge: Harvard University Press, 2005).

3. Paul A. C. Koistinen, *Beating Plowshares into Swords: The Political Economy of American Warfare, 1606–1865* (Lawrence: University Press of Kansas, 1996), 53–93; Erna Risch, *Quartermaster Support of the Army: A History of the Corps* (Washington, D.C.: Office of the Quartermaster General, 1962), 145–47, 202, 251–57; Merritt Roe Smith, *Harpers Ferry Armory and the New Technology: The Challenge of Change* (Ithaca: Cornell University Press, 1977); Cynthia Ann Miller, "The United States Army Logistics Complex, 1818–1845: A Case Study of the Northern Frontier" (Ph.D. diss., Syracuse University, 1991), 359–430. On the navy yards, see also Kurt Hackemer, *The U.S. Navy and the Origins of the Military-Industrial Complex, 1847–1883* (Annapolis: Naval Institute Press, 2001); Thomas R. Heinrich, *Ships for the Seven Seas: Philadelphia Shipbuilding in the Age of Industrial Capitalism* (Baltimore: Johns Hopkins University Press, 1997).

4. Edward George Everett, "Pennsylvania's Mobilization for War, 1860–1861" (Ph.D. diss., University of Pittsburgh, 1954), 228; Nathan Rosenberg, ed., *The American System of Manufactures* (Edinburgh: Edinburgh University Press, 1969), 365; "Miscellaneous Summary," *Scientific American* 9, no. 12 (19 Sept. 1863): 179; W. F. G. Shanks, "The Brooklyn Navy Yard," *Harper's New Monthly Magazine* 42 (Dec. 1870): 1–13; Richard E. Winslow III, *Constructing Munitions of War: The Portsmouth Navy Yard Confronts the Confederacy, 1861–1865* (Portsmouth, N.H.: Peter E. Randall, 1995).

5. Donald A. MacDougall, "The Federal Ordnance Bureau, 1861–1865" (Ph.D. diss., University of California at Berkeley, 1951), 177–78; Everett, "Pennsylvania's Mobilization for War," 238–39; James V. Murray and John Swantek, eds., *The Watervliet Arsenal, 1813–1993: A Chronology of the Nation's Oldest Arsenal* (Watervliet, N.Y., 1993), 91–108; James J. Farley, *Making Arms in the Machine Age: Philadelphia's Frankford Arsenal, 1816–1870* (University Park: Pennsylvania State University Press, 1994), 21–25, 53–64, 75–86. A terrible accident at the Allegheny Arsenal in September 1862, which killed seventy-eight ammunition plant workers, mostly girls and young women in their teens, has been described by Alan Becer in "An Appalling Disaster: The Allegheny Arsenal and the Great Explosion of 1862" (unpublished paper presented at the Twenty-sixth Annual Duquesne University History Forum, 14 Oct. 1993).

6. "The Government Bakery," *Scientific American* 9, no. 11 (12 Sept. 1863): 167. For an account of a public bakery in Philadelphia with twenty-five workers, see "Another Government Enterprise," *North American and U.S. Gazette*, 29 Jan. 1864.

7. "Government Laboratory," *Scientific American* 9, no. 4 (25 July 1863): 51; "The National Laboratory—a Great Public Medicine Factory," *North American and U.S. Gazette*, 26 Jan. 1864; George Winston Smith, *Medicines for the Union Army: The United States Army Laboratories during the Civil War* (Madison, Wisc.: American Institute for the History of Pharmacy, 1962).

8. Risch, *Quartermaster Support*, 253–56; Miller, "United States Army Logistics Complex," 359–437; G. H. Crosman to [Meigs], 30 July 1864, vol. 23, e. 999, RG 92, NA.

9. "Report of the Chief of Ordnance," in 1866 *ARSW*; James Biser Whisker, *The United States Armory at Springfield, 1795–1865* (Lewiston, N.Y.: Edwin Mellen Press, 1997), 170–71. For a more detailed discussion of the small arms industry, see Carl L. Davis, *Arming the Union: Small Arms in the Civil War* (Port Washington, N.Y.: Kennikat Press, 1973); Mark R. Wilson, "The Business of Civil War: Military Enterprise, the State, and Political Economy in the United States, 1850–1880" (Ph.D. diss., University of Chicago, 2002), 349–66; James Biser Whisker, *U.S. and Confederate Arms and Armories during the American Civil War*, 4 vols. (Lewiston, N.Y.: Edwin Mellen Press, 2002).

10. "Report of the Chief of Ordnance," 1866 *ARSW*. See also Berkeley R. Lewis, *Small Arms and Ammunition in the United States Service* (Washington, D.C.: Smithsonian Institution, 1956), and Dean S. Thomas, *Round Ball to Rimfire: A History of Civil War Small Arms Ammunition*, pts. 1–3 (Gettysburg, Pa.: Thomas Publications, 1997–2003).

11. Hackemer, *The U.S. Navy and the Origins of the Military Industrial Complex*, 116; Koistinen, *Beating Plowshares into Swords*, 173. See also William H. Roberts, *Civil War Ironclads: The U.S. Navy and Industrial Mobilization* (Baltimore: Johns Hopkins University Press, 2002).

12. A. J. Perry report, 19 Oct. 1865, in "Report of the Quartermaster General," 1865 *ARSW*.

13. Dunan payrolls, e. 238, 2nd subseries (oversize), 1864, RG 92, NA; "Hurtt Court Martial," pp. 63, 106, House Exec. Doc. 225, 43rd Congr., 1st Sess., ser. 1614; Russell F. Weigley,

Quartermaster General of the Union Army: A Biography of M. C. Meigs (New York: Columbia University Press, 1959), 235.

14. Crane payrolls, e. 238, second subseries (oversize), 1864, RG 92, NA. On the U.S. military railroads, see Thomas Weber, *The Northern Railroads in the Civil War, 1861–1865* (New York: King's Crown Press, 1952), 134–219, and Robert G. Angevine, *The Railroad and the State: War, Politics, and Technology in Nineteenth-Century America* (Stanford: Stanford University Press, 2004), 152–54.

15. "Report of Employees," e. 1163, RG 92, NA; E. J. Strang personal narrative reports for 1864 and 1865, e. 1128, RG 92, NA; Ira Berlin et al., eds., *Freedom: A Documentary History of Emancipation, 1861–1867,* ser. 1, vol. 2, *The Wartime Genesis of Free Labor: The Upper South* (Cambridge: Cambridge University Press, 1993), 19–23, 250–56, 375–78.

16. For an accounting that is generally careful but understates Ordnance Department employment, see Paul P. Van Riper and Keith A. Sutherland, "The Northern Civil Service: 1861–1865," *Civil War History* 11 (1965): 351–69.

17. E. S. Sibley to J. K. Morehead [sic], 3 May 1861, vol. 55, roll 35, M745; George H. Pendleton to Cameron, 29 July 1861, box 6, e. 22, RG 92, NA.

18. "Distribution of War Money," *Missouri Democrat,* 2 July 1861; "Army Clothing—Neglect of the West," *Cincinnati Daily Commercial,* 19 July 1861; Cincinnati City Council Resolution, 11 Dec. 1861, copies in box 7, e. 22, RG 92, NA, and roll 30, Lincoln Papers, LC. Weeks later, in February 1862, the Ohio state legislature sent Congress a similar petition. See "Branch Clothing Bureau," House Misc. Doc. 46, 37 Congr., 2nd Sess., ser. 1141.

19. "An Enthusiastic Meeting of Workingmen," *Cincinnati Daily Enquirer,* 12 Jan. 1862; "The Removal of Government Work from Cincinnati," *Cincinnati Daily Commercial,* 20 Jan. 1862; "Removal of Government Work from Cincinnati," *Cincinnati Daily Enquirer,* 21 Jan. 1862. See also Clinton W. Terry, "'The Most Commercial of People': Cincinnati, the Civil War, and the Rise of Industrial Capitalism, 1861–1865" (Ph.D. diss., University of Cincinnati, 2002), 74.

20. H. M. Armstrong and John G. Piper to [Yates], 21 Aug. 1862, Richard Yates Correspondence, RG 101.013, Illinois State Archives; Yates to Stanton, 15 Aug. 1862, box 7, Richard Yates Papers, Illinois State Historical Library.

21. Meigs to McClellan, 20 June 1861, p. 460, vol. 55, roll 35, M745; Meigs to Stanton, 1 Feb. 1862, reprinted in *Missouri Democrat,* 1 Mar. 1862; Meigs to various state governors, 13 Aug. 1862, vol. 19, e. 999, RG 92, NA.

22. Meigs endorsement on Wisconsin tailors' petition, 18 May 1862, box 7, e. 22, RG 92, NA.

23. Charles C. Nazro et al. to Cameron, 18 Oct. 1861, and Meigs to Cameron, 22 and 30 Oct. 1861, all in *OR,* ser. 3, 1:582–86, 1:608–9. See also Risch, *Quartermaster Support,* 355.

24. P. H. Watson to Meigs, 14 Mar. 1862, box 7, e. 22, RG 92, NA; Meigs to Stanton, 22 Aug. 1862, roll 1, M494.

25. James Rusling to Meigs, 28 Jan. and 10 June 1865, vol. 1, box 1, e. 1092, RG 92, NA; Donaldson report for fiscal 1865, box 515 e. 225, RG 92, NA.

26. James M. Moore, Personal Narrative Report for fiscal 1865, e. 1128, RG 92, NA; Henry C. Symonds, *Report of a Commissary of Subsistence, 1861–1865* (Sing Sing, N.Y., 1888); Palmer H. Boeger, "The Great Kentucky Hog Swindle of 1864," *Journal of Southern History* 28 (1962): 59–70.

27. *Philadelphia Daily Evening Bulletin,* 24 Apr. 1861; Crosman and George W. Martin 1864 payrolls, in e. 238, second subseries (oversize), 1864, RG 92, NA; Crosman to [Meigs], 30 July 1864, vol. 23, e. 999, RG 92, NA. Erna Risch has suggested that the Schuylkill Arsenal employed as many as eight thousand to ten thousand sewing women at once, but the source for this estimate remains unclear. Risch, *Quartermaster Support,* 348.

28. Justus McKinstry to Charles Thomas, 30 Apr. and 1 May 1861, reprinted in *Vindication of Brig. Gen. J. McKinstry* (St. Louis, 1862), 10–11; GovCon 1, 843–50; Testimony of Thomas A. Simpson, 8 Dec. 1862, in Court-Martial Case File LL 21, RG 153, NA; "Observations in the Quartermaster's Department," *Cincinnati Daily Commercial,* 2 Dec. 1861.

29. GovCon 2, 741–43; Abbott testimony, 18 May 1863, "Court of Inquiry, Case of J. H. Dickerson," box 504, e. 225, RG 92, NA; Terry, "The Most Commercial of People," 179; R. S. Hart, Personal Narrative Report for fiscal 1864, e. 1128, RG 92, NA.

30. George Crosman to military storekeeper C. M. Siter, 11 Feb. 1862, vol. 4, e. 2291, RG 92, National Archives and Records Administration–Mid-Atlantic Region (Philadelphia).

31. S. E. Meigs to M. C. Meigs, 5 Jan. 1862, roll 12, Meigs Papers, LC; M. C. Meigs to Frank P. Blair, 6 Jan. 1862, vol. 19, e. 999, RG 92, NA.

32. M. C. Meigs to Robert Allen, 1 Feb. 1862, p. 153, vol. 58, roll 37, M745.

33. "Mass Meeting of Mechanics," *Missouri Democrat,* 3 Feb. 1862; Allen to Meigs, 3 Feb. 1862, in Records of Quartermaster William Myers, vol. 1, e. 381, RG 92, NA.

34. "Meeting of the Workingmen and Women," *Missouri Democrat,* 6 Feb. 1862; F. P. Blair to Meigs, 20 Feb. 1862, box 963, e. 225, RG 92, NA; Scott to Stanton, 9 Feb. 1862, vol. 3, roll 2, Stanton Papers, LC.

35. Meigs to Frank P. Blair, 26 Feb. 1862, and Meigs to Crosman and Vinton, 26 Feb. 1862, vol. 19, e. 999, RG 92, NA; Charles Thomas to Meigs, 19 Nov. 1862, box 963, e. 225, RG 92, NA; R. S. Hart, Annual Report for Fiscal 1863, e. 1127, RG 92, NA; Hart 1864 payrolls, e. 238, second subseries (oversize), 1864, RG 92, NA; A. J. Perry Report, 19 Oct. 1865, in "Report of the Quartermaster General," 1865 *ARSW.*

36. C. E. Bliven Report, 30 Sept. 1865, box 316, e. 225, RG 92, NA.

37. Meigs to Swords, 23 July 1862, vol. 21, e. 999, RG 92, NA; David Costigan, "A City in Wartime: Quincy, Illinois and the Civil War" (D.A. thesis, Illinois State University, 1994).

38. Meigs to Swords, 23 July 1862, vol. 21, e. 999, RG 92, NA; Meigs to Mary Andrews, 17 Apr. 1863, vol. 21, e. 999, RG 92, NA; *North American and U.S. Gazette,* 27 July 1863; J. Matthew Gallman, *Mastering Wartime: A Social History of Philadelphia during the Civil War* (New York: Cambridge University Press, 1990), 243; "Philadelphia Seamstresses—Meeting of Women Employed in the U.S. Arsenal," *Fincher's Trades Review,* 8 Aug. 1863.

39. Swords and Jenkins annual reports for fiscal 1863, vols. 5 and 3, e. 1127, RG 92, NA.

40. The standard allowance varied slightly over the course of the war. See George Patten, *Patten's Army Manual* (New York: J. W. Fortune, 1862), 38; Roeliff Brinkerhoff, *The Volunteer Quartermaster* (New York: D. Van Nostrand, 1865), 133; Risch, *Quartermaster Support,* 444–45.

41. The best recent study of this extraordinarily important antebellum industry is Michael Zakim, *Ready-made Democracy: A History of Men's Dress in the American Republic, 1760–1860* (Chicago: University of Chicago Press, 2003). See also William E. Devlin, "Shrewd Irishmen: Irish Entrepreneurs and Artisans in New York's Clothing Industry, 1830–1880," in *The New York Irish,* ed. Ronald H. Bayor and Timothy J. Meagher (Baltimore: Johns Hopkins University Press, 1996), 169–92, 616–19.

42. Zakim, *Ready-made Democracy*, 63–68, 100, 127–56; Christine Stansell, "The Origins of the Sweatshop: Women and Early Industrialization in New York City," in *Working-Class America: Essays on Labor, Community, and American Society*, ed. Michael H. Frisch and Daniel J. Walkowitz (Urbana: University of Illinois Press, 1983), 78–103.

43. On labor in antebellum New York, see Sean Wilentz, *Chants Democratic: New York City and the Rise of the American Working Class, 1788–1850* (New York: Oxford University Press, 1984); Christine Stansell, *City of Women: Sex and Class in New York, 1789–1860* (Urbana: University of Illinois Press, 1987); Richard B. Stott, *Workers in the Metropolis: Class, Ethnicity, and Youth in Antebellum New York City* (Ithaca Cornell University Press, 1990).

44. "Meeting of Tailors," *New York Tribune*, 20 July 1861; "Strike in the Navy-Yard," *New York Tribune*, 3 Jan. 1862; "Strike at the Navy Yard—Excitement among the Workmen," *Philadelphia Daily Evening Bulletin*, 17 Jan. 1862; "Another Strike in the Navy Yard," *New York Sun*, 3 July 1863; "Strike in the Boston Navy Yard," *New York Sun*, 11 Nov. 1863; "The Charlestown Navy Yard Strike: The Masses and Uncle Sam," *Fincher's Trades Review*, 21 Nov. 1863; David R. Roediger and Philip S. Foner, *Our Own Time: A History of American Labor and the Working Day* (New York: Greenwood Press, 1989), 84.

45. David Montgomery, *Beyond Equality: Labor and the Radical Republicans, 1862–1872* (New York: Alfred A. Knopf, 1967), 90–101; Philip Shaw Paludan, *"A People's Contest": The Union and Civil War, 1861–1865* (New York: Harper & Row, 1988), 170–97; Grace Palladino, *Another Civil War: Labor, Capital, and the State in the Anthracite Regions of Pennsylvania* (Urbana: University of Illinois Press, 1990); Iver Bernstein, *The New York City Draft Riots: Their Significance for American Society and Politics in the Age of the Civil War* (New York: Oxford University Press, 1990), 75–124; Louis S. Gerteis, *Civil War St. Louis* (Lawrence: University Press of Kansas, 2001), 256–59.

46. Zakim, *Ready-made Democracy*, 157–84; Mari Jo Buhle, "Needlewomen and the Vicissitudes of Modern Life: A Study of Middle-class Construction in the Antebellum Northeast," in *Visible Women: New Essays on American Activism*, ed. Nancy A. Hewitt and Suzanne Lebsock (Urbana: University of Illinois Press, 1993), 145–65.

47. Gallman, *Mastering Wartime*, 240–47; Seidman, "Beyond Sacrifice," 130–31, 158–71.

48. Hannah Rose et al. to Cameron, 1 Aug. 1861, box 798, e. 225, RG 92, NA; *North American and U.S. Gazette*, 4 Sept. 1861. This paragraph assumes that the "Mrs. Yeaker" mentioned by the newspaper was Mary Yeager, who would be involved in subsequent protests.

49. *North American and U.S. Gazette*, 5 Sept. 1861; Anna Long et al. to Stanton, 29 July 1862 or 1863 (?), box 798, e. 225, RG 92, NA. Unfortunately, the year of the Long petition is unclear. It is possible that these seamstresses were among those dismissed in 1863 when Quartermaster George Crosman began to enforce a policy of employing only those women with close male relations in the Union armies. This paragraph assumes that the "Mrs. Long" referred to in 1861 by Quartermaster Thomas was the same Anna Long who was chief signatory of the 1862 petition.

50. Leonard Myers to Meigs, 27 Apr. 1863, box 114, e. 20, RG 92, NA; William D. Kelley et al. to Stanton, 13 May 1863, box 9, e. 22, RG 92, NA.

51. George Crosman to [Meigs], 3 Nov. 1863, box 122; David Vinton to Meigs, 20 Nov. 1863, box 130; Crosman to Meigs, 14 Apr. 1864 box 132; M. S. Miller to Meigs, 30 Apr. 1864, box 133, all in e. 20, RG 92, NA; Thomas to Meigs, 25 Jan. 1864, pp. 172–73, vol. 74-A, roll 45, M745.

52. *New York Sun*, 2 July and 6 Aug. 1863.

53. *New York Sun*, 9 Dec. 1863; Philip S. Foner, *Women and the American Labor Movement: From Colonial Times to the Eve of World War I* (New York: Free Press, 1979), 114–17. By the beginning of July 1864, some seamstresses had joined a "Sewing Women's Union #1 of Philadelphia." Seidman, "Beyond Sacrifice," 154.

54. "The Working Women—Great Mass Meeting—White Slavery to Be Abolished," *New York Sun*, 13 Nov. 1863; "Women's Work, and Their Struggle for Better Compensation," *New York Times*, 15 Nov. 1863; "Another Great Mass Meeting of Working Women," *New York Sun*, 19 Nov. 1863; "The Workingwomen's Union," *New York Times*, 19 Nov. 1863; "Working Women of New York," *Fincher's Trades Review*, 21 Nov. 1863.

55. "Sad Story of a Poor Girl," *New York Sun*, 17 Nov. 1863; "Who Pockets the Difference?" *Fincher's Trades Review*, 5 Sept. 1863; "A Word for Our Starving Seamstresses," *Fincher's Trades Review*, 12 Dec. 1863.

56. "Meeting of Sewing Women," *Fincher's Trades Review*, 23 Apr. 1864; "The Wrongs of the Sewing Women," *North American and U.S. Gazette*, 27 Apr. 1864; "The Sewing Women's Mass Meeting" and "Meetings of the Sewing Women," *Fincher's Trades Review*, 14 May 1864; "The Sewing Women Again," *North American and U.S. Gazette*, 19 May 1864; "Working-Women's Meeting," *Fincher's Trades Review*, 28 May 1864. See also Seidman, "Beyond Sacrifice," 148–54.

57. Petition attached to Crosman to Meigs, 6 June 1864, box 439, e. 225, RG 92, NA; Gallman, *Mastering Wartime*, 245.

58. Crosman to Meigs, 6 June 1864, box 439, e. 225, RG 92, NA.

59. J. J. Dana to Swords and W. Myers, 10 June 1864, p. 267, vol. 77, roll 46, M745; Vinton to Meigs, 20 June 1864, box 439, e. 225, RG 92, NA.

60. For a last-minute request for an investigation and report, see Stanton to Meigs, 27 July 1864, p. 104, vol. 3, roll 2, M421. On the meetings with Stanton and Lincoln, see "The Working Women," *North American and U.S. Gazette*, 4 Aug. 1864.

61. Stanton to Meigs, 18 Aug. 1864, p. 183, vol. 3, roll 2, M421; *New York Times*, 26 Aug. 1864; "The Sewing Women and the War Department—Additional Pay Granted," *New York Sun*, 29 Aug. 1864; *Fincher's Trades Review*, 10 Sept. 1864. See also Seidman, "Beyond Sacrifice," 155–56.

62. "Fair Wages for Workingwomen," *New York Sun*, 3 Sept. 1864; "Workingwomen's Petition," *New York Sun*, 4 Sept. 1864; "Wrongs of Sewing Women," *New York Sun*, 22 Sept. 1864; *New York Sun*, 22 Oct. 1864.

63. "Wrongs of Sewing Women," *New York Sun*, 22 Sept. 1864; reprint of *Philadelphia Daily Evening Bulletin* article in "Women's Wages," *New York Sun*, 21 Sept. 1864.

64. On the January 1865 meeting with Lincoln, see "The President and the Arsenal Women," *Fincher's Trades Review*, 4 Feb. 1865, and Seidman, "Beyond Sacrifice," 152–57. On a subsequent meeting in Philadelphia, see "The Sewing Women and Arsenal Work," *Fincher's Trades Review*, 11 Feb. 1865. For Cincinnati, see *Cincinnati Daily Commercial*, 3 Feb. 1865; "From the Sewing Women," *Cincinnati Daily Commercial*, 4 Mar. 1865. This Cincinnati petition is reproduced in John R. Commons et al., eds., *A Documentary History of American Industrial Society* (New York: Russell & Russell, 1958), 9:72–73.

65. Meigs to Thomas, 9 July 1861, vol. 18, e. 999, RG 92, NA. For very similar comments that Meigs made slightly later to New York City depot chief David Vinton, see Meigs to Vinton, 9 Sept. 1861 and 2 Dec. 1861, ibid., and Risch, *Quartermaster Support*, 353–54.

66. Parsons annual report for fiscal 1863, e. 1127, RG 92, NA; Meigs to Wilson, 20 Feb. 1864, pp. 516–17, vol. 74-B, roll 45, M745; Risch, *Quartermaster Support*, 351–52, 409.

67. Allen to Meigs, 23 Apr. 1863, box 111, e. 20, RG 92, NA; Donaldson to Meigs, 19 Nov. 1863, box 127, and 18 Apr. 1864, box 132, both e. 20, RG 92, NA; Perry to Meigs, 5 Sept. 1864, box 2, e. 27, RG 92, NA; "Many Working Women" to Stanton, 17 Oct. 1864, box 963, e. 225, RG 92, NA.

68. Meigs to Stanton, 7 Nov. 1862, roll 9, M494; Eddy to Meigs, 15 Sept. 1864, box 136, e. 20, RG 92, NA.

69. Meigs to Brown, 4 Mar. 1864, roll 51, M494.

70. Jane Hasler to Lincoln, 22 Aug. 1864, attached to Meigs to Stanton, 13 Oct. 1864, roll 63, M494.

71. Meigs to McKim, 1 Sept. 1864, p. 285, vol. 79, roll 47, M745; McKim to Meigs, 2 Sept. 1864, and Moulton to McKim, 17 Sept. 1864, both enclosed in Meigs to Stanton, 13 Oct. 1864, roll 63, M494.

72. Meigs to Stanton, 13 Oct. 1864, roll 63, M494.

73. Perry to McKim, 4 Jan. 1865, vol. 24, e. 999, RG 92, NA.

CHAPTER 4. THE TROUBLE WITH CONTRACTING

1. Meigs to Stanton, 13 Aug. 1862, roll 1, and Meigs to Stanton, 2 Feb. 1864, roll 37, both in M494; A. J. Perry to Crosman, Swords, and Vinton, 3 Aug. 1864, vol. 23, e. 999, RG 92, NA; Russell F. Weigley, *Quartermaster General of the Union Army: A Biography of M. C. Meigs* (New York: Columbia University Press, 1959), 252–53.

2. Mark R. Wilson, "The Business of Civil War: Military Enterprise, the State, and Political Economy in the United States, 1850–1880" (Ph.D. diss., University of Chicago, 2002), 260.

3. *U.S. Statutes at Large* 12 (1863), 220; *Revised United States Army Regulations of 1861* (Washington, 1863), 155–56.

4. E. S. Sibley to Charles Thomas, 21 May and 30 May 1861, and Meigs to Thomas, 14 June 1861, vol. 55, roll 35, M745.

5. "Army Supplies," *New York Tribune*, 4 June 1861; "Hurtt Court Martial," pp. 123, 209, House Exec. Doc. 255, 43rd Congr., 1st Sess., ser. 1614.

6. Abstract of proposals in Ekin to Meigs, 10 Apr. 1863, box 113, e. 20, RG 92, NA.

7. W. W. Northrup testimony, 15 May 1863, in transcript of "Court of Inquiry, Case of J. H. Dickerson," box 504, e. 225, RG 92, NA.

8. Abstract of proposals received, 2 May 1863, book 52 B-C, box 112, e. 20, RG 92, NA.

9. Abstract of proposals in Ekin to Meigs, 10 Apr. 1863, box 113, e. 20, RG 92, NA.

10. Meigs to Stanton, 13 Aug. 1862, roll 1, M494.

11. Susan Hoffman, *Politics and Banking: Ideas, Public Policy, and the Creation of Financial Institutions* (Baltimore: Johns Hopkins University Press, 2001), 67–68, 91; Heather Cox Richardson, *The Greatest Nation of the Earth: Republican Economic Policies during the Civil War* (Cambridge: Harvard University Press, 1997), 38–65.

12. John Dickerson to Meigs, 8 Oct. 1861, box 316, e. 225, RG 92, NA; Robert Allen to Wilson, Childs & Co., 2 Nov. 1861, p. 113, vol. 1, e. 381, RG 92, NA; Allen to Meigs, 14 Jan. 1862, box 27, e. 225, RG 92, NA.

13. Allen testimony, 25 Oct. 1862, in Court-Martial Case File LL 21, RG 153, NA.

14. Cash Book, 1859–1865, box 13, and Day Book, 1861–1871, box 27, William Whitaker & Sons Papers, accession 1471, Hagley Museum and Library, Wilmington, Del.

15. On commercial paper during this era, see Harold D. Woodman, *King Cotton and His Retainers: Financing and Marketing the Cotton Crop of the South, 1800–1925* (Lexington: University of Kentucky Press, 1968); Glenn Porter and Harold C. Livesay, *Merchants and Manufacturers: Studies in the Structure of Nineteenth-Century Marketing* (Baltimore: Johns Hopkins University Press, 1971).

16. Thomas Scott to Stanton, 9 Feb. 1862, vol. 3, roll 2, Stanton Papers, LC; Meigs to Stanton, 11 Feb. 1865, roll 72, M494.

17. John Dickerson to Meigs, 3 May 1863, box 113, e. 20, RG 92, NA; *Cincinnati Daily Commercial*, 13 Jan. 1864.

18. *New York Tribune*, 4 Mar. 1862; "United States Government Securities," *Bankers' Magazine and Statistical Register* 11 (new ser.), no. 12 (June 1862): 964–65; Albert S. Bolles, *The Financial History of the United States*, vol. 3, *From 1861 to 1885*, 2nd ed. (New York: D. Appleton & Co., 1894), 92–93. After April 1863, both interest and principal for these certificates were paid in greenbacks; before then, in some cases at least, interest was payable in gold. See Richard Roll, "Interest Rates and Price Expectations during the Civil War," *Journal of Economic History* 32 (1972): 487–89.

19. Thomas Swords to Meigs, 24 Mar. and 7 Apr. 1862, and Swords to Jones Bros. & Co., 7 Apr. 1862, vol. 1, e. 3554, RG 393 (pt. 1), NA.

20. Dickerson to Meigs, 3 May 1863, box 113, e. 20, RG 92, NA; N. J. T. Dana to Myers, 13 June 1864, p. 309, vol. 77, roll 44, M745.

21. Davis Rich Dewey, *Financial History of the United States*, 10th ed. (New York: Longmans, Green & Co., 1928), 308; *Banker's Magazine and Statistical Register* 14 (new ser.), no. 3 (Sept. 1864): 172, no. 6 (Dec. 1864): 511; Ellis Paxson Oberholtzer, *Jay Cooke, Financier of the Civil War* (Philadelphia: G. W. Jacobs & Co., 1907), 1:658, 2:1. At the very end of the war, starting in March or April 1865, the Treasury began to allow contractors to exchange vouchers for bonds (including a new batch of the 7.3%, three-year bonds known as "seven-thirties") instead of certificates. See Oberholtzer, *Jay Cooke*, 1:519–23, 569, 658.

22. *New York Tribune*, 25 Mar. 1862, 5 Apr. 1862, 14 Apr. 1862; John Wiegard testimony, pp. 3629–40, box 1125A, Court-Martial Case File MM 2250, RG 153, NA.

23. Jay Cooke & Co. to Stanton, 4 Apr. 1862, box 7, e. 22, RG 92, NA; Oberholtzer, *Jay Cooke*, 1:194–95, 523; "United States Bonds and Notes," *Banker's Magazine and Statistical Register* 12 (new ser.), no. 10 (Apr. 1863): 785.

24. A. G. Sanford to James Donaldson, 7 Feb. 1865, enclosed in Donaldson to [Meigs], 8 Feb. 1865, box 137, e. 20, RG 92, NA.

25. This conclusion contradicts the findings of some of the best studies of business enterprise during the Civil War—although perhaps not as much as it might seem at first glance. In their studies of Philadelphia, both Philip Scranton and Matthew Gallman have emphasized the participation of relatively small-scale manufacturers and merchants in the war contracting business. See Philip Scranton, *Proprietary Capitalism: The Textile Manufacture at Philadelphia, 1800–1885* (New York: Cambridge University Press, 1982); J. Matthew Gallman, *Mastering Wartime: A Social History of Philadelphia during the Civil War* (New York: Cambridge University Press, 1990); and Gallman, "Entrepreneurial Experiences in the Civil War: Evidence from Philadelphia," in *American Development in Historical Perspective*, ed. Thomas Weiss

and Donald Schaefer (Stanford: Stanford University Press, 1994), 205–22, 299–303. Lacking the kind of comprehensive analysis of prime contracting that was undertaken for this book, Scranton and Gallman tend to underestimate the concentration of prime contracts in the hands of a relatively small group of merchants and manufacturers. On the other hand, by discussing subcontractors at greater length, they succeed in showing, correctly, that hundreds of firms of various kinds were involved in the war supply business.

26. J. Leander Bishop, *A History of American Manufactures from 1608 to 1860* (1869; reprint, New York: Augustus M. Kelley, 1966), 3:252–53; Henry Hall, *America's Successful Men of Affairs* (New York: N.Y. Tribune, 1896), 2:129–31.

27. Meigs to Stanton, 7 June 1863, box 3, e. 27, RG 92, NA.

28. E. S. Sibley to Burden & Co., 20 Dec. 1861, p. 456, vol. 57, roll 36, M745.

29. L. M. Kollack, Treasurer Gosnold Mills to [Meigs], 11 Apr. 1863, box 113, e. 20, RG 92, NA. For evidence of a machine-made horseshoe factory in southern Massachusetts, which may have been owned by Gosnold Mills, selling to the army in 1863, see "Machine-Made Horse Shoes," *Scientific American*, 9, no. 18 (31 Oct. 1863): 280. On Stone, Chisholm & Jones and other manufacturers, see Ann Norton Greene, "Harnessing Power: Industrializing the Horse in Nineteenth-Century America" (Ph.D. diss., University of Pennsylvania, 2004), 136–38.

30. Cameron to [Meigs], 1 Nov. 1851, *OR*, ser. 3, 1:615. See also Mike Woshner, *India Rubber and Gutta-Percha in the Civil War Era* (Alexandria, Va.: O'Donnell Publications, 1999).

31. The data are from Quartermaster's Department registers for 1864: vols. 14–15, e. 1238, RG 92, NA, and "Contracts Made by the Quartermaster's Department," House Exec. Doc. 84, 38th Congr., 2nd Sess., ser. 1230. Although this case suggests that Providence India Rubber could have saved the North large sums, other contracts show smaller differences in prices offered by the two firms.

32. A. O. Brown to Herman Biggs, 19 Dec. 1864, and Meigs to Biggs, 22 Dec. 1864, vol. 24, e. 999, RG 92, NA; *Goodyear v. Providence Rubber Co.*, 10 Fed. Cas. 712 (1864). After the war, the U.S. Supreme Court upheld the district court's wartime ruling in favor of Union India Rubber. See *Providence Rubber Co. v. Goodyear*, 6 Wall. 153 (1868); *Providence Rubber Co. v. Goodyear*, 9 Wall. 788 (1870). For more details, see Wilson, "Business of Civil War," 334–49.

33. Arthur Pine van Gelder and Hugo Schlatter, *History of the Explosives Industry in America* (New York: Columbia University Press, 1927), 115–16; Harold B. Hancock and Norman B. Wilkinson, "A Manufacturer in Wartime: Du Pont, 1860–1865," *Business History Review* 40 (1966): 213–36; Mark R. Wilson, "Gentlemanly Price-Fixing and Its Limits: Collusion and Competition in the U.S. Explosives Industry during the Civil War Era," *Business History Review* 77 (2003): 207–34. On Hazard in particular, see J. Hammond Trumbull, *Memorial History of Hartford County, Connecticut, 1633–1884* (Boston: Edward L. Osgood, 1886), 2:158–62; Berkeley R. Lewis, *Small Arms and Ammunition in the United States Service* (Washington: Smithsonian Institution, 1956), 35–36.

34. Charles B. Dew, *Ironmaker to the Confederacy: Joseph R. Anderson and the Tredegar Iron Works* (New Haven: Yale University Press, 1966).

35. U.S. census of manufactures, 1860, manuscript returns, National Archives Microfilm Publications T1157, roll 6; Bishop, *History of American Manufactures*, 3:98–106; "Cannon Manufacture at Pittsburgh," *Scientific American* 6, no. 11 (15 Mar. 1862): 163; "A Great Cannon Foundry," *Scientific American* 11, no. 11 (10 Sept. 1864): 165. For more on technological change in artillery production before and during the Civil War, see Edward Clinton Ezell, "The

Development of Artillery for the United States Land Service before 1861: With Emphasis on the Rodman Gun" (M.A. thesis, University of Delaware, 1963), esp. 111–58, and Alexander L. Holley, *A Treatise on Ordnance and Armor* (New York: D. Van Nostrand, 1865).

36. U.S. census of manufactures, 1860, manuscript returns, New York State Archives (microfilm); J. H. French, *Gazeteer of the State of New York* (Syracuse: R. Pearsall Smith, 1860), 542; "Manufacture of Heavy Ordnance for the Government," *Scientific American* 5, no. 5 (3 Aug. 1861): 67; untitled item on Hotchkiss, *Scientific American* 6, no. 28 (28 June 1862): 404; "Extreme Activity in Machine Works," *Scientific American* 9, no. 17 (24 Oct. 1863): 263; "Heavy Ordnance," in *Report of the Joint Committee on the Conduct of the War at the Second Session, Thirty-Eighth Congress* (Washington, 1865); Bishop, *History of American Manufactures*, 3:485–88; Orra L. Stone, "The Industries of Metropolitan Boston," in *Metropolitan Boston: A Modern History*, ed. Albert P. Langtry (New York: Lewis Historical Publishing Co., 1929), 2:651–52; Colin T. Naylor Jr., *Civil War Days in a Country Village* (Peekskill, N.Y.: Highland Press, 1961), 14–15. For more details, see Wilson, "Business of Civil War," 317–27. Specialized studies of Civil War artillery and projectiles include Warren Ripley, *Artillery and Ammunition of the Civil War* (New York: Van Nostrand Reinhold Co., 1970); Thomas S. Dickey and Peter C. George, *Field Artillery Projectiles of the American Civil War* (Atlanta: Arsenal Press, 1980); Spencer Tucker, *Arming the Fleet: U.S. Navy Ordnance in the Muzzle-Loading Era* (Annapolis: Naval Institute Press, 1989); and Jack Bell, *Civil War Heavy Explosive Ordnance: A Guide to Large Artillery Projectiles, Torpedoes, and Mines* (Denton: University of North Texas Press, 2003).

37. *Manufactures of the United States in 1860* (Washington, 1865). The 40,000 figure for wagons and ambulances purchased in 1860 is a low estimate based on data for the second half of the war, provided in B. C. Card's report in "Report of the Quartermaster General," 1865 *ARSW*.

38. For evidence of one small Philadelphia shop that supplied Henry Simons & Co., a leading prime contractor, see Franklin & Co. Papers, Vols. 1–5 (1861–65), Historical Society of Pennsylvania, Philadelphia. In keeping with their usual concern to prevent monopoly, supply officers tried deliberately to spread orders for wheeled vehicles. In June 1861, for instance, the Office of the Quartermaster General ordered Washington depot chief Morris Miller to acquire wagons with "a judicious distribution" of contracts with firms in New Jersey and Pennsylvania. Erna Risch, *Quartermaster Support of the Army: A History of the Corps* (Washington, D.C.: Office of the Quartermaster General, 1962), 373.

39. Edwin T. Freedley, *Philadelphia and Its Manufactures . . . in 1857* (Philadelphia: Edward Young, 1859), 447–49; U.S. census of manufactures, 1860, manuscript returns, National Archives Microfilm Publications T1157, roll 8; Pennsylvania, vol. 132, pp. 431, 285d, Dun; *Philadelphia Daily Evening Bulletin*, 2 Sept. 1861; "From Philadelphia," *New York Tribune*, 30 Sept. 1861; Bishop, *History of American Manufactures*, 3:86–87; Charles Robson, *The Manufactories and Manufacturers of Pennsylvania in the Nineteenth Century* (Philadelphia: Galaxy Publishing Co., 1875), 277–79.

40. Ohio, vol. 79, p. 325; Ohio, vol. 80, pp. 135–38; and Ohio, vol. 82, p. 187, all in Dun; M. Joblin & Co., *Cincinnati Past and Present, or Its Industrial History . . .* (Cincinnati, 1872), 349–53; *Biographical Encyclopaedia of Ohio of the Nineteenth Century* (Cincinnati: Galaxy Publishing Co., 1876), 419–20.

41. Quartermaster's Department report of 14 Nov. 1863 on various wagon manufacturers, copy in O. Cross to Charles Thomas, 1 Jan. 1864, box 127, e. 20, RG 92, NA.

42. For more details, see Wilson, "Business of Civil War," 349–66.

43. U.S. census of manufactures, 1860, manuscript returns, National Archives Microfilm Publications T1204, roll 15; U.S. census of manufactures, 1860, manuscript returns, New York State Archives (microfilm); Benjamin Hobart, *History of the Town of Abington, Plymouth County, Massachusetts* (Boston: T. H. Carter & Son, 1866), 291–95; Seth Bryant, *Shoe and Leather Trade of the Last Hundred Years* (Boston, 1891); Blanche Evans Hazard, *The Organization of the Boot and Shoe Industry in Massachusetts before 1875* (Cambridge: Harvard University Press, 1921); Frank W. Norcross, "Wholesale Shoe Trade of Suffolk County," in *Professional and Industrial History of Suffolk County, Massachusetts*, vol. 3 (Boston: Boston History Co., 1894); Norcross, *A History of the New York Swamp* (New York: Chiswick Press, 1901), 219–22; William Hall testimony, in "Report of the Majority of the Select Committee on . . . the Transactions of the State Military Board," doc. 194, *Documents of the Assembly of the State of New York*, 85th sess., vol. 7 (Albany, 1862), 401–4; Ross Thomson, *The Path to Mechanized Shoe Production in the United States* (Chapel Hill: University of North Carolina Press, 1989). For evidence of the use of McKay stitchers in New York City by 1863, possibly at a Benedict, Hall & Co. factory, see "Manufacture of Boots and Shoes by Machinery," *Scientific American* 9, no. 20 (14 Nov. 1863): 312. Mundell apparently used stitching machinery designed by Eugene LeMercier of Paris, the rights to which he purchased after it was patented in the United States in March 1863. See Charles H. McDermott, *A History of the Shoe and Leather Industries of the United States* (Boston: John W. Denehy & Co., 1920), 78–79. On Simpkinson & Co. and Pomeroy & Co., see Ohio, vol. 79, pp. 188, 223; vol. 80, pp. 122, 172, 241, 316; vol. 84, pp. 101, 273, Dun. John Simpkinson is profiled in *Biographical Cyclopaedia and Portrait Gallery of . . . Ohio* (Cincinnati: Western Biographical Publishing Co., 1883), 1:255–57. For more details, see Wilson, "Business of Civil War," 377–406.

44. Woodman, *King Cotton and His Retainers*; Porter and Livesay, *Merchants and Manufacturers*; Alfred D. Chandler Jr., *The Visible Hand: The Managerial Revolution in American Business* (Cambridge: Harvard University Press, 1977); Howard Bodenhorn, *A History of Banking in Antebellum America: Financial Markets and Economic Developments in an Era of Nation-Building* (New York: Cambridge University Press, 2000), 107.

45. On Lord and Dobson, among other Philadelphia manufacturers, see Scranton, *Proprietary Capitalism*, 63, 245–46.

46. Benjamin Bullock & Co. to Charles I. Du Pont & Co., 27 Nov. 1855, incoming correspondence, box 3; Charles I. Du Pont & Co. Records, accession 500, Hagley Museum and Library.

47. W. C. Houston and Houston & Robinson to Charles I. Du Pont & Co., various letters dated 1845–50, incoming correspondence, box 10, Charles I. Du Pont & Co. Records; Pennsylvania, vol. 137, pp. 492, 518; vol 141, pp. 114, 117, Dun.

48. Benjamin Bullock & Co. to Charles I. Du Pont & Co., series of letters on wool operations written in 1855, incoming correspondence, box 3, Charles I. Du Pont & Co. Records.

49. Pennsylvania, vol. 137, pp. 492, 500, 766, 851, Dun; Stephen N. Winslow, *Biographies of Successful Philadelphia Merchants* (Philadelphia: James K. Simon, 1864), 213–17; *Biographical Encyclopaedia of Pennsylvania of the Nineteenth Century*, 26–27; Bishop, *History of American Manufactures*, 3:48–50; Scranton, *Proprietary Capitalism*, 119–20; *North American and U.S. Gazette*, 23 Oct. 1861 and 2 Feb. 1863; *Report of the Commissioner of Agriculture for the Year 1864* (Washington, D.C., 1865), 512–13.

50. Mack and Glaser testimony in GovCon 2, 765–67, 918–20; Ohio, vol. 78, pp. 164, 307, 334, 371; vol. 79, pp. 77, 123, 285, 290–91; vol. 81, p. 310; vol. 82, p. 60, Dun; Joblin & Co., *Cincinnati Past and Present* (Cincinnati, 1872), 243–46, 317–18, 401–4; *Biographical Encyclopaedia of Ohio of the Nineteenth Century*, 408–9, 439; Steven Mostov, "A 'Jerusalem' on the Ohio: The Social and Economic History of Cincinnati's Jewish Community, 1840–1875" (Ph.D. diss., Brandeis University, 1981), 74–76, 109; Michael W. Rich, "Henry Mack: An Important Figure in Nineteenth-Century American Jewish History," *American Jewish Archives* 48 (1995): 261–79.

51. U.S. census of manufactures, 1860, manuscript returns, National Archives Microfilm Publications T1157, roll 8; *North American and U.S. Gazette*, 5 Feb. 1862; Winslow, *Biographies of Successful Philadelphia Merchants*, 91–95.

52. Massachusetts, vol. 104, pp. 532, 679, Dun. C. W. Freeland & Co. signed large army contracts not only in the names of Charles and James Freeland but also in the names of other partners and clerks, such as Lyman L. Harding and Alvin Rose.

53. New York, vol. 198, pp. 102, 111–12, 116, 132A, 183; vol. 203, pp. 700H–700I; vol. 204, pp. 721, 800R, Dun; William C. Browning, "Clothing and Furnishing Trade," in *One Hundred Years of American Commerce*, ed. Chauncey M. Depew (New York: D. O. Haynes & Co., 1895), 561–65; William R. Bagnall, "Sketches of Manufacturing Establishments in New York City, and of Textile Establishments in the United States," ed. Victor S. Clark (microfiche edition of 1908 typescript, Merrimack Valley Textile Museum, North Andover, Mass., 1977), 1:315–34; Egal Feldman, *Fit for Men: A Study of New York's Clothing Trade* (Washington, D.C.: Public Affairs Press, 1960), 43–51. For biographical sketches of Martin, see Henry W. B. Howard, ed., *The Eagle and Brooklyn . . . History of the City of Brooklyn* (Brooklyn: Brooklyn Daily Eagle, 1893), 274–75, 787–89; Hall, *America's Successful Men of Affairs*, 1:428; *National Cyclopaedia of American Biography* (New York: James T. White & Co., 1900), 8:419.

54. On the importance of textile merchants in the mid-nineteenth-century economy, see Robert Greenlough Albion, *The Rise of New York Port, 1815–1860* (New York: Charles Scribner's Sons, 1939), 55–63, 275–82; Frank L. Walton, *Tomahawks to Textiles: The Fabulous Story of Worth Street* (New York: Appleton-Century-Crofts, 1953); Stephen N. Elias, *Alexander T. Stewart: The Forgotten Merchant Prince* (Westport, Conn.: Praeger, 1992).

55. Lewis, Boardman & Wharton to H. Biggs, 16 Dec. 1864, roll 63, M494; Mark R. Wilson, "The Extensive Side of Nineteenth-Century Military Economy: The Tent Industry in the Northern United States during the Civil War," *Enterprise and Society* 2 (2001): 297–337; Frederick C. Gaede, *The Federal Civil War Shelter Tent* (Alexandria, Va.: O'Donnell Publications, 2001).

56. Charles Bush Estate Papers, boxes 1–2, accession 753, Hagley Museum and Library; Porter and Livesay, *Merchants and Manufacturers*, 84–97, 106–8; Lucius F. Ellsworth, "The Delaware Leather Industry in the Mid-Nineteenth Century," *Delaware History* 11 (1965): 264–65.

57. Ellsworth, "Delaware Leather Industry," 270–74.

58. Charles Bush Estate Papers, H. S. McComb Account Books 1 and 2, box 1, accession 753; H. S. McComb Papers, box 2, accession 474, all at Hagley Museum and Library.

59. Meigs to David Vinton, 16 Nov. 1861, p. 264, vol. 57, roll 36, M745.

60. Stevens testimony, 8 Mar. 1862, in GovCon 2, 911; Elleard testimony, 30 Sept. and 3 Oct. 1862, in Court-Martial Case File LL 21, RG 153, NA.

61. *Argument of F. Carroll Brewster, Esq., on Behalf of William B. N. Cozens, Delivered at Philadelphia, June 12, 1865* (Philadelphia: King & Baird, 1865), 30.

62. P. H. Watson to Charles Thomas, 31 Oct. 1863, pp. 272–73, vol. 1, roll 1, M421.

63. Herman Boker & Co. to Stanton, 13 Mar. 1862, pp. 79–80, in "Report of the Commission on Ordnance and Ordnance Stores," Senate Exec. Doc. 72, 37th Congr., 2nd Sess., ser. 1123.

64. Wilson, "Gentlemanly Price-Fixing," 228.

65. Charles Thomas to H S. McComb, 12 July 1861 and 16 July 1861, vol. 2, e. 2194, RG 92, National Archives and Records Administration–Mid Atlantic Region (Philadelphia).

66. Ledgers, 1863–1869, box 54, William Whitaker & Sons Papers.

67. Kunkel, Hall & Co. to C. W. Moulton, 26 Mar. 1864, enclosed in Swords to [Meigs], 31 Mar. 1864, box 133, e. 20, RG 92, NA.

68. William L. Hodge to Stanton, 12 Feb. 1862, box 6, e. 22, RG 92, NA.

69. F. B. Loomis to Stanton, 3 Feb. and 8 Feb. 1862, box 6, e. 22, RG 92, NA.

70. "Government Contracts," *New York Tribune*, 26 Oct. 1861.

71. Tyler, Stone & Co. to Meigs, 24 Jan. 1863, and same to A. Boyd, 11 Mar. 1863, book 51 T-Z, box 111, e. 20, RG 92, NA.

72. Read, Drexel & Co. to Stanton, 3 June 1862, box 7, e. 22, RG 92, NA.

73. Henry D. Moore to Thomas, 20 Nov. 1863, box 128, e. 20, RG 92, NA.

74. D. W. McClung to Thomas, 3 Dec. 1863, box 128, e. 20, RG 92, NA.

75. Oberholtzer, *Jay Cooke*, 1:429–30; Roll, "Interest Rates and Price Expectations," 490–97.

76. Lewis, Boardman & Wharton et al. to Crosman, 8 July 1864, enclosed in Crosman to [Meigs], 9 July 1864, box 134, e. 20, RG 92, NA; for a similar complaint, see copy of Thomas Murphy to Stanton, 28 July 1854, in Charles Thomas to D. Vinton, 9 Sept. 1864, box 1, e. 2153, RG 92, National Archives and Records Administration–Northeast Regional Branch (New York City).

77. Kunkel, Hall & Co. to Meigs, 9 July 1864, box 134, e. 20, RG 92, NA.

78. Wilson, "Gentlemanly Price-Fixing," 230.

79. Meigs to Stanton, 29 Aug. 1864, roll 37, M494.

80. William C. Browning, "The Clothing and Furnishing Trade," in *One Hundred Years of American Commerce*, ed. Chauncey M. Depew (New York: D. O. Haynes & Co., 1895), 561–65.

81. Philip Shaw Paludan, *"A People's Contest": The Union and Civil War, 1861–1865* (New York: Harper & Row, 1988), 144–49; Thomas Weber, *The Northern Railroads in the Civil War, 1861–1865* (New York: King's Crown Press, 1952), 43.

82. For a legislative history, see Richardson, *Greatest Nation of the Earth*, 120–33.

83. *The Income Record: A List Giving the Taxable Income for the Year 1863 of Every Resident of New York* (New York: American News Co., 1865), 139; *Income Tax of the Residents of Philadelphia and Bucks County for the Year Ending April 30, 1865* (Philadelphia, 1865), 6–7, 11. For Cincinnati income tax data, see *Cincinnati Daily Commercial*, 12 Dec. 1864 and 1 July 1865; for revenue data, see *Cincinnati Daily Commercial*, 11 July 1865.

84. *Income Tax of the Residents of Philadelphia*, 5–6; Harold B. Hancock, "The Income and Manufacturers' Tax of 1862–1872 as Historical Source Material," *Delaware History* 14 (1971): 255–61; Charles Bush Estate Papers, H. S. McComb Account Book 1; *Incomes of the*

Citizens of Boston and Other Cities and Towns in Massachusetts (Boston: A. Williams & Co., 1866); *Incomes of the Citizens of Boston and Other Cities and Towns in Massachusetts* (Boston: Worthington, Flanders & Co., 1867).

85. Thomas A. Scott to E. S. Sibley, 12 July 1861, *OR*, ser. 3, 1:325–26; Meigs to Stanton, 1 May 1862, *OR*, ser. 3, 2:839–41. See also Weber, *Northern Railroads in the Civil War*, 127–33; Robert G. Angevine, *The Railroad and the State: War, Politics, and Technology in Nineteenth-Century America* (Stanford: Stanford University Press, 2004), 130–64.

86. Vinton testimony, 2 Jan. 1862, GovCon 2, 286–91.

87. *U.S. Statutes at Large* 12 (1863), 411; Risch, *Quartermaster Support*, 346–47. A copy of the June 1862 act was appended to the 1863 edition of the *Revised United States Army Regulations of 1861*, 533.

88. Meigs to Crosman, 17 Apr. 1863, vol. 21, e. 999, RG 92, NA.

89. Crosman to Cameron, 18 Mar. 1863, roll 9, Cameron Papers, LC.

90. John Dickerson testimony, 30 Dec. 1861, GovCon 2, 743–45.

91. Allen to Meigs, 2 Sept. 1864, and Meigs to War Department Solicitor General, 17 Sept. 1864, both in roll 63, M494. For the congressional legislation of the summer of 1864 that restated the need for formal contracting, see *U.S. Statutes at Large* 13 (1865): 394.

92. Meigs to Stanton, 13 Aug. 1862, *OR*, ser. 3, 2:372; D. W. McClung to Thomas, 3 Dec. 1863, box 128, e. 20, RG 92, NA; Chapman personal narrative report for fiscal 1864, e. 1128, RG 92, NA.

93. Samuel L. Brown to Charles Thomas, 23 Dec. 1863, box 126, e. 20, RG 92, NA; Meigs to Brown, 4 Mar. 1864, roll 51, M494; Brown to Meigs, 22 Oct. 1864, box 136, e. 20, RG 92, NA.

94. Vinton testimony, GovCon 2, 857; Meigs to Stanton, 13 June 1862, in "Letter of the Secretary of the Navy on Contracts, etc.," Senate Misc. Doc. 105, 37th Congr., 2nd Sess., ser. 1124; War Department General Order 69 of 1862, reproduced in Walworth Jenkins, *Q. M. D., or a Book of Reference for Quartermasters* (Louisville: John P. Morton & Co., 1865), 34–35.

95. Accurate, comprehensive figures on the purchase of horses and mules are difficult to find. This estimate is based on a combination of statistics from a variety of sources, including Quartermaster's Department correspondence and Agriculture Department reports. For more details, see Wilson, "Business of Civil War," 466–98, and Greene, "Harnessing Power," 114–71.

96. Abstract of proposals in Van Vliet to Meigs, 29 Apr. 1863, box 116, e. 20, RG 92, NA.

97. Court-Martial Case Files MM 1169 and NN 893, RG 153, NA.

98. Calder testimony, 23 Nov. 1861 and 17 Apr. 1862, GovCon 2, 155–59 and 1387–97; William Henry Egle, *History of the Counties of Dauphin and Lebanon in the Commonwealth of Pennsylvania* (Philadelphia: Evarts & Peck, 1883), 473.

99. Charles M. Elleard testimony, 29 Apr. 1864, in Court-Martial Case File MM 1426; 9 May 1864, in Court-Martial Case File NN 1720; 3 June 1864, in Court-Martial Case File NN 1865, all in RG 153, NA.

100. Testimony in Court-Martial Case Files MM 1426 and NN 1865, RG 153, NA.

101. "Government Horse Contracts," *Missouri Democrat*, 18 Apr. 1863.

102. James M. Hopkins to Stanton, 29 May 1862, box 7, e. 22, RG 92, NA.

103. James M. Hopkins to Stanton, 3 Sept. 1863, and Charles Thomas to Stanton, 15 Sept. 1863, box 10, e. 22, RG 92, NA.

104. Meigs to Wilson, 20 Feb. 1864, pp. 516–17, vol. 74b, roll 45, M745.

105. George Stoneman to Stanton, 15 Oct. 1863, *OR*, ser. 3, 3:884–86.

106. Thomas Swords's annual report for fiscal 1863, e. 1127, RG 92, NA.

107. *Congressional Globe*, 38th Congr., 1st Sess. (28 June 1864), 3355.

108. C. A. Dana to Ekin, 13 Apr. 1864, *OR*, ser. 3, 4:228; C. W. Tolles to Meigs, 26 Aug. 1864, box 907, e. 225, RG 92, NA; disbursement notices for September 1864 on pp. 349–428, vol. 79, roll 47, M745.

109. Abstracts of proposals in Donaldson to Meigs, 11 June 1863, box 118, and McClung to Thomas, 10 Nov. 1863, box 128, both in e. 20, RG 92, NA.

CHAPTER 5. THE MIDDLEMAN ON TRIAL

1. Among the many studies that suggest such a narrative are James M. McPherson, *Battle Cry of Freedom: The Civil War Era* (New York: Oxford University Press, 1988), 323–25; Paul A. C. Koistinen, *Beating Plowshares into Swords: The Political Economy of American Warfare, 1606–1865* (Lawrence: University Press of Kansas, 1996), 132–38; and Stuart D. Brandes, *Warhogs: A History of War Profits in America* (Lexington: University Press of Kentucky, 1997), 67–107.

2. "A Word to Stock-Jobbers," *New York Tribune*, 2 May 1861; "Army Contractors!" *Vanity Fair* 3 (8 June 1861): 265; "Government Contracts," *New York Tribune*, 26 Oct. 1861.

3. "Shoddy," *United States Economist* 4 (10 Dec. 1853): 129–30; "Devil's Dust," *Chambers's Journal* 15 (16 Feb. 1861): 103–5; Brandes, *Warhogs*, 72–73; "Song of the Shoddy," *Vanity Fair* 4 (21 Sept. 1861): 142.

4. Ronald Schultz, "Small Producer Thought in Early America," pt. 1–2, *Pennsylvania History* 54 (1987): 115–47, 197–229; Albert Saboul, *The Sans-Culottes: The Popular Movement and Revolutionary Government, 1793–1794* (Princeton: Princeton University Press, 1980); Marvin Meyers, *The Jacksonian Persuasion: Politics and Belief* (Stanford: Stanford University Press, 1957); Eric Foner, *Free Soil, Free Labor, Free Men: The Ideology of the Republican Party before the Civil War* (New York: Oxford University Press, 1970); John Ashworth, *"Agrarians" and "Aristocrats": Party Political Ideology in the United States, 1837–1846* (London: Royal Historical Society, 1983); Victoria C. Hattam, *Labor Visions and State Power: The Origins of Business Unionism in the United States* (Princeton: Princeton University Press, 1993); Martin J. Burke, *The Conundrum of Class: Public Discourse on the Social Order in America* (Chicago: University of Chicago Press, 1995).

5. "Army Peculators," *New York Tribune*, 25 May 1861; Meigs to Charles Thomas, 12 Aug. 1861, pp. 202–3, vol. 56, roll 36, M745. For circulation figures, see *New York Tribune*, 1 Oct. 1861.

6. *Chicago Tribune*, 23 Dec. 1861; Brandes, *Warhogs*, 100; "Resolutions of the State of New York in Relation to Reported Frauds in Supplies for Our Armies," House Misc. Doc. 34, and "Resolutions of the Legislature of Ohio, in Relation to Fraud and Corruption by Agents and Officers of the Government," House Misc. Doc. 44, both in 37th Congr., 2nd Sess., ser. 1141. For one example of a reprinting of the *Tribune's* "Army Peculators" editorial, see *Cincinnati Daily Commercial*, 31 May 1861.

7. The Van Wyck committee and the content of its reports have been described often. See esp. Fred Nicklason, "The Civil War Contracts Committee," *Civil War History* 17 (1971): 232–44; Allan G. Bogue, *The Congressman's Civil War* (New York: Cambridge University Press, 1989), 81–88; and Brandes, *Warhogs*, 73–83, 105–6. For the major reports, see GovCon 1 and GovCon 2. The committee issued short concluding reports in 1863: "Government

Contracts," Final Report, and "Government Contracts: Views of the Minority," House Reports 49–50, 37th Congr., 3rd Sess., ser. 1173. A small number of unpublished committee files, mainly letters complaining of abuses or praising the committee's work, may be found in "Select Committee on Government Contracts," HR37A-E21.1, RG 233 (Records of the U.S. House of Representatives), NA.

8. GovCon 1, report, 31–34, and testimony, 278; Dexter A. Hawkins to Washburne, 6 Jan. 1862, vol. 21, Washburne Papers, LC.

9. GovCon 1, report, 68, and testimony, 416.

10. See William E. Smith, "The Blairs and Frémont," *Missouri Historical Review* 23 (1929): 214–60; Allan Nevins, *Frémont: Pathmarker of the West* (New York: D. Appleton-Century Co., 1939), 473–549; Bruce Tap, "Reconstructing Emancipation's Martyr: John C. Frémont and the Joint Committee on the Conduct of the War," *Gateway Heritage* 14, no. 4 (1994): 36–53.

11. Washburne to Lincoln, 17 Oct., 19 Oct., and 21 Oct. 1861, roll 28, Lincoln Papers, LC; F. P. Blair Jr. to Simon Cameron, 19 May 1861, box 641, e. 225, RG 92, NA; Blair to Van Wyck committee, undated 1861 memo, folder 8, HR37A-E21.1, RG 233, NA; Louis S. Gerteis, *Civil War St. Louis* (Lawrence: University Press of Kansas, 2001), 147.

12. GovCon 1, report, 78, 83–84.

13. "Case of Frémont," *Chicago Tribune*, 6 Nov. 1861; *Cincinnati Daily Commercial*, 10 Jan. 1862; J. W. Jones to Dawes, 14 Jan. 1862, folder 3, HR37A-E21.1, RG 233, NA; "Shake Em Up, Charlie!" [cartoon], *Harper's Weekly* 6 (1 Mar. 1862): 144.

14. *Missouri Republican*, 21 Dec. 1861; *Congressional Globe*, 37th Congr., 2nd Sess. (21–29 Apr. 1862), 1746–53, 1848–53, 1862–70; Nicklason, "Civil War Contracts Committee"; Tap, "Reconstructing Emancipation's Martyr." Thaddeus Stevens had been linked indirectly to the procurement scandals in the case of a small arms sale to Frémont's department by Simon Stevens, who had read law under the congressman. Because this small transaction was financed by the young J. P. Morgan, it received a great deal of attention in the years after the war. See R. Gordon Wasson, *The Hall Carbine Affair: A Study in Contemporary Folklore* (New York: Pendick, 1948).

15. GovCon 1, testimony, 489, 695.

16. GovCon 2, vii.

17. GovCon 1, report, 83; *Congressional Globe*, 37th Congr., 2nd Sess. (7 Feb. 1862 and 29 Apr. 1862), 710–15, 1862–70.

18. David Davis to Sarah Davis, 16 Nov. 1861, box 10, David Davis Papers, Chicago Historical Society; "War Claims at St. Louis," House Exec. Doc. 94, 37th Congr., 2nd Sess., ser. 1135.

19. "War Claims at St. Louis," 5–16. For data on the disposition of claims, see J. S. Fullerton to David Davis, 5 May 1862, box 11, Davis Papers.

20. "Report of the Commission on Ordnance and Ordnance Stores," Senate Exec. Doc. 72, 37th Congr., 2nd Sess., ser. 1123; Brandes, *Warhogs*, 86–87.

21. See Richard Bruce Winders, *Mr. Polk's Army: The American Military Experience in the Mexican War* (College Station: Texas A&M University Press, 1997), 17.

22. Van Vliet to Stanton, 27 Sept. 1862, vol. 9, roll 4, Stanton Papers, LC; Van Vliet to Samuel Barlow, 10 Feb. 1862 and 21 July 1862, box 44, Samuel Barlow Papers, Huntington Library, San Marino, Calif.; Van Vliet to Simon Cameron, 20 Nov. 1862 and 11 Nov. 1864, rolls 8–9, Cameron Papers, LC.

23. Jay Cooke to John Sherman, 12 Apr. 1864, roll 1, John Sherman Papers, LC.

24. Meigs to Stanton, 21 June 1864, roll 51, M494.

25. Parsons to Meigs, 26 Feb. 1864, box 583, e. 225, RG 92, NA; Meigs to Allen, 31 Mar. 1864, pp. 438–39, vol. 75, roll 45, M745; Allen, annual report for fiscal 1864, e. 1128, RG 92, NA.

26. Jos. G. Blunt to S. R. Curtis, 30 Mar. 1863; Meigs to Halleck, 8 Apr. 1863; Easton to Meigs, 18 Apr. 1863, all in box 9, e. 22, RG 92, NA.

27. Meigs to Dennison, 26 Aug. 1861, p. 277, vol. 56, roll 36, M745; *Cincinnati Daily Enquirer*, 1 Mar. 1862; *Missouri Democrat*, 3–4 Mar. 1862; Meigs to Dickerson, 11 Mar. 1862, p. 383, vol. 58, roll 37, M745; Dickerson to Meigs, 4 Apr. 1862, box 504, e. 225, RG 92, NA.

28. "The Commercial and Quartermaster Dickerson," *Cincinnati Daily Enquirer*, 2 Aug. 1862; Stanton to Meigs, 24 Apr. 1863, Burnside to Stanton, 27 Apr. 1863, transcript, "Court of Inquiry, Case of J. H. Dickerson, Capt. & A.Q.M.," and Dickerson to Meigs, 28 Nov. 1864, all in box 504, e. 225, RG 92, NA; "Dickerson, J. H.," D101 CB 1864, Letters Received by the Commission Branch of the Adjutant General's Office (RG 94), roll 84, National Archives Microfilm Publication M1064; "Hurtt Court-Martial," House Exec. Doc. 255, 43rd Congr., 1st Sess., ser. 1614., pp. 289, 317, 321. For a reference to Dickerson's activities as a contractors' agent and commission merchant in Cincinnati and St. Louis between his 1864 resignation and his death at age fifty in 1872, see George W. Cullum, *Biographical Register of the Officers and Graduates of the U.S. Military Academy*, 3rd ed. (Boston: Houghton Mifflin & Co., 1891), 2:313.

29. Crosman testimony, 6 Mar. 1862, GovCon 2, 885–86; Crosman to Meigs, 18 Mar. 1863 and 12 May 1863, box 112, e. 20, RG 92, NA; O'Neill to Stanton, 13 May 1863, box 116, e. 20, RG 92, NA. It was evidently not unusual for Quartermaster's Department civilian employees, like other government employees, to swear or sign loyalty oaths. See, e.g., Van Vliet to Meigs, 29 Oct. 1864, vol. 1, e. 2140, RG 92, National Archives and Records Administration–Northeast Regional Branch (New York).

30. Clinton W. Terry, "'The Most Commercial of People': Cincinnati, the Civil War, and Industrial Capitalism, 1861–1865" (Ph.D. diss., University of Cincinnati, 2002), 180–82.

31. P. H. Watson to Charles Thomas, 18 Dec. 1863, p. 339, vol. 1, roll 1, M421; Brown to Meigs, 5 Mar. 1864, roll 51, M494.

32. Swords to Joseph Grannon, 19 June 1862, vol. 1, e. 3554, RG 393 (pt. 1), NA.

33. Henry D. Bacon to Samuel L. M. Barlow, 3 Dec. 1861, 4 Dec. 1861, 3 Mar. 1862; Mendes Cohen to Barlow, 2 Mar. 1862, Barlow Papers.

34. David A. Schlueter, "The Court-Martial: An Historical Survey," *Military Law Review* 87 (1980): 129–67; *The Army Lawyer: A History of the Judge Advocate General's Corps, 1775–1975* (Washington, D.C.: GPO, 1975).

35. "Turner, Horace," box 531, LL 981; "Latshaw, H. J.," box 966, MM1015; "Black, Samuel," box 592, LL1525, all in RG 153, NA.

36. This is leaving aside the case of Surgeon General William Hammond, who was dismissed in 1864 after a military trial on charges that he had ordered his subordinates to purchase from certain "favorite" contractors, including Wyeth & Co. in Philadelphia. For the trial transcript and a published version of Hammond's defense, see "Hammond, W. A.," MM 1430, RG 153, NA, and *A Statement of the Causes which Led to the Dismissal of Surgeon-General William A. Hammond, with a Review of the Evidence Adduced before the Court* (New York, 1864). Among the several studies of the Hammond case is Henry C. Friend,

"Abraham Lincoln and the Court Martial of Surgeon General Hammond," *Commercial Law Journal* 62 (1957): 71–80.

37. For McKinstry's antebellum trials (not the unpublished Civil War court-martial transcript), see "Courts-Martial in Case of Major McKinstry," House Exec. Doc. 144, 37th Congr., 2nd Sess., ser. 1138.

38. "Gen. Frémont and His Policy," *New York Tribune*, 19 Aug. 1861; *Missouri Republican*, 17 Sept. 1861.

39. "The Case of Gen. McKinstry," *New York Tribune*, 22 Jan. 1862; "The Van Wyck Committee," *New York Tribune*, 12 Feb. 1862.

40. *Vindication of Brig. Gen. J. McKinstry* (St. Louis, 1862), 5, 20.

41. Ibid., 8–9, 18.

42. The trial transcript of more than fifteen hundred manuscript pages may be found under "McKinstry, J.," LL 21, RG 153, NA. For more details, see Mark R. Wilson, "The Business of Civil War: Military Enterprise, the State, and Political Economy in the United States, 1850–1880" (Ph.D. diss., University of Chicago, 2002), 675–83.

43. A published version of the McKinstry court-martial verdict, along with the Van Wyck committee's comments on it, may be found in "Government Contracts" (final report), House Report 49, 37th Congr., 3rd Sess., ser. 1173. For a report of the "copperhead" speech, see *Cincinnati Daily Commercial*, 4 Sept. 1863. For a brief note on McKinstry's subsequent career, which remains mysterious, see Cullum, *Biographical Register of the Officers and Graduates of the U.S. Military Academy*, 1:727.

44. Meigs to Belger, 20 June and 24 June 1861, vol. 55, roll 35; Meigs to T. H. Hicks, 26 and 30 Aug. 1861, pp. 284–85 and 385, vol. 56, roll 36, all in M745; Belger to Meigs, 9 Sept. and 1 Dec. 1861, box 131, e. 225, RG 92, NA; "The Baltimore Quartermaster's Department," *New York Tribune*, 29 Aug. 1861.

45. Meigs to Belger, 9 and 29 July 1862, NN 893; Thomas to Meigs, 26 Aug. 1862, box 1597, NN 784, RG 153, NA. Records related to Belger's court-martial case are split between these two files.

46. "Employment of Transport Vessels," Senate Report 84, 37th Congr., 3rd Sess., ser. 1151, pp. 1–59, 114, 152.

47. "Belger, James," box 1597, NN 784.

48. *Argument of Judge Advocate Maj. W. L. Marshall* (Baltimore: Henry A. Robinson, 1865); copy in NN 784.

49. *Defence of Col. James Belger, Quartermaster U.S.A.* (Baltimore: Henry A. Robinson, 1865), 35–36, 64; copy in NN 784.

50. Ibid., 55–56.

51. P. H. Watson to Meigs, 5–6 Mar. 1863, box 9, e. 22, RG 92, NA.

52. *Defence of Col. James Belger*, 79; Stanton order, 30 Nov. 1863, in NN 893.

53. GovCon 2, 73, 102–6, 118–19, 299–300; Brandes, *Warhogs*, 77; Meigs to P. H. Watson, 21 Jan. 1864, roll 37, M494; Crosman testimony, 31 May 1865, p. 3921, box 1125b, MM 2250, RG 153, NA; W. W. McKim to Meigs, 26 Dec. 1864, box 316, e. 225, RG 92, NA.

54. For testimony on the early efforts of Wilson, Childs & Co., see GovCon 2, 1591–1610. For one Gilbert Hubbard & Co. proposal accompanied by Congressman Arnold's recommendation, see Meigs to Stanton, 19 Jan. 1863, roll 17, M494. For evidence of one senator's efforts to secure contracts for his constituents and speed up their settlement, see Theodore

Calvin Pease and James G. Randall, eds., *The Diary of Orville Hickman Browning*, vol. 1, *1850–1864* (Springfield: Illinois State Historical Library, 1925), 499–534, 650–76.

55. For an early example, see Meigs to Edward H. Rollins, 16 July 1861, pp. 76–77, vol. 56, roll 36, M745.

56. GovCon 2, 155–59, 1372–97; M. Hall Stanton to Cameron, 22 Mar. 1865, roll 9, Cameron Papers; "Mr. M. Hall Stanton," *Harper's Weekly* 20 (12 Feb. 1876): 127.

57. New York, vol. 200, pp. 360, 364, Dun; *The Great Libel Case: Opdyke vs. Weed* (New York: American News Co., 1865), 70–71, 103–4. On Opdyke's antebellum career, see Michael Zakim, *Ready-made Democracy: A History of Men's Dress in the American Republic, 1760–1860* (Chicago: University of Chicago Press, 2003), 65–66.

58. "Workingmen of Cincinnati, Attention," *Cincinnati Daily Enquirer*, 4 Apr. 1863; "Shoddy to Be Made to Bleed," *Cincinnati Daily Enquirer*, 2 Sept. 1864; "Shoddy Contractor & Co.," *Cincinnati Daily Enquirer*, 26 Sept. 1864; Mark W. Summers, "The Spoils of War," *North and South* 6 (2003): 86; Bernard F. Reilly Jr., *American Political Prints, 1766–1876: A Catalog of the Collections in the Library of Congress* (Boston: G. K. Hall & Co., 1991), 519–20.

59. Harry J. Carman and Reinhard H. Luthin, *Lincoln and the Patronage* (New York: Columbia University Press, 1943), 290–92.

60. Pennsylvania, vol. 135, pp. 123, 320L, Dun; H. Biggs to M. C. Meigs, 5 Nov. 1864, roll 13, Meigs Papers.

61. Crane circular, 19 Oct. 1864, and Crane to William Browning, 26 Oct. 1864, roll 11, Andrew Johnson Papers, LC.

62. McComb to W. P. Jones, 18 June 1864, roll 10, Johnson Papers; McComb to Lincoln, 27 Sept. 1864, roll 82, Lincoln Papers; Meigs to Stanton, 20 Apr. 1863, roll 17, M494.

63. On antebellum contracting, see Wilson, "Business of Civil War," chap. 1. For an obituary of Martin that claimed that he "belonged to no social or political clubs, and was never identified with any political party," see "Death of John T. Martin," *New York Times*, 12 Apr. 1897.

64. Henry Morford, *The Days of Shoddy: A Novel of the Great Rebellion in 1861* (Philadelphia: T. B. Peterson & Bros., 1863), 23, 174–95.

65. Stephen V. Ash, "Civil War Exodus: The Jews and Grant's General Orders No. 11," *Historian* 44 (1982): 505–23; Tyler Ambinder, "Ulysses S. Grant, Nativist," *Civil War History* 43 (1997): 119–41; Gary L. Bunker and John Appel, "'Shoddy,' Anti-Semitism, and the Civil War," *American Jewish History* 82 (1994): 43–71.

66. Iver Bernstein, *The New York City Draft Riots: Their Significance for American Society and Politics in the Age of the Civil War* (New York: Oxford University Press, 1990), 35–38, 298; Brandes, *Warhogs*, 86.

67. *The Great Libel Case*, 3–7, quotation at 150.

68. *New York Sun* editorial of mid-January 1865, reproduced in *The Opdyke Libel Suit: A Full Metrical, Juridicial, and Analytical Report of the Extraordinary Suit for Libel of George Opdyke "Verses" Thurlow Weed* (New York, 1865), 53.

69. Morford, *Days of Shoddy*, 478; Robert Tomes, "The Fortunes of War: How They Are Made and Spent," *Harper's New Monthly Magazine* 29 (July 1864): 227–31.

70. "The Contractors and the Public," *North American and U.S. Gazette*, 6 Jan. 1863; "Corruption and Extravagance," *North American and U.S. Gazette*, 10 Mar. 1863; Tomes, "Fortunes of War," 227–28.

71. "Official Corruption," *Newark Daily Advertiser*, 25 June 1864.

72. *Congressional Globe*, 37th Congr., 2nd Sess. (12 June 1862), 2684–85; *U.S. Statutes at Large* 12 (1863): 594.

73. Frank L. Klement, *Dark Lanterns: Secret Political Societies, Conspiracies, and Treason Trials in the Civil War* (Baton Rouge: Louisiana State University Press, 1984); Mark Neely, *The Fate of Liberty: Abraham Lincoln and Civil Liberties* (New York: Oxford University Press, 1991).

74. *U.S. Statutes at Large* 12 (1863): 696. Some historians, while noting the long-lasting effects of the 1863 frauds statute, have concluded mistakenly that contractors were not subjected to military law during the Civil War. See Brandes, *Warhogs*, 101.

75. *Congressional Globe*, 37th Congr., 3rd Sess. (14 Feb. 1863). 952–58.

76. I estimate that between twenty-five and fifty Union contractors were tried by court-martial in 1863–65. Manuscript indexes of army courts-martial are available in microcopy 1105, RG 153, NA. These indexes sometimes refer to defendants as contractors, but in other cases contractors are listed only under the more general category of civilians. This makes it difficult to arrive at an exact total. Navy suppliers were also subject to court-martial proceedings. For published documents and a brief secondary account of the case of navy contractor Franklin W. Smith, see "The Prosecution of Franklin W. Smith by the United-States Navy Department," in Boston Board of Trade, *Twelfth Annual Report* (Boston: Geo. C. Rand & Avery, 1866); Curtis Dahl, "Lincoln Saves a Reformer," *American Heritage* 23, no. 6 (Oct. 1972): 74–78.

77. "Punishment of Army Frauds" and "Findings of the Court Martial," *Cincinnati Daily Commercial*, 24 Nov. 1863; "Hasty Generalization—Fraudulent Contracts—The Case of Stettler [sic]," *Cincinnati Daily Enquirer*, 12 Dec. 1863; Stetler Court-Martial Case File MM 1107, RG 153, NA.

78. Such routine practices are evident from Quartermaster's Department correspondence. For example, Crosman to Meigs, [no day] Dec. 1863, box 126, e. 20, RG 92, NA; Crosman to Meigs, 30 Apr. 1864, e. 29, RG 92, NA; Crosman testimony, 31 May 1865, pp. 3914–15, box 1125B, MM 2250, RG 153, NA. See also chap. 4.

79. William H. White Court-Martial Case File MM 1192, RG 153, NA.

80. For examples of acquittals, see D. W. Whitney and Henry W. Scott Court-Martial Case Files MM 1161 and MM 1229, RG 153, NA.

81. White Court-Martial Case File MM 1192.

82. Benjamin C. Evans Court-Martial Case File NN 3036, RG 153, NA.

83. Hall and Smith Court-Martial Case Files MM 1169 and MM 893, RG 153, NA.

84. Wilmore Court-Martial Case File NN 327, RG 153, NA.

85. Hall Court-Martial Case File MM 1169.

86. Smith Court-Martial Case File MM 893.

87. Olcott to James A. Hardie, 4 and 10 Oct. 1863; Brown & Co. to P. H. Watson, 23 Oct. 1863, all in roll 28, M494.

88. Swords to Meigs, 26 Feb. 1864; J. H. Wilson to Meigs, 6 Mar. 1864, both in box 1, e. 29, RG 92, NA; Wilson testimony of 23 Apr. 1864, in MM 1426, RG 153, NA.

89. Spicer, Smoot, and Wormer Court-Martial Case Files MM 1426, NN 1720, NN 1865, RG 153, NA.

90. Roeliff Brinkerhoff, *The Volunteer Quartermaster* (New York: D. Van Nostrand, 1865), 180.

CHAPTER 6. THE UNACKNOWLEDGED MILITARIZATION OF AMERICA

1. Virtually none of the recent literature on memories of the war considers the question of the legacies of large-scale economic mobilization. See David W. Blight, *Race and Reunion: The Civil War in American Memory* (Cambridge: Harvard University Press, 2001); Matthew J. Grow, "The Shadow of the Civil War: A Historiography of Civil War Memory," *American Nineteenth Century History* 4 (2003): 77–103; Alice Fahs and Joan Waugh, eds., *The Memory of the Civil War in American Culture* (Chapel Hill: University of North Carolina Press, 2004).

2. On the Civil War as an impetus to new ways of thinking among Northern elites about large-scale organization and nationalism, see esp. George M. Fredrickson, *The Inner Civil War: Northern Intellectuals and the Crisis of the Union* (New York: Harper & Row, 1965); Fredrickson has sounded this theme more recently in "Nineteenth-Century American History," in *Imagined Histories: American Historians Interpret the Past*, ed. Anthony Molho and Gordon S. Wood (Princeton: Princeton University Press, 1998), 177–79. For similar arguments, see Allan Nevins, "A Major Result of the Civil War," *Civil War History* 5 (1959): 237–50; Morton Keller, *Affairs of State: Public Life in Late Nineteenth Century America* (Cambridge: Harvard University Press, 1977), 1–48; Wilfred M. McClay, *The Masterless: Self and Society in Modern America* (Chapel Hill: University of North Carolina Press, 1994), 25–26. For a stimulating discussion of the legacies of wartime financial innovations, see Richard Franklin Bensel, *Yankee Leviathan: The Origins of Central State Authority in America, 1859–1877* (New York: Cambridge University Press, 1990). For a historiographical essay that situates Bensel's book in the context of a long-running debate about the effects of the Civil War on the political and economic powers of Northern business, see Phillip Shaw Paludan, "What Did the Winners Win? The Social and Economic History of the North during the Civil War," in *Writing the Civil War: The Quest to Understand*, ed. James M. McPherson and William J. Cooper Jr. (Columbia: University of South Carolina Press, 1998), 174–200.

3. W. W. McKim to Lewis, Boardman, & Wharton, 13 Apr. 1865, vol. 7, e. 2195, RG 92, National Archives and Records Administration–Mid-Atlantic Region (Philadelphia); War Department order in *OR*, ser. 3, 4:618–19.

4. "Report of the Quartermaster General," 1866 *ARSW*.

5. "The War Ended," *Scientific American* 12 (24 June 1865): 407.

6. William B. Holberton, *Homeward Bound: The Demobilization of the Union and Confederate Armies, 1865–66* (Mechanicsburg, Penn.: Stackpole Books, 2001), 35; Robert M. Utley, *Frontier Regulars: The United States Army and the Indian, 1866–1891* (New York: Macmillan, 1973), 10–16; Edward M. Coffman, *The Old Army: A Portrait of the American Army in Peacetime, 1784–1898* (New York: Oxford University Press, 1986), 215. The U.S. Navy followed a different trajectory: before World War I, it became a powerful force that relied upon ships developed through higher levels of integration between the military and industry. See Benjamin Franklin Cooling, *Gray Steel and Blue Water Navy: The Formative Years of America's Military-Industrial Complex, 1881–1917* (Hamden, Conn.: Archon Books, 1979); Paul A. C. Koistinen, *Mobilizing for Modern War: The Political Economy of American Warfare, 1865–1919* (Lawrence: University Press of Kansas, 1997), 19–57; Dirk Bönker, "Militarizing the Western World: Navalism, Empire, and State-Building in Germany and the United States before World War I" (Ph.D. diss., Johns Hopkins University, 2002).

7. John Niven, *Connecticut for the Union: The Role of the State in the Civil War* (New Haven: Yale University Press, 1965), 435.

8. J. D. Crittenden to Meigs, 15 Aug. 1865, box 583; J. L. Donaldson report for fiscal 1865 and Donaldson to Meigs, 13 Dec. 1865, box 515; D. H. Rucker, report for fiscal 1866, box 944, all in e. 225, RG 92, NA; Felicia Johnson Deyrup, *Arms Makers of the Connecticut Valley* (Northampton, Mass.: Smith College, 1948), 245.

9. "Report of the Quartermaster General," 1869 *ARSW;* "Army—Staff Organization," House Report 74, 42nd Congr., 3rd Sess., ser. 1576, p. 222.

10. "Public Property Sold by the War Department," House Exec. Doc. 200, 40th Congr., 2nd Sess., ser. 1513.

11. "Report of the Quartermaster General," 1866 *ARSW;* Erna Risch, *Quartermaster Support of the Army: A History of the Corps* (Washington, D.C.: Office of the Quartermaster General, 1962), 458; Robert G. Angevine, *The Railroad and the State: War, Politics, and Technology in Nineteenth-Century America* (Stanford: Stanford University Press, 2004), 157. Navy sales are calculated from "Sale of Public Vessels," House Exec. Doc. 282, 40th Congr., 2nd Sess., ser. 1343.

12. "Public Property Sold by the War Department"; "Sales of Ordnance Stores," House Report 46, 42nd Congr., 2nd Sess., ser. 1528; "Sale of Arms by the Ordnance Department," Senate Report 183, 42nd Congr., 2nd Sess., ser. 1497.

13. Meigs testimony, 3 Feb. 1874, pp. 289-92, in "Reduction of the Military Establishment," House Report 384, 43rd Congr., 1st Sess., ser. 1624; National Association of Wool Manufacturers, *Bulletin . . . for the Year 1872* (Boston, 1872), 236-43; Edward Stanwood, *American Tariff Controversies in the Nineteenth Century* (Boston: Houghton, Mifflin, & Co., 1903), 2:169; Risch, *Quartermaster Support,* 492.

14. Rucker annual report for fiscal 1866, box 944, e. 225, RG 92, NA.

15. *Nashville Dispatch,* 10 June 1865; *Cincinnati Daily Enquirer,* 14 June 1865.

16. A. J. Perry to R. Allen, 8 Mar. 1866, vol. 26, e. 999, RG 92, NA; Meigs to R. Brinkerhoff, 24 July 1866, vol. 92, roll 54, M745; "Report[s] of the Quartermaster General," in 1868 and 1869 *ARSW;* "The Cost of Moths and Mildew," *Scientific American* 34 (8 Apr. 1876): 225. Some surplus clothing, tents, and food were issued to suffering settlers on the Great Plains and other refugees during the 1870s and 1880s. See Gilbert C. Fite, "The United States Army and Relief to Pioneer Settlers, 1874-1875," *Journal of the West* 6 (1967): 99-107; Mark R. Wilson, "The Extensive Side of Nineteenth-Century Military Economy: The Tent Industry in the United States during the Civil War," *Enterprise and Society* 2 (2001): 330.

17. "Government Claims," *American Law Review* 1 (July 1867): 653-54.

18. Kyle S. Sinisi, *Sacred Debts: State Civil War Claims and American Federalism, 1861-1880* (New York: Fordham University Press, 2003).

19. "Claims of Loyal Citizens Growing Out of the Late War," House Exec. Doc. 121, 43rd Congr., 1st Sess., ser. 1607; "Reduction of the Military Establishment," House Report 384, 43rd Congr., 1st Sess., ser. 1624, p. 286; *The Law of Claims against Governments, Including the Mode of Adjusting Them and the Procedure Adopted in Their Investigation* (Washington, D.C.: GPO, 1875), 301; Frank W. Klingberg, *The Southern Claims Commission* (Berkeley and Los Angeles: University of California Press, 1955).

20. William W. Wiecek, "The Origin of the United States Court of Claims," *Administrative Law Review* 20 (1968): 387-406; Wilson Cowen, Philip Nichols Jr., and Marion T. Bennett, *The United States Court of Claims: A History,* pt. 2, *Origin—Development—Jurisdiction, 1855-1978* (Washington, D.C.: Committee on Bicentennial of Independence and the Constitution of the Judicial Conference of the United States, 1978).

21. "Government Contracts," *American Law Review* 4 (Oct. 1869): 16–17.

22. *Brandeis v. U.S.*, 3 Ct. Cl. 99 (1868); *Cobb v. U.S.*, 7 Ct. Cl. 470 (1872); *Cobb v. U.S.*, 9 Ct. Cl. 291 (1874); *Thompson v. U.S.*, 9 Ct. Cl. 187 (1874).

23. *Reeside v. U.S.*, 2 Ct. Cl. 1 (1867); *Mowry v. U.S.*, 2 Ct. Cl. 68 (1867); *Burton v. U.S.*, 2 Ct. Cl. 223 (1867); *Brady v. U.S.*, 3 Ct. Cl. 203 (1868).

24. *Stevens v. U.S.*, 2 Ct. Cl. 95 (1867); *Ramsdell v. U.S.*, 2 Ct. Cl. 508 (1867).

25. *Livingston v. U.S.*, 3 Ct. Cl. 131 (1868); *Child v. U.S.*, 4 Ct. Cl. 176 (1869).

26. Smoot and Spicer Court-Martial Case Files NN 1720 and MM 1426, RG 153, NA.

27. *Spicer v. U.S.*, 1 Ct. Cl. 316 (1866).

28. *Ex Parte Milligan*, 4 Wall. 2 (1866); W. Winthrop, *Digest of Opinions of the Judge Advocate General of the Army*, 3rd ed. (Washington, D.C.: GPO, 1868), 118.

29. *Wormer v. U.S.*, 4 Ct. Cl. 258 (1869); *Smoot v. U.S.*, 5 Ct. Cl. 490 (1870).

30. *U.S. v. Adams*, 7 Wall. 463 (1869); *U.S. v. Adams*, 9 Wall. 554 (1870); Robert B. Murray, "The Fremont-Adams Contracts," *Journal of the West* 5 (1966): 517–24; *Child v. U.S.*, 6 Ct. Cl. 44 (1871); *U.S. v. Child*, 12 Wall. 232 (1371); *U.S. v. Justice*, 14 Wall. 535 (1872); *Smith v. U.S.*, 5 Ct. Cl. 496 (1870) and 8 Ct. Cl. 67 (1873); *Mason v. U.S.*, 6 Ct. Cl. 57 (1871) and 8 Ct. Cl. 125 (1873); *Haskell v. U.S.*, 9 Ct. Cl. 410 (1874).

31. *Smoot's Case*, 15 Wall. 36 (1873) at 44. See also *U.S. v. Wormer*, 13 Wall. 25 (1872).

32. E. B. Washburne to C. C. Washburn, 5 Feb. 1870, box 1, C. C. Washburn Papers, SHSW.

33. *U.S. v. Reeside*, 8 Wall. 38 (1869); Murray, "Fremont-Adams Contracts"; Adams award discussed in *Congressional Globe*, 42nd Congr., 3rd Sess. (8 and 15 Jan. 1873), 396–409, 608–9.

34. J. Holt to U. S. Grant, 12 Nov. 1867, folder 1; U.S. War Department Office of Board of Claims memo, 13 Apr. 1869, folder 3; Cozens account book, folder 4, all in Cozens Papers, Historical Society of Pennsylvania. See also J. Matthew Gallman, *Mastering Wartime: A Social History of Philadelphia during the Civil War* (New York: Cambridge University Press, 1990), 290; Wilson, "Extensive Side of Nineteenth-Century Military Economy," 325–30.

35. Box 592, Court-Martial Case File LL 1525, RG 153, NA; B 1066 CB 1864, roll 72, National Archives Microform Publication M1054 [Letters Received by the Commission Branch of the Adjutant General's Office, RG 94].

36. "James Belger," House Exec. Doc. 72, 41st Congr., 2nd Sess., ser. 1417; "James Belger," Senate Report 330, 51st Congr., 1st Sess., ser. 2704. On McKinstry, see 1959 ACP 1872, RG 94, NA; Mexican War Certificate 11146, Military Pension Files, NA.

37. *OR*, ser. 3, 5:577.

38. Keller, *Affairs of State*, 107.

39. *Congressional Globe*, 39th Congr., 2nd Sess. (19 Feb. 1867), 1355.

40. Meigs to W. Myers, 8 May 1865, p. 339, vol. 84, roll 50; Meigs to Sherman, 26 Aug. 1865, p. 238, vol. 86, roll 51; Meigs to Sheridan, 2 Nov. 1865, pp. 412–413, vol. 87, roll 51, all in M745; "Army Reduction," *Galaxy* 21 (Feb. 1876): 260.

41. Ibid., 264; James E. Sefton, *The United States Army and Reconstruction, 1865–1877* (Baton Rouge: Louisiana State University Press, 1967), 207–8, 260–62; Bensel, *Yankee Leviathan*, 380, 390.

42. On the West during the war, see Alvin M. Josephy Jr., *The Civil War in the American West* (New York: Random House, 1991).

43. For detailed accounts of postwar military-economic activity in particular areas, see

Thomas T. Smith, *The U.S. Army and the Texas Frontier Economy, 1845–1900* (College Station: Texas A&M University Press, 1999); William A. Dobak, *Fort Riley and Its Neighbors: Military Money and Economic Growth, 1853–1895* (Norman: University of Oklahoma Press, 1998); and Darlis A. Miller, *Soldiers and Settlers: Military Supply in the Southwest, 1861–1885* (Albuquerque: University of New Mexico Press, 1989).

44. William Morris Hoge Jr., "The Logistical System of the U.S. Army during the Indian Wars, 1866–1889" (M.A. thesis, University of Washington, 1968), 25.

45. Dobak, *Fort Riley and Its Neighbors*, 80–86, 101–2; William E. Connelley, ed., *A Standard History of Kansas and Kansans* (Chicago: Lewis Publishing Co., 1918), 3:1219; W. N. Davis Jr., "The Sutler at Fort Bridger," *Western Historical Quarterly* 2 (1971): 37–54; David Michael Delo, *Peddlers and Post Traders: The Army Sutler on the Frontier* (Salt Lake City: University of Utah Press, 1992), 144–47.

46. William E. Lass, *From the Missouri to the Great Salt Lake: An Account of Overland Freighting* (Lincoln: Nebraska State Historical Society, 1972); Henry Pickering Walker, *The Wagonmasters: High Plains Freighting from the Earliest Days of the Santa Fe Trail to 1880* (Norman: University of Oklahoma Press, 1966); Gary S. Freedom, "Moving Men and Supplies: Military Transportation on the Northern Great Plains, 1866–1891," *South Dakota History* 14 (1984): 114–33. For more details, see Mark R. Wilson, "The Business of Civil War: Military Enterprise, the State, and Political Economy in the United States, 1850–1880" (Ph.D. diss., University of Chicago, 2002), 811–40.

47. *Congressional Globe*, 42nd Congr., 3rd Sess. (25 Feb. 1873), 1766.

48. "Report of the Quartermaster General," 1879 *ARSW*.

49. Angevine, *Railroad and the State*, 186–87; *Sinking Fund Cases*, 99 U.S. 700 (1878), at 755.

50. Heather Cox Richardson, *The Greatest Nation of the Earth: Republican Economic Policies during the Civil War* (Cambridge: Harvard University Press, 1997), 170–208.

51. "Railroad Routes to the Pacific," Senate Report 219, 40th Congr., 3rd Sess., ser. 1362; *U.S. v. Union Pacific R.R. Co.*, 91 U.S. 72 (1875), at 80.

52. Robert G. Athearn, *Union Pacific Country* (Chicago: Rand McNally, 1971); Maury Klein, *Union Pacific: Birth of a Railroad, 1863–1893* (Garden City, N.Y.: Doubleday, 1987).

53. Because of the semipublic status of these railroad companies, only about a quarter of these amounts were paid in cash; the remainder went toward the companies' public debts. See "Report of the Quartermaster General," 1879 *ARSW*; *Pacific R.R. Co. v. U.S.*, 13 Ct. Cl. 401 (1877), at 461; Risch, *Quartermaster Support*, 478–79; Angevine, *Railroad and the State*, 184.

54. "Report[s] of the Quartermaster General," in 1866–80 *ARSW*; George W. Cullum, *Biographical Register of the Officers and Graduates of the U.S. Military Academy* (Boston: Houghton, Mifflin & Co., 1891), 1:652–53, 2:195, 2:502. On Ekin, see *A Biographical History of Eminent and Self-made Men of the State of Indiana* (Cincinnati: Western Biographical Publishing Co., 1880), 1:13–16.

55. Potter died of heart failure in 1888, at the age of seventy-two, at home in Painesville, Ohio. See Potter to Meigs, 22 Feb. 1867; W. F. Howell to Rucker, 12 Sept. 1867; Potter to Rucker, 21 Dec. 1867; Potter to Meigs, 1 Sept. 1868; Potter to Adjutant General, 30 June 1868; Howard W. Potter to Adjutant General, 25 Apr. 1888, all in boxes 839–40, e. 225, NA; 3607 ACP 1877, RG 94, NA; Civil War Certificate no. 249,246, Civil War Pension Files, NA.

56. Cullum, *Biographical Register*, 1:284, 316, 437, 632.

57. For one study of the postwar careers of a handful of top Confederate and Union officers, see Mark Van Rhyn, "Beyond the Battlefield: Post-War Careers of Middle Rank Civil War Generals" (Ph.D. diss., University of Nebraska, 2003).

58. Russell F. Weigley, *Quartermaster General of the Union Army: A Biography of M. C. Meigs* (New York: Columbia University Press, 1959), 343–61.

59. Officer personnel files 278 ACP 1875, 1047 ACP 1877, 800 ACP 1879, RG 94, NA; miscellaneous materials in box 515, e. 225, RG 92, NA; Cullum, *Biographical Register*, 1:637–41, 1:652–53, 2:195.

60. Ezra J. Warner, *Generals in Blue: Lives of the Union Commanders* (Baton Rouge: Louisiana State University Press, 1964), 360–61; *Biographical Encyclopedia of Ohio of the Nineteenth Century* (Cincinnati: Galaxy Publishing Co., 1876), 28–29; C. W. Moulton, *The Review of General Sherman's Memoirs Examined* (Cincinnati: Robert Clarke & Co., 1875); "General Roeliff Brinkerhoff, 1828–1911," *Ohio Archaeological and Historical Publications* 20 (1911): 353–67; Van Rhyn, "Beyond the Battlefield," 342–66.

61. Van Vliet to Thomas William Sweeny, 30 Sept. 1891, Sweeny Papers, SW 714, Huntington Library; Warner, *Generals in Blue*, 524.

62. Olivier Zunz, *Making America Corporate, 1870–1920* (Chicago: University of Chicago Press, 1990), 39. Seminal "organizational synthesis" studies include Robert H. Wiebe, *The Search for Order, 1877–1920* (New York: Hill & Wang, 1967); Louis Galambos, "The Emerging Organizational Synthesis in Modern American History," *Business History Review* 44 (1970): 279–90; Alfred D. Chandler Jr., *The Visible Hand: The Managerial Revolution in American Business* (Cambridge: Harvard University Press, 1977); Louis Galambos, "Technology, Political Economy, and Professionalization: Central Themes of the Organizational Synthesis," *Business History Review* 57 (1983): 471–93. More recent discussions include Brian Balogh, "Reorganizing the Organizational Synthesis: Federal-Professional Relations in Modern America," *Studies in American Political Development* 5 (1991): 126; Kenneth Cmiel, "Destiny and Amnesia: The Vision of Modernity in Robert Wiebe's *The Search for Order*," *Reviews in American History* 21 (1993): 352–68.

63. Several excellent studies have questioned whether the large industrial corporations were as efficient or representative as Chandler suggested. For example, Gerald Berk, *Alternative Tracks: The Constitution of American Industrial Order, 1865–1917* (Baltimore: Johns Hopkins University Press, 1994); Philip Scranton, *Endless Novelty: Specialty Production and American Industrialization, 1865–1925* (Princeton: Princeton University Press, 1997); William G. Roy, *Socializing Capital: The Rise of the Large Industrial Corporation in America* (Princeton: Princeton University Press, 1997). For one of the few direct dissents from Chandler's narrative of the development of bureaucracy in America, see Richard R. John, "Private Enterprise, Public Good? Communications Deregulation as a National Political Issue, 1839–1851," in *Beyond the Founders: New Approaches to the Political History of the Early American Republic*, ed. Jeffrey L. Pasley, Andrew W. Robertson, and David Waldstreicher (Chapel Hill: University of North Carolina Press, 2004), 332–33.

64. William E. Nelson, *The Roots of American Bureaucracy, 1830–1900* (Cambridge: Harvard University Press, 1982); JoAnne Yates, *Control through Communication: The Rise of System in American Management* (Baltimore: Johns Hopkins University Press, 1989); Kenneth Lipartito, "The Utopian Corporation," in *Constructing Corporate America: History, Politics, Culture*, ed. Kenneth Lipartito and David B. Sicilia (New York: Oxford University Press, 2004), 94–119.

65. For efforts to distinguish between militarization and militarism, see John Gillis, "Introduction," and Michael Geyer, "The Militarization of Europe, 1914–1945," in *The Militarization of the Western World,* ed. John R. Gillis (New Brunswick, N.J.: Rutgers University Press, 1989), 1–10, 65–102; Patrick M. Regan, *Organizing Societies for War: The Process and Consequences of Societal Militarization* (Westport, Conn.: Praeger, 1994).

66. Carl Russell Fish, *The Civil Service and the Patronage* (New York: Longmans, Green & Co., 1905); Paul P. Van Riper, *History of the United States Civil Service* (Evanston, Ill.: Row, Peterson & Co., 1958); Ari Hoogenboom, *Outlawing the Spoils: A History of the Civil Service Reform Movement, 1865–1883* (Urbana: University of Illinois Press, 1961); John G. Sproat, *"The Best Men": Liberal Reformers in the Gilded Age* (New York: Oxford University Press, 1968); Thomas L. Haskell, *The Emergence of Professional Social Science: The American Social Science Association and the Nineteenth-Century Crisis of Authority* (Urbana: University of Illinois Press, 1977), 115–21; Stephen Skowronek, *Building a New American State: The Expansion of National Adminstrative Capacities, 1877–1920* (Cambridge: Cambridge University Press, 1982), 43–53, 78–83; Daniel P. Carpenter, *The Forging of Bureaucratic Autonomy: Reputations, Networks, and Policy Innovation in Executive Agencies, 1862–1928* (Princeton: Princeton University Press, 2001), 10–11, 45–47; Nancy Cohen, *The Reconstruction of American Liberalism, 1865–1914* (Chapel Hill: University of North Carolina Press, 2002), 134–35; Andrew L. Slap, "Transforming Politics: The Liberal Republican Movement and the End of Civil War Era Political Culture" (Ph.D. diss., Pennsylvania State University, 2002), 103–40.

67. Kirk H. Porter and Donald Bruce Johnson, comp., *National Party Platforms, 1840–1964* (Urbana: University of Illinois Press, 1966), 41–48; Leonard D. White, *The Republican Era, 1869–1901: A Study in Administrative History* (New York: Macmillan, 1958), 281; Sean M. Theriault, "Patronage, the Pendleton Act, and the Power of the People," *Journal of Politics* 65 (2003): 55.

68. *Congressional Globe,* 39th Congr., 1st Sess. (28 June 1866), 3450, and 39th Congr., 2nd Sess. (29 Jan. 1867 and 6 Feb. 1867), 837–41, 1034; White, *The Republican Era,* 280.

69. Julius Bing, "Civil Service of the United States," *North American Review* 105 (Oct. 1867): 488.

70. Jacob D. Cox, "The Civil-Service Reform," *North American Review* 112 (1871): 97; A. R. Macdonough, "Civil Service Reform," *Harper's New Monthly Magazine* 40 (Mar. 1870): 546–56.

71. Charles Eliot Norton, ed., *Orations and Addresses of Charles William Curtis* (New York: Harper & Bros., 1894), 2:1–28; "A Vital Reform," *Harper's Weekly* 12 (28 Nov. 1868): 754; "General Pleasonton and the Civil Service," *Harper's Weekly* 15 (21 Jan. 1871): 50; "Another Glimpse of the Public Service," *Harper's Weekly* 21 (13 Jan. 1877): 23; "Civil Service Clubs," *Harper's Weekly* 21 (3 Nov. 1877): 858; "The Administration and Reform," *Harper's Weekly* 21 (10 Nov. 1877): 878; "Party and Patronage," *Harper's Weekly* 21 (29 Dec. 1877): 1022–23.

72. E. L. Godkin, "Commercial Immorality and Political Corruption," *North American Review* 107 (July 1868): 265–66.

73. Carpenter, *The Forging of Bureaucratic Autonomy,* 76–102. See also Daniel P. Carpenter, "The Corporate Metaphor and Executive Department Centralization in the United States, 1888–1928," *Studies in American Political Development* 12 (1998): 162–203.

74. "Civil Service of the United States," House Report 47, 40th Congr., 2nd Sess. (1868), p. 52.

75. Ibid., 19, 34, 81–88.

76. Dorman B. Eaton, "A New Phase of the Reform Movement," *North American Review* 132 (June 1881): 548; Albion W. Tourgée, "Reform versus Reformation," *North American Review* 132 (Apr. 1881): 316–17. In his own very brief discussion of the importance of the military model for the civil service movement, George Fredrickson uses Eaton's essay as his primary example. Fredrickson, *The Inner Civil War,* 209.

77. *Congressional Globe,* 39th Congr., 2nd Sess. (6 Feb. 1867), 1034.

78. *Congressional Globe,* 40th Congr., 3rd Sess. (8 Jan. 1869), 262–66.

79. *Congressional Globe,* 42nd Congr., 2nd Sess. (18 Jan. 1872), 454.

80. *Congressional Globe,* 41st Congr., 3rd Sess. (23 Jan. 1871), 674.

81. "An Imaginary Conversation," *Putnam's Magazine* 3 (Mar. 1869): 358.

82. Keller, *Affairs of State,* 106.

83. *Report of the Select Committee to Investigate the Alleged Credit Mobilier Bribery* (Washington, D.C.: GPO, 1873), x.

84. For overviews that stress the cynical origins and limited influence of these innovations, see Keller, *Affairs of State,* 289–318; Skowronek, *Building a New American State,* 121–62; Mark Wahlgren Summers, *Party Games: Getting, Keeping, and Using Power in Gilded Age Politics* (Chapel Hill: University of North Carolina Press, 2004), 229–37.

85. Gustavus Myers, *History of the Great American Fortunes* (1909; reprint, New York: Modern Library, 1936), 291–93, 400; Matthew Josephson, *The Robber Barons: The Great American Capitalists, 1861–1901* (New York: Harcourt, Brace & Co., 1934), 59–102. For a partial roster of the early corporate leaders, see Frances W. Gregory and Irene D. Neu, "The American Industrial Elite in the 1870s: Their Social Origins," in *Men in Business: Essays in the History of Entrepreneurship,* ed. William Miller (Cambridge: Harvard University Press, 1952), 193–211.

86. John F. Stover, "Colonel Henry S. McComb, Mississippi Railroad Adventurer," *Journal of Mississippi History* 17 (1955): 177–90; Fred Carstensen, "Deering, William," in *American National Biography,* ed. John A. Garraty and Mark C. Carnes (New York: Oxford University Press, 1999), 6:334–35; Stuart Seely Sprague, "The Economic Impact of the Civil War: The Case of Cincinnati," *Essays in Business and Economic History* 10 (1992): 12–28; Norman B. Wilkinson, *Lammot Du Pont and the American Explosives Industry, 1850–1884* (Charlottesville: University Press of Virginia, 1984).

87. New York, vol. 196, pp. 931, 1000LL; Pennsylvania, vol. 137, pp. 492, 500, 766, 851; Massachusetts, vol. 70, pp. 960–61, all in Dun; William R. Bagnall, "Sketches of Manufacturing Establishments in New York City, and of the Textile Establishments of the United States" (microfiche of typescript, Merrimack Valley Textile Museum, 1977), 315–34. J. Matthew Gallman, "Entrepreneurial Experiences in the Civil War: Evidence from Philadelphia," in *American Development in Historical Perspective,* ed. Thomas Weiss and Donald Schaefer (Stanford: Stanford University Press, 1994), 219–20.

88. William Henry Engle, *History of the Counties of Dauphin and Lebanon in the Commonwealth of Pennsylvania* (Philadelphia: Evarts & Peck, 1883), 473; Michael W. Rich, "Henry Mack: An Important Figure in Nineteenth-Century American Jewish History," *American Jewish Archives* 48 (1995): 261–79; *Biographical Cyclopaedia and Portrait Gallery of . . . Ohio* (Cincinnati: Western Publishing Co., 1883), 1:255–57.

89. Henry W. B. Howard, ed., *The Eagle and Brooklyn . . . History of the City of Brooklyn* (Brooklyn: Brooklyn Daily Eagle, 1893), 274–75, 787–89; "Obituary: John T. Martin," *New*

York Tribune, 12 Apr. 1897. However much pleasure the art collection gave the Martins, as a financial investment it had mixed results. When the collection was sold at auction in 1909, Millet's "Going to Work-Dawn of Day" sold for $50,000, three times what Martin had paid in 1882. On the other hand, Ludwig Knaus's "The Christening," which reportedly cost Martin $50,000, brought only $8,900. See "Millet Painting Sells for $50,000," *New York Times,* 17 Apr. 1909.

90. Charles F. Adams Jr., "An Erie Raid," *North American Review* 112 (Apr. 1871): 241–42.

91. James A. Ward, "Image and Reality: The Railway Corporate-State Metaphor," *Business History Review* 55 (1981): 491–516.

92. Angevine, *Railroad and the State,* 175.

93. Thomas C. Cochran, *Railroad Leaders, 1845–1890: The Business Mind in Action* (Cambridge: Harvard University Press, 1953), 84, 429–30.

94. Stuart Morris, "Stalled Professionalism: The Recruitment of Railway Officials in the United States, 1855–1940," *Business History Review* 47 (1973): 317.

95. Cochran, *Railroad Leaders,* 405.

96. Allan Nevins, *Study in Power: John D. Rockefeller, Industrialist and Philanthropist* (New York: Charles Scribner's Sons, 1953), 1:93–94.

97. For historical studies that suggest this was the case, see esp. Berk, *Alternative Tracks;* Roy, *Socializing Capital;* David A. Hounshell, *From the American System to Mass Production, 1800–1932: The Development of Manufacturing Technology in the United States* (Baltimore: Johns Hopkins University Press, 1984).

98. Charles F. Adams Jr., "Railway Problems in 1869," *North American Review* 110 (Jan. 1870): 147. On businessmen's support for regulation and public enterprise, see Lee Benson, *Merchants, Farmers, and Railroads: Railroad Regulation and New York Politics, 1850–1887* (Cambridge: Harvard University Press, 1955); Richard R. John, "Unnatural Monopoly: The Political Economy of Telegraphy in the Civil War Era," unpublished paper presented to MIT Science, Technology, and Society Program, Nov. 2004. The growing influence of laissez-faire ideas by the 1880s is described in Keller, *Affairs of State,* 162–96.

99. Cochran, *Railroad Leaders,* 341, 436.

100. Charles McArthur Destler, "The Opposition of American Businessmen to Social Control during the 'Gilded Age,'" *Mississippi Valley Historical Review* 39 (1953): 641–72; Edward Chase Kirkland, *Dream and Thought in the Business Community, 1860–1900* (Ithaca: Cornell University Press, 1956), 16–24, 121–23; Sidney Fine, *Laissez Faire and the General Welfare State: A Study of Conflict in American Thought, 1865–1901* (Ann Arbor: University of Michigan Press, 1957), 96–120; Jack Blicksilver, *Defenders and Defense of Big Business in the United States, 1880–1900* (New York: Garland, 1985), 226–33.

101. Alfred L. Thimm, *Business Ideologies in the Reform-Progressive Era, 1880–1914* (n.p.: University of Alabama Press, 1976), 134–35.

102. Edward C. Kirkland, *Industry Comes of Age: Business, Labor, and Public Policy, 1860–1897* (New York: Holt, Rinehart & Winston, 1961), 99.

103. Lawrence Goodwyn, *Democratic Promise: The Populist Moment in America* (New York: Oxford University Press, 1976); Leon Fink, *Workingmen's Democracy: The Knights of Labor and American Politics* (Urbana: University of Illinois Press, 1983); Steve Leikin, *The Practical Utopians: American Workers and the Cooperative Movement in the Gilded Age* (Detroit: Wayne State University Press, 2005).

104. Charles McArthur Destler, "Western Radicalism, 1865–1901: Concepts and Origins," *Mississippi Valley Historical Review* 31 (1944): 354–55; Kim Voss, *The Making of American Exceptionalism: The Knights of Labor and Class Formation in the Nineteenth Century* (Ithaca: Cornell University Press, 1993), 72–101; Elizabeth Sanders, *Roots of Reform: Farmers, Workers, and the American State, 1877–1917* (Chicago: University of Chicago Press, 1997), esp. 397–400.

105. Richard R. John, "Recasting the Information Infrastructure for the Industrial Age," in *A Nation Transformed by Information*, ed. Alfred D. Chandler Jr. and James Cortada (New York: Oxford University Press, 2000), 81.

106. Martin J. Burke, *The Conundrum of Class: Public Discourse on the Social Order in America* (Chicago: University of Chicago Press, 1995), 133, 141.

107. Gladden quoted in Wiebe, *The Search for Order*, 63. For an international perspective, see Michael Mann, *The Sources of Social Power*, vol. 2, *The Rise of Classes and Nation States, 1760–1914* (Cambridge: Cambridge University Press, 1993), 644.

108. Jerry M. Cooper, *The Army and Civil Disorder: Federal Military Intervention in Labor Disputes, 1877–1900* (Westport, Conn.: Greenwood Press, 1980); Cooper, *The Rise of the National Guard* (Lincoln: University of Nebraska Press, 1997); David Montgomery, *Citizen Worker: The Experience of Workers in the United States with Democracy and the Free Market during the Nineteenth Century* (New York: Cambridge University Press, 1993), 89–104; Larry Isaac, "To Counter 'The Very Devil' and More: The Making of Independent Capitalist Militia in the Gilded Age," *American Journal of Sociology* 108 (2002): 353–405. On military interventions on behalf of employers during the Civil War, see esp. David Montgomery, *Beyond Equality: Labor and the Radical Republicans, 1862–1872* (New York: Alfred A. Knopf, 1967), and Grace Palladino, *Another Civil War: Labor, Capital, and the State in the Anthracite Regions of Pennsylvania, 1840–68* (Urbana: University of Illinois Press, 1990).

109. See John F. Kasson, *Civilizing the Machine: Technology and Republican Values in America, 1776–1900* (New York: Penguin, 1977), 183–234; Alan Trachtenberg, *The Incorporation of America: Culture and Society in the Gilded Age* (New York: Hill & Wang, 1982), 48–52; Neil Harris, "Utopian Fiction and Its Discontents," in *Cultural Excursions: Marketing Appetites and Cultural Tastes in Modern America* (Chicago: University of Chicago Press, 1990), 150–73.

110. Ignatius Donnelly, *Caesar's Column: A Story of the Twentieth Century*, ed. Walter B. Rideout (Cambridge: Harvard University Press, 1960). More openly anti-Semitic than most Civil War criticisms of profiteering, the novel deliberately places Jewish bankers among the imagined global elite.

111. Edward Bellamy, "Progress of Nationalism in the United States," *North American Review* 154 (June 1892): 743.

112. John L. Thomas, *Alternative America: Henry George, Edward Bellamy, Henry Demarest Lloyd, and the Adversary Tradition* (Cambridge: Harvard University Press, 1983), 30–31, 271–73; McClay, *The Masterless*, 74–103; Peter Karsten, "Militarization and Rationalization in the United States, 1870–1914," in Gillis, *Militarization of the Western World*, 37–38.

113. Edward Bellamy, "First Steps toward Nationalism," *Forum* 10 (Oct. 1890): 174–84; Bellamy, "'Looking Backward' Again," *North American Review* 150 (Mar. 1890): 354. On Bellamy's influence among Populists, see Destler, 'Western Radicalism, 1865–1901," 350–51; Robert C. McGrath Jr., *American Populism: A Social History, 1877–1898* (New York: Hill & Wang, 1993), 113–14.

114. Donnelly, *Caesar's Column*, 300–303.

115. Bellamy, "'Looking Backward' Again," 357.

116. For a recent article that challenges common assumptions about the quietness of Bellamy's home town but does not discuss the potential significance of nearby Springfield, see John Robert Mullin, "Bellamy's Chicopee: A Laboratory for Utopia?" *Journal of Urban History* 29 (2003): 133–50. See also Daphne Patai, ed., *Looking Backward, 1988–1888: Essays on Edward Bellamy* (Amherst: University of Massachusetts Press, 1988). For a recent intellectual history that locates Bellamy between older producerist thinking and a newer consumerist agenda, see Kathleen G. Donohue, *Freedom from Want: American Liberalism and the Idea of the Consumer* (Baltimore: Johns Hopkins University Press, 2003), 51–65.

117. Edward Bellamy, "The Programme of the Nationalists," *Forum* 17 (Mar. 1894): 88.

118. Edward Bellamy, "What 'Nationalism' Means," *Eclectic Magazine of Foreign Literature* 52 (Sept. 1890): 297.

119. For suggestions that an exceptionally nonmilitarist modernity was attributed to the United States for ideological reasons in the wake of World War II, see Daniel T. Rodgers, "Exceptionalism," in Molho and Wood, *Imagined Histories*, 21–40; Irmgard Steinisch, "Different Path to War: A Comparative Study of Militarism and Imperialism in the United States and Imperial Germany, 1871–1914," in *Anticipating Total War: The German and American Experiences, 1871–1914*, ed. Manfred F. Boemke, Roger Chickering, and Stig Förster (Cambridge: Cambridge University Press, 1999), 29–53.

120. Karsten, "Militarization and Rationalization," 42–44.

Essay on Sources

This study is based mainly on primary sources, many of them unpublished. The most important single archive for the purposes of this book is the large body of records left by the wartime Quartermaster's Department, including correspondence, contract registers, and payrolls. These are held in Record Group 92 (Office of the Quartermaster General of the Army) at the National Archives in Washington. A smaller amount of RG 92 material is held in regional branches of the National Archives in New York City and Philadelphia. Another large set of records held at the National Archives that proved especially useful for this book are the Civil War court-martial case files, in RG 153 (Office of the Judge Advocate General of the Army). For a recent general discussion of these records, see Thomas P. Lowry, "Research Note: New Access to a Civil War Resource," *Civil War History* 49 (2003): 52–63.

The Library of Congress holds the largest collection of personal papers of leading officials in the Civil War era. For this study, among the most important Library of Congress manuscript holdings were the papers of Presidents Abraham Lincoln and Andrew Johnson, Secretaries of War Simon Cameron and Edwin Stanton, and Quartermaster General Montgomery Meigs. Many of these collections are available in microfilm editions.

Some national government documents used in this study are available in published form. One essential published source is the large multivolume collection, *War of the Rebellion . . . Official Records of the Union and Confederate Armies* (Washington, D.C.: GPO, 1880–1901). Although the *Official Records* are devoted largely to War Department correspondence regarding military operations in the field, they contain some documents pertaining to procurement and logistics, especially in series 3. No less important is the U.S. Serial Set, the indispensable collection of executive and legislative documents printed for Congress. Among the Serial Set documents from the Civil War era that proved most useful for this book were lists of military contracts, Treasury Department "Receipts and Expenditures" reports, and the reports of the wartime investigative committees. The Serial Set is also one source for the *Annual Report[s] of the Secretary of War,* which for the Civil War years also appear in the *Official Records;* it is also possible to find these reports bound separately. The essential source for congressional debates and proceedings during this era is the *Congressional Globe;* in 1873, it became the *Congressional Record.*

ESSAY ON SOURCES

Of the several dozen published legal cases involving Civil War era military contractors that were identified and used for this study, nearly all may be found in the reports of the U.S. Court of Claims and the Supreme Court. Future researchers may well uncover more rich material in the surviving records of lower courts, scattered across the country, as well as in the manuscript Court of Claims case files, held at the National Archives.

State government documents also served as important sources for this study. Many state archives hold unique manuscript collections that describe state-level procurement efforts; these include governors' papers and supply department records. During the research for this book, I examined such collections at the Archives Division of the State Historical Society of Wisconsin, the Illinois State Archives and Illinois State Historical Library, and the Indiana State Archives. No less rich are the many published state government documents that speak to the mobilization effort, including adjutant generals' reports, executive documents, legislative journals, and committee reports In many Northern states, these published documents contain rich sources of information, including state procurement officers' annual reports, lists of purchases of military supplies, and testimony gathered by legislative committees investigating procurement practices.

Unfortunately, surviving business records from the Civil War era are rare. As readers will gather from the notes to this book, there are a only a few collections, many of them fragmentary, that contain the business papers of Northern war contractors. All the more valuable, then, is the remarkable set of manuscript credit reports in the R. G. Dun & Company Collection, held at Baker Library, Harvard Business School. Compiled and updated regularly throughout the Civil War era for hundreds of individuals and firms in counties across the United States, these reports contain valuable information about many businesses, including war contractors. Also useful are the microlevel data on manufacturing firms recorded in manuscript returns compiled by U.S. Census enumerators for the 1860 and 1870 censuses. In many cases, these returns are available on microfilm, either through state archives or the National Archives. Other useful sources of information about contractors and the general business environment include local board of trade or chamber of commerce reports, local business directories, published lists of local or national income tax assessments, and county histories and biographical dictionaries, many of which were published in the late nineteenth century. An important source of biographical information about army officers is George W. Cullum, *Biographical Register of the Officers and Graduates of the U.S. Military Academy*, 3rd ed. (Boston: Houghton, Mifflin, 1891).

For this book, as for so many historical research projects, published periodicals proved to be a rich source both for specific details on the war economy and as a guide to the broader context of wartime public discourse. In many Northern cities, there were two influential daily newspapers, each of which had a distinctive editorial position; in some cases, there was at least one paper associated with each of the major political parties. For this reason, I examined two or more dailies for the war years for several of the Union's urban supply hubs: for Saint Louis, the *Missouri Democrat* and *Missouri Republican;* for Cincinnati, the *Commercial* and the *Enquirer;* for Philadelphia, the *Daily Evening Bulletin* and the *North American and U.S. Gazette;* and for New York City, the *Sun* the *Times,* and the *Tribune.* The most important labor paper for the war years was *Fincher's Trades Review,* based in Philadelphia. Among the weekly and monthly periodicals that provide rich visual illustrations or unique coverage of special topics are the *Army and Navy Journal, Bankers' Magazine and Statistical Register, Frank*

Leslie's Illustrated Newspaper, Harper's Weekly, Prairie Farmer, Scientific American, U.S. Service Magazine, and *Vanity Fair.*

IN THE SECONDARY LITERATURE, the subject of the economic mobilization for the Civil War received the most attention in the years following the First World War and Second World War, when industrial mobilization became recognized as an essential aspect of modern warfare. Thus it was no accident that the 1920s saw the publication of two of the first serious academic studies to discuss the Union's administration of material mobilization: Fred Albert Shannon, *The Organization and Administration of the Union Army,* 2 vols. (Cleveland: Arthur C. Clark, 1928), and A. Howard Meneely, *The War Department, 1861: A Study in Mobilization and Administration* (New York: Columbia University Press, 1928).

Scholarly interest in the economic aspects of the Civil War peaked during the 1950s and 1960s, in the wake of the Second World War. In his multivolume synthetic history of the Civil War, *The War for the Union,* 4 vols. (New York: Charles Scribner's Sons, 1959–71), Allan Nevins suggested that the challenges of economic mobilization pushed Americans to practice large-scale organization. At the same time, Russell F. Weigley published what would become the standard biography of Montgomery C. Meigs, which included a substantial discussion of Union logistics and procurement: *Quartermaster General of the Union Army: A Biography of M. C. Meigs* (New York: Columbia University Press, 1959). Soon after, army historian Erna Risch published *Quartermaster Support of the Army: A History of the Corps* (Washington, D.C.: Office of the Quartermaster General, 1962), an outstanding survey, based on extensive archival research, of Quartermaster's Department history up to World War II. Other significant studies of Union procurement and logistics completed during these years included Donald A. MacDougall, "The Federal Ordnance Bureau, 1861–1865" (Ph.D. diss., University of California, Berkeley, 1951); George Winston Smith, *Medicines for the Union Army: The United States Army Laboratories during the Civil War* (Madison, Wisc.: American Institute of the History of Pharmacy, 1962); Thomas Weber, *The Northern Railroads in the Civil War, 1861–1865* (New York: King's Crown, 1952); and an often-overlooked but excellent study of the army's food supply effort by Palmer Henry Boeger, "Hardtack and Coffee: The Commissary Department, 1861–1865" (Ph.D. diss., University of Wisconsin, 1953). The widespread interest in the organizational dimension of the Civil War was also reflected in George M. Fredrickson, *The Inner Civil War: Northern Intellectuals and the Crisis of the Union* (New York: Harper & Row, 1965). Like Nevins before him, Fredrickson pointed to the work of the United States Sanitary Commission, a wartime voluntary group that monitored military hospitals and procured and delivered medical supplies, as evidence that the Civil War had exposed American elites to the problems and possibilities of large-scale, national organization.

Many of the now-classic studies of Confederate procurement and logistics also date from the early cold war era. These include Frank E. Vandiver, *Ploughshares into Swords: Josiah Gorgas and Confederate Ordnance* (Austin: University of Texas Press, 1952); Charles B. Dew, *Ironmaker to the Confederacy: Joseph R. Anderson and the Tredegar Iron Works* (New Haven: Yale University Press, 1966); and Richard D. Goff, *Confederate Supply* (Durham, N.C.: Duke University Press, 1969). For important recent additions to this literature, see Mary A. DeCredico, *Patriotism for Profit: Georgia's Urban Entrepreneurs and the Confederate War Effort*

(Chapel Hill: University of North Carolina Press, 1990); Harold S. Wilson, *Confederate Industry: Manufacturers and Quartermasters in the Civil War* (Jackson: University Press of Mississippi, 2002); and Chad Morgan, "The Public Nature of Private Industry in Confederate Georgia," *Civil War History* 50 (2004): 27–46.

In the 1960s, as the wave of broad scholarly interest in Civil War mobilization crested, specialists in economic history debated the question of whether the Civil War had indeed boosted American industrial growth, as many older histories had implied. The question was raised in a provocative article by Thomas C. Cochran, who argued that the Civil War had actually stunted economic growth: "Did the Civil War Retard Industrialization?" *Mississippi Valley Historical Review* 48 (1961): 197–210. Drawing upon new estimates of aggregate national product for the nineteenth century, economic historians mostly confirmed Cochran's thesis. Among the important works along these lines were David T. Gilchrist and W. David Lewis, eds., *Economic Change in the Civil War Era* (Greenville, Del.: Eleutherian Mills–Hagley Foundation, 1965); Stanley L. Engerman, "The Economic Impact of the Civil War," *Explorations in Entrepreneurial History* 3 (Spring 1966): 176–99; Ralph L. Andreano, ed., *The Economic Impact of the American Civil War* (Cambridge: Schenckman, 1967); Claudia Goldin and Frank D. Lewis, "The Economic Cost of the American Civil War: Estimates and Implications," *Journal of Economic History* 35 (1975): 299–326.

For much of the 1970s and 1980s, as many historians concentrated on writing new social and cultural histories, relatively little was done to build upon the previous generation's work on the economic dimensions of the Civil War. This is evident in the best recent overview of Union and Confederate mobilization efforts, Paul A. C. Koistinen's *Beating Plowshares into Swords: The Political Economy of American Warfare, 1606–1865* (Lawrence: University Press of Kansas, 1996), which makes admirable use of published primary sources but is compelled to rely heavily upon secondary sources published before 1970. In his discussion of state-level mobilization efforts, for example, Koistinen reaches back to classic studies such as William B. Weeden, *War Government: Federal and State in Massachusetts, New York, Pennsylvania and Indiana, 1861–1865* (Boston: Houghton Mifflin, 1906); William B. Hesseltine, *Lincoln and the War Governors* (New York: Alfred A. Knopf, 1948); and James A. Rawley, *Edwin D. Morgan, 1811–1883: Merchant in Politics* (New York: Columbia University Press, 1955).

Historians' interest in the economic dimensions of the Civil War revived in the late 1980s, when social historians began to focus on the war. Perhaps the clearest announcement of this shift came in Maris Vinovskis, ed., *Toward a Social History of the American Civil War: Exploratory Essays* (New York: Cambridge University Press, 1990). By this time, there was already one fine new survey of the Union home front, Phillip Shaw Paludan, *"A People's Contest": The Union and the Civil War, 1861–1865* (New York: Harper & Row, 1988); another, J. Matthew Gallman, *The North Fights the Civil War: The Home Front* (Chicago: Ivan R. Dee, 1994), would soon follow. Suddenly, readers interested in the Northern home front could find several outstanding specialized studies, including J. Matthew Gallman, *Mastering Wartime: A Social History of Philadelphia during the Civil War* (New York: Cambridge University Press, 1990); Iver Bernstein, *The New York City Draft Riots: Their Significance for American Society and Politics in the Age of the Civil War* (New York: Oxford University Press, 1990); Grace Palladino, *Another Civil War: Labor, Capital, and the State in the Anthracite Regions of Pennsylvania, 1840–1868* (Urbana: University of Illinois Press, 1990); Ira Berlin et al., eds., *Freedom: A Documentary History of Emancipation*, ser. 1, vol. 2, *The Wartime Genesis of Free Labor: The Upper South* (New York:

Cambridge University Press, 1993); Rachel Seidman, "Beyond Sacrifice: Women and Politics on the Pennsylvania Home Front during the Civil War" (Ph.D. diss., Yale University, 1995); and Jeanie Attie, *Patriotic Toil: Northern Women and the American Civil War* (Ithaca: Cornell University Press, 1998). These were among the first books to add substantially to the labor history of the Civil War, which had been synthesized previously in David Montgomery's still-valuable *Beyond Equality: Labor and the Radical Republicans, 1862–1872* (New York: Alfred A. Knopf, 1967). That the social history of the Civil War continues to be of considerable interest is suggested by two recent collections edited by Paul A. Cimbala and Randall M. Miller: *Union Soldiers and the Northern Home Front: Wartime Experiences, Postwar Adjustments* (New York: Fordham University Press, 2002) and *An Uncommon Time: The Civil War and the Northern Home Front* (New York: Fordham University Press, 2002).

Although much of this new work concentrated on the experiences of ordinary Americans on the Civil War home front, it inevitably revived questions about the activities of business and government in the shaping of the war economy. Indeed, both of the authors of the two new surveys of the Union home front, Phillip Paludan and Matthew Gallman, soon re-examined the older debate over the war's economic impact: Stanley L. Engerman and J. Matthew Gallman, "The Civil War Economy: A Modern View," in *On the Road to Total War: The American Civil War and the German Wars of Unification, 1861–1871*, ed. Stig Förster and Jörg Nagler (New York: Cambridge University Press, 1997), 217–47; Phillip Shaw Paludan, "What Did the Winners Win? The Social and Economic History of the North during the Civil War," in *Writing the Civil War: The Quest to Understand*, ed. James M. McPherson and William J. Cooper Jr. (Columbia: University of South Carolina Press, 1998), 174–200. Although the older consensus about the war's failure to jump-start American industrialization at the aggregate level appeared sound, the experiences of individual localities or business firms could vary tremendously. This was evident from the few available studies of Northern military contractors, including Harold B. Hancock and Norman B. Wilkinson, "A Manufacturer in Wartime: Du Pont, 1860–1865," *Business History Review* 40 (1966): 213–36, and Philip Scranton, *Proprietary Capitalism: The Textile Manufacture at Philadelphia, 1800–1885* (New York: Cambridge University Press, 1985), a pioneering new business history that argued that smaller Philadelphia textile firms adjusted more easily to war production than did their larger New England counterparts. Even Gallman, whose careful work on the Union home front tended to stress social and economic continuity rather than change, suggested in a study of Philadelphia contractors that economic historians would benefit from moving down from the aggregate impact story to explore the great variance in microlevel developments: J. Matthew Gallman, "Entrepreneurial Experiences in the Civil War: Evidence from Philadelphia," in *American Development in Historical Perspective*, ed. Thomas Weiss and Donald Schaefer (Stanford: Stanford University Press, 1994), 205–22, 299–303.

More recent studies of Civil War contracting, which build on the work of Scranton and Gallman, include Clinton W. Terry, "'The Most Commercial of People': Cincinnati, the Civil War, and Industrial Capitalism, 1861–1865" (Ph.D. diss., University of Cincinnati, 2002); Mark R. Wilson, "Gentlemanly Price Fixing and Its Limits: Collusion and Competition in the U.S. Explosives Industry during the Civil War Era," *Business History Review* 77 (2003): 207–34; and Wilson, "The Extensive Side of Nineteenth-Century Military Economy: The Tent Industry in the Northern United States during the Civil War," *Enterprise and Society* 2 (June 2001): 297–337. In the latter essay, I suggest that it is a mistake for students of Civil

War contracting to focus only on manufacturers because mercantile firms were equally important. For two classic studies in business history that emphasize the role of specialized mercantile enterprises during the mid-nineteenth century, see Glenn Porter and Harold Livesay, *Merchants and Manufacturers: Studies in the Structure of Nineteenth-Century Marketing* (Baltimore: Johns Hopkins Press, 1971), and Alfred D. Chandler Jr., *The Visible Hand: The Managerial Revolution in American Business* (Cambridge: Harvard University Press, 1977). For more details about contractors for the various states and the U.S. Army, some researchers may want to consult my doctoral dissertation, Mark R. Wilson, "The Business of Civil War: Military Enterprise, the State, and Political Economy in the United States, 1850–1880" (Ph.D. diss., University of Chicago, 2002). This dissertation also includes a bibliography that some specialists may find useful as a supplement to this essay on sources.

For all of their contributions, most of the new social, labor, and business histories failed to add much to what was already known about the wartime activities of governmental organizations and officers. This weakness became increasingly glaring in light of a new wave of historical studies by sociologists and political scientists, who transformed the field of political history by turning away from elections and party ideology to focus on state institutions. For the United States, among the most important of the new studies were Stephen Skowronek, *Building a New American State: The Expansion of National Administrative Capacities, 1877–1920* (New York: Cambridge University Press, 1982); Richard Franklin Bensel, *Yankee Leviathan: The Origins of Central State Authority in America, 1859–1877* (New York: Cambridge University Press, 1990); and Theda Skocpol, *Protecting Soldiers and Mothers: The Political Origins of Social Policy in the United States* (Cambridge: Harvard University Press, 1992). Strangely, although the bulk of the original research and sometimes brilliant analysis featured in these studies described the years after the Civil War, they did not hesitate to proclaim that the early American state, to use Skowronek's influential phrase, was a "state of courts and parties."

The "courts and parties" label, adopted easily by social scientists but without any serious historical investigation of the actual operations of the early American state, turned out to be far from adequate. This was pointed out most effectively by Richard R. John, the author of *Spreading the News: The American Postal System from Franklin to Morse* (Cambridge: Harvard University Press, 1995) Two review essays by John demonstrate that the "courts and parties" label fails to do enough to describe a variety of important early American state institutions: "Governmental Institutions as Agents of Change: Rethinking American Political Development in the Early Republic, 1787–1835," *Studies in American Political Development* 11 (1997): 347–80, and "Farewell to the 'Party Period': Political Economy in Nineteenth-Century America," *Journal of Policy History* 16 (2004): 117–25.

Perhaps the greatest problem with the "courts and parties" label was its implication that military institutions were an insignificant part of the early American state. This was a characterization of American state formation that set the United States far apart from its counterparts in Europe, if not the entire world. Many leading social scientists had recently emphasized the military dimensions of state action and the primary role played by the pressures of organizing for war in the creation of national states: Charles Tilly, ed., *The Formation of National States in Western Europe* (Princeton: Princeton University Press, 1975); Tilly, *Coercion, Capital, and European States, AD 990–1992*, rev. ed. (Cambridge, Mass.: Blackwell, 1992); Anthony Giddens, *A Contemporary Critique of Historical Materialism*, vol.

2, *The Nation-State and Violence* (Berkeley and Los Angeles: University of California Press, 1987); Michael Mann, *The Sources of Social Power*, vol. 2, *The Rise of Classes and Nation States, 1760–1914* (Cambridge: Cambridge University Press, 1993).

Only very recently have the social scientists interested in early American political development begun to consider the military. For one of the first calls to do so, see Ira Katznelson, "Flexible Capacity: The Military and Early American Statebuilding," in *Shaped by War and Trade: International Influences on American Political Development*, ed. Katznelson and Martin Shefter (Princeton: Princeton University Press, 2002), 82–110. Fortunately, political historians interested in the early military can turn to the several excellent studies of the antebellum army that have emerged in the last two decades. These include Edward M. Coffman, *The Old Army: A Portrait of the American Army in Peacetime, 1784–1898* (New York: Oxford University Press, 1986); William B. Skelton, *An American Profession of Arms: The Army Officer Corps, 1784–1861* (Lawrence: University Press of Kansas, 1992); Charles F. O'Connell Jr., "The Corps of Engineers and the Rise of Modern Management, 1827–1856," in *Military Enterprise and Technological Change: Perspectives on the American Experience*, ed. Merritt Roe Smith (Cambridge: MIT Press, 1985), 88–116; James L. Morrison Jr., *The Best School in the World: West Point, the Pre–Civil War Years, 1833–1866* (Kent, Ohio: Kent State University Press, 1986); Cynthia Ann Miller, "The United States Army Logistics Complex, 1818–1845: A Case Study of the Northern Frontier" (Ph.D. diss., Syracuse University, 1991); Samuel J. Watson, "Professionalism, Social Attitudes, and Civil-Military Accountability in the United States Army Officer Corps, 1815–1846" (Ph.D. diss., Rice University, 1996); and Robert G. Angevine, *The Railroad and the State: War, Politics, and Technology in Nineteenth-Century America* (Stanford: Stanford University Press, 2004).

Among historians of the American West, the military institutions of the national state have long been recognized as important. In fact, there is already a remarkably fine body of work on the specific subject of the relationship between army procurement and regional development. Francis Paul Prucha, *Broadax and Bayonet: The Role of the United States Army in the Development of the Northwest, 1815–1860* (Madison: State Historical Society of Wisconsin, 1953), established the genre. More recent works in this field include Robert W. Frazer, *Forts and Supplies: The Role of the Army in the Economy of the Southwest, 1846–1861* (Albuquerque: University of New Mexico Press, 1983); Darlis A. Miller, *Soldiers and Settlers: Military Supply in the Southwest, 1861–1885* (Albuquerque: University of New Mexico Press, 1989); William A. Dobak, *Fort Riley and Its Neighbors: Military Money and Economic Growth, 1853–1895* (Norman: University of Oklahoma Press, 1998); and Thomas T. Smith, *The U.S. Army and the Texas Frontier Economy, 1845–1900* (College Station: Texas A&M University Press, 1999). For the standard history of the greatest military contractor in antebellum America, a western transport firm, see Raymond W. Settle and Mary Lund Settle, *War Drums and Wagon Wheels: The Story of Russell, Majors, and Waddell* (Lincoln: University of Nebraska Press, 1966).

Like students of the history of the western United States, students of the history of technology have been sensitive to the importance of military institutions. For two important studies of the army's influence on early American manufacturing, see Merritt Roe Smith, *Harpers Ferry Armory and the New Technology: The Challenge of Change* (Ithaca: Cornell University Press, 1977), and David A. Hounshell, *From the American System to Mass Production, 1800–1932: The Development of Manufacturing Technology in the United States* (Baltimore: Johns Hopkins University Press, 1984). The essays in Merritt Roe Smith, ed., *Military Enterprise and*

Technological Change: Perspectives on the American Experience (Cambridge: MIT Press, 1985), are also rewarding. Historians of technology have been particularly interested in small arms manufacture and shipbuilding, where the challenges of innovation were especially high and government enterprise especially important. On the public and private manufacture of small arms and ammunition in the Civil War era, see especially Felicia Johnson Deyrup, *Arms Making in the Connecticut Valley: A Regional Study of the Economic Development of the Small Arms Industry, 1788–1870* (Northampton, Mass.: Smith College Studies in History, 1948); Carl L. Davis, *Arming the Union: Small Arms in the Civil War* (Port Washington: Kennikat Press, 1973); Smith, *Harpers Ferry Armory* and James J. Farley, *Making Arms in the Machine Age: Philadelphia's Frankford Arsenal, 1816–1870* (University Park: Pennsylvania State University Press, 1994). On shipbuilding and navy procurement, there are several interesting recent studies, including Richard E. Winslow III, *Constructing Munitions of War: The Portsmouth Navy Yard Confronts the Confederacy, 1861–1865* (Portsmouth, N.H.: Peter E. Randall, 1995); Thomas R. Heinrich, *Ships for the Seven Seas: Philadelphia Shipbuilding in the Age of Industrial Capitalism* (Baltimore: Johns Hopkins University Press, 1997); Kurt Hackemer, *The U.S. Navy and the Origins of the Military-Industrial Complex, 1847–1883* (Annapolis: Naval Institute Press, 2001); and William H. Roberts, *Civil War Ironclads: The U.S. Navy and Industrial Mobilization* (Baltimore: Johns Hopkins University Press, 2002). Although the long-run historical development of U.S. procurement law remains poorly understood, one valuable survey is James F. Nagle, *A History of Government Contracting*, 2nd ed. (Washington, D.C.: George Washington University, 1999).

Readers who seek more information about army logistics in the field may consult Edward Hagerman, *The American Civil War and the Origins of Modern Warfare: Ideas, Organization, and Field Command* (Bloomington: Indiana University Press, 1988), and Charles R. Shrader, "Field Logistics in the Civil War," in *The U.S. Army War College Guide to the Battle of Antietam: The Maryland Campaign of 1862*, ed. Jay Luvaas and Harold W. Nelson (New York: Harper & Row, 1988), 255–84. For a more general survey, which says little about the Civil War, see Martin van Creveld, *Supplying War: Logistics from Wallenstein to Patton* (New York: Cambridge University Press, 1977).

For a fine synthetic account that surveys the older historical literature on corruption and profiteering in the Civil War economy, see Stuart D. Brandes, *Warhogs: A History of War Profits in America* (Lexington: University Press of Kentucky, 1997), 67–107. For studies of antebellum and postwar corruption that are based on prodigious research in newspapers and personal papers, consult Mark W. Summers, *The Plundering Generation: Corruption and the Crisis of the Union, 1849–1861* (New York: Oxford University Press, 1987), and Summers, *The Era of Good Stealings* (New York: Oxford University Press, 1993). One of the most original political historians of the Civil War is Mark E. Neely Jr., whose book *The Fate of Liberty: Abraham Lincoln and Civil Liberties* (New York: Oxford University Press, 1991) provides important context for the discussion of the operations of military courts in this book.

There is no shortage of works that address the broader context of politics in the wartime North, which served as the background to the political struggles over procurement. For a recent study of the intellectual and political origins of the economic legislation passed by the Republican-dominated Congress, see Heather Cox Richardson, *The Greatest Nation of the Earth: Republican Economic Policies during the Civil War* (Cambridge: Harvard University Press, 1997). Among the most original recent studies of politics in the wartime North are Melinda

Lawson, *Patriot Fires: Forging a New American Nationalism in the Civil War North* (Lawrence: University Press of Kansas, 2002), and Mark E. Neely Jr., *The Union Divided: Party Conflict in the Civil War North* (Cambridge: Harvard University Press, 2002).

Although this book makes some new claims about the importance of producerist ideas in the wartime struggles over military economy, historians have long been aware of the producerist tone of much popular political economy in nineteenth-century America (as well as Europe). Among the many studies that address this subject are Selig Perlman, *A Theory of the Labor Movement* (New York: Macmillan, 1928); Marvin Meyers, *The Jacksonian Persuasion: Politics and Belief* (Stanford: Stanford University Press, 1957); Sean Wilentz, *Chants Democratic: New York City and the Rise of the American Working Classes, 1788–1850* (New York: Oxford University Press, 1984); Victoria C. Hattam, *Labor Visions and State Power: The Origins of Business Unionism in the United States* (Princeton: Princeton University Press, 1993); Martin J. Burke, *The Conundrum of Class: Public Discourse on the Social Order in America* (Chicago: University of Chicago Press, 1995); and Iorwerth Prothero, *Radical Artisans in England and France, 1830–1870* (Cambridge: Cambridge University Press, 1996).

On the military, business, politics, and culture in the years that followed the Civil War, the secondary literature is so vast that only a tiny fraction of it may be mentioned here. That the bulk of the postwar army quickly returned to the West, well before the end of Reconstruction, is shown clearly in James E. Sefton, *The United States Army and Reconstruction, 1865–1877* (Baton Rouge: Louisiana State University Press, 1967). For a fine recent book that describes one part of the huge Civil War claims problem, see Kyle S. Sinisi, *Sacred Debts: State Civil War Claims and American Federalism, 1861–1880* (New York: Fordham University Press, 2003). For a classic synthetic study of the postwar polity, see Morton Keller, *Affairs of State: Public Life in Late Nineteenth Century America* (Cambridge: Harvard University Press, 1977); from a cultural history perspective, one influential survey is Alan Trachtenberg, *The Incorporation of America: Culture and Society in the Gilded Age* (New York: Hill & Wang, 1982). On civil service reform in particular, the standard study remains Ari A. Hoogenboom, *Outlawing the Spoils: A History of the Civil Service Reform Movement, 1865–1883* (Urbana: University of Illinois Press, 1961). An important newer study of the late-nineteenth-century national state is Daniel P. Carpenter, *The Forging of Bureaucratic Autonomy: Reputations, Networks, and Policy Innovation in Executive Agencies, 1862–1928* (Princeton: Princeton University Press, 2001). The standard account of the rise of big business is Chandler, *The Visible Hand;* for a provocative revisionist view, see William G. Roy, *Socializing Capital: The Rise of the Large Industrial Corporation in America* (Princeton: Princeton University Press, 1997). Although more work remains to be done in fleshing out the history of the ideological origins of the modern industrial corporation, there are some studies that address this subject, including Thomas C. Cochran, *Railroad Leaders: The Business Mind in Action* (Cambridge: Harvard University Press, 1953); Edward Chase Kirkland, *Dream and Thought in the Business Community, 1860–1900* (Ithaca: Cornell University Press, 1956); a 1955 Northwestern University Ph.D. dissertation by Jack Blicksilver, published unaltered thirty years later as *Defenders and Defense of Big Business in the United States, 1880–1900* (New York: Garland, 1985); Alfred L. Thimm, *Business Ideologies in the Reform-Progressive Era, 1880–1914* (University of Alabama Press, 1976), and James A. Ward, "Image and Reality: The Railway Corporate-State Metaphor," *Business History Review* 55 (1981): 491–516.

Index

Page numbers in *italics* refer to figures.

A. & G. A. Arnoux, 20
A. T. Stewart & Co., 18, 22, 23, 222
Abbott, E. F., 87
Adams, Charles Francis, 216–18
Adams, Nathanial, 204
Adams, Theodore, 200–201
Adonijah Peacock & Son, 232
advertising, 109–10
African Americans, 78
Albert Jewett & Co., 122, 233, 235
Alden, Alonzo, 212
Alfred Jenks & Son, 120, 231
Alger. *See* Cyrus Alger & Co.
Allegheny Arsenal, 74, 255n. 5
Allen, Robert: antebellum career of, 45, 55, 62; background of, 42; on cash shortages, 102, 111, 130; postwar career and death of, 205, 207; purchasing practices of, 138; as St. Louis and Louisville depot chief, 68–69; and Saint Louis clothing halls, 88–89; self-promotion by, 61–62, 161; on vouchers, 112
ambulances. *See* wagons
American Law Review, 198, 200
Ames Manufacturing Co., 232
ammunition, 1, 13–14, 76, 79
Andrew, John, 15, 31, 83
Anspach & Stanton, 134, 176, 178, 235
anti-Semitism, 103, 180
Arnold, Isaac N., 175

Arnoux. *See* A. & G. A. Arnoux
auctions, 192, 194
Austin Baldwin & Co., 195

Babbitt, Edwin B., 41, 50, 52, 69
Bailhache, William, 60
bakeries, 31, 75
Baldwin. *See* Austin Baldwin & Co.; F. B. Baldwin & Co.
Baltimore, 76–77, 170–72
Bandini, Ysidora, 167
banks, 111–12, 115
Batcheller. *See* T. & E. Batcheller
Beach, Moses, 95
Belger, James, 45, 170–73, 179, 201
Bellamy, Edward, 222–24
Benedict, Hall & Co., 120, 215, 233
Benjamin Bullock's Sons & Co., 122–23, 130–31, 134, 215, 234
Betts & Nichols, 21
Bickta, Joseph, 233
bids, advertising for, 109–10
big business, 4, 116, 206–9, 214–19
Biggs, Herman, 178
Bing, Julius, 211
Black, Samuel, 166, 201
blacksmiths, 85
Blair family, 63, 153–54
Blair, Frank P., Jr., 88–89, 154, 167–68, 182
blankets, 1, 112, 121, 128–29, 235

Blunt, James G., 162
Boker & Co., 122, 128, 195, 231
bonds. *See* government bonds
bonuses and bounties, 31
boots and shoes, 44, 53, 120–21, 233
Boston Board of Trade, 83
Boyd, Augustus, 69
Boylan, James B., 122, 235
Boylan, John, 122, 235
Brady, S. P., 158
Braidwood, T. W., 96
Branch, Lawrence, 54
Brand, William, 132
Brandeis & Crawford, 198
Brandes, Stuart, 149
Brent, Thomas, 52
bribery, 21, 165, 174–75, 186
Brinkerhoff, Roeliff, 70, 207
Brooks Brothers, 19–20, 92, 180
Brown, B. Gratz, 210
Brown, Moses, 187–88
Brown, Samuel L., 59, 69, 103, 139, 164–65
Browning, Orville H., 89, 151
Browning, William C., 122, 124, 132, 215 235
Bryant, J. M., 187–88
Bryant, Seth, 233
Buchanan, James, 51, 53
Buckley. *See* Samuel Buckley & Co.
Bull Run, first battle of, 7, 26
Bullock. *See* Benjamin Bullock's Sons & Co.
Bullock, George, 178, 185
Burbridge, Stephen, 93
Burden & Co., 116–17
bureaucracy, 1, 46–51, 56, 60, 191–92, 208–9
Burnett, Henry L., 186–87
Burnham, H. B., 185–86
Burnside, Ambrose, 163
Burnside Rifle Co., 120
Bush, Charles, 125
Button, Worthington B., 122, 125, 235

Caesar's Column, 221
Calder, William, 141, 146, 175–76, 215
Calhoun, John C., 37, 39
Cameron, Simon, 27, 58, 137, 154, 175; Anspach & Stanton's connection with, 178; criticized for mismanagement, 152–53, 155–56; end to Illinois buying ordered by, 30–31; governors' correspondence with, 8–10; Meigs and, 65; named ambassador to Russia, 153; and patronage, 34; Pendleton's appeal to, 81
Campbell, Hugh, 157
cannon. *See* ordnance
Carhart, Thomas, 122, 176, 180, 235
Carnegie, Andrew, 116, 216
Carpenter, Daniel, 211
Carpenter, Matthew Hale, 212
Cassell, Martin, 22
Cavalry Bureau, 143
cavalry depots, 78, 80. *See also* horses and mules
certificates of indebtedness, 102–3, 113–15, 130–32. *See also* government bonds; Treasury Department
Chandler, Alfred D., Jr., 208
Chapin, George, 122, 235
Chapman, Edmund, 138–39
Charles W. Freeland & Co., 122, 124–25, 215, 234–35
Chattanooga, Tenn., 78, 80
Cheney, Benjamin, 18
Chicago Tribune, 151, 155
Child, Pratt, & Fox, 154–55, 158, 169, 199–200
children, 14, 74, 76
Childs, John. *See* Wilson, Childs & Co.
Cincinnati: clothing and tent-making halls at, 75, 86, 89, 104–5; depot at, 65, 77, 83; residents' protests in, 81–82
Cincinnati Daily Commercial, 26, 155
Cincinnati Enquirer, 163, 176
City Point, Va., 78
civil service reform, 4, 209–14, 223–24
claims, 192, 196–201. *See also* United States Court of Claims
Clark, Amos, 235
Clary, Robert E., 41
clothing industry, 91–92. *See also* uniforms
Cobb, Oliver P., 198
coffee, 1, 75, 184

Cohen, Mendes, 165
Cole & Hopkins, 22
collusion, 118, 138
Colman, Moses, 18
Colt Patent Fire Arms Co., 120, 122–23, 194, 231
Commissary General of Purchases, 37, 44
Commissary General of Subsistence. *See* Gibson, George; Subsistence Department
Confederacy, 5, 62
Congress, 2, 147, 202; and army officers, 61; and contracting rules, 136; and contractors, 175; military establishment redefined by, 182–83; and postwar army, 193; and postwar economic regulation, 214; and postwar ordnance sales, 195; and Union Pacific Railroad, 205; Van Wyck committee reports debated by, 156. *See also* Grimes committee; Joint Select Committee on Retrenchment; Van Wyck committee
Connecticut, 13
Connecticut Yankee in King Arthur's Court, 221
Constitution, 7
Cooke, Henry D., 115
Cooke, Jay, 111, 131, 161
corruption, 24–26, 159–60, 165–66, 169, 175, 182
courts-martial, 148, 160; of contractors, 182–89, 273n. 76; of supply officers, 165–74, 179
Couts, Cave J., 167
Cowen, Edgar, 183
Cozens, William, 127–28, 201
Crane, John C., 76–77, 178
Crimean War, 1, 51, 242n. 2
Cronin, Hurxthal & Sears, 234
Crosman, George H., 161, 178; antebellum career of, 41, 52, 55; and contractors, 131; on inspectors, 175; and open-market buying, 136–37; as Philadelphia depot chief, 68–69; political leanings of, 62, 163–64; retirement of, 206; rubber blankets purchased by, 117; and seamstresses, 86, 90, 96; and wages, 94, 97
Cross, Osborn, 41

Crowninshield, Francis B., 15
Cummings, Alexander, 153
Currier & Ives, 176–77
Curtin, Andrew, 9, 25
Curtis, George William, 211
Cyrus Alger & Co., 118–19, 122, 232

Davis, David, 157–58, 200
Davis, Jefferson, 49, 62, 68
Davis-Holt commission, 157–58, 168–69, 199–200
Dawes, Henry L., 155, 157, 182
Days of Shoddy, 179–81
Deering, William, 215, 235
Delano, Columbus, 11
delivery schedules, 110
demobilization, 1, 191–96, 203
Democratic Party, 60–64, 150, 162, 176–77, 210
Dennison, William, 11, 26, 162
depot-arsenal system, 1, 74, 78, 82, 108
Detroit, 18, 77
Devlin, Hudson & Co., 20, 92
Dickerson, John H.: antebellum career of, 52; on cash shortages, 111; and certificates of indebtedness, 114; controversies surrounding, 162–64, 166; as public entrepreneur, 86–87, 89; purchasing practices of, 110, 137–38; resignation of, 163; in Utah expedition, 52–53, 55–56; vouchers of, 113; as wartime depot officer in Cincinnati, 68–69
Dickerson, Julia, 162–63
Dimick, J. W., 234
discounting, 112–13, 129–32
Dobson, John, 121, 234
Dodge, Grenville, 217
Dom, Philip, 232
Donaldson, James L., 42, 69, 84, 102, 207
Donnelly, Ignatius, 221–23
Donner Party, 46
Douglas, Stephen, 167
Downs, H. S., 233
drug laboratories, 75
Dunan, Sylvester, 76
Du Pont de Nemours (E. I.) Co., 118, 122, 127, 215

E. P. Fenton & Co., 233
E. Remington & Sons, 120, 122, 195, 232
E. S. Higgins & Co., 234
E. Tracy & Co., 122, 178, 235
Easton, Langdon C., 50, 162
Eaton, Dorman B., 212
economic inequalities, 107
Eddy, Asher: antebellum career of, 35, 55; on cash shortages, 103; in clash with Illinois officials, 29–30, 60; as Memphis depot chief, 103; postwar career and death of, 205–7
Ekin, James, 69; antebellum career of, 59; contracting methods of, 109–10; at Indianapolis, 28; postwar career of, 206; in Washington, 144
Elleard, Charles M., 127, 141
Enfield rifle, 15
equipment (infantry and cavalry), 7, 20–21, 91, 125–26
Erie Canal, 10
Espensheid, Louis, 232
Evans, Benjamin C., 185–86
Evans, John R., 235
Ex parte Milligan, 200
expenditures: state-level, 12–13, 31; U.S., 1, 36, 38, 203

F. B. Baldwin & Co., 20
Farmers' Alliance, 219, 221, 223
federalism, 5, 244–45n. 3
feed, animal. *See* forage
Fenton. *See* E. P. Fenton & Co.
Fessenden, William P., 177
Field, David Dudley, 180–81
Field, Stephen, 205
Fincher, Jonathan, 96
Fincher's Trades Review, 96–97
Floyd, John B., 40, 54, 64, 66
food. *See* coffee; hardtack; pork; Subsistence Department
footwear. *See* boots and shoes
forage, 103, 113, 131, 138–39, 164–65
Fort Laramie, 52
Fort Leavenworth, 45, 47–48

Fort Pitt Works foundry, 118–20, 232
Fort Sumter, 7
Foudray & Evans, 109
Fox & Polhemus, 122, 125, 179
France, 1, 36, 49
Franco-Prussian War, 195
Frankford Arsenal, 74
Freeland, Charles W. *See* Charles W. Freeland & Co.; James H. Freeland & Co.
Frémont, John, 153–55, 157–58, 167, 182
Frowenfield & Bros., 24

Galaxy, 203
Gallman, J. Matthew, 93
Geer, J. C., 212
Geisendorff, J. C., 21
George, Milton, 219
Gettysburg, 171, 173
Gibson, George, 40
Gilbert Hubbard & Co., 175
Girard House Hotel (Philadelphia), 12
Gladden, Washington, 220
Glaser Bros., 17, 124
Godkin, E. L., 211
Godwin, Joseph, 21
Gosnold Mills, 117
government bonds, 9, 11, 132–33. *See also* certificates of indebtedness
grain. *See* forage
Grangers, 219
Grant, Ulysses S.: antebellum career of, 43, 45; anti-Semitic order of, 180; City Point supply base of, 78; as Civil War general, 73, 213; as president, 194, 207, 210
Great Britain, 1, 36
greenbacks, 111, 114–15
Grimes committee, 171
gunpowder, 117–18, 128
guns. *See* ordnance; small arms

H. Robinson & Co., 123
Hadden, H. G., 122, 133
Hagner, Peter V., 158
Hale, R. C., 12, 17
Hall, Amasa C., 171

Hall, Charles W., 186–87
Hammond, William A., 75, 270n. 36
Hanford, John E., 92, 122, 124, 235
hardtack, 7, 75
Harkness Bros., 96
Harmer. *See* James B. Harmer & Co.
Harpers Ferry Armory, 47, 74
Harper's Monthly, 181
Harper's Weekly, 155, 211
Hart, R. S., 87
Hartley, J. F., 212
Hartley, Marcellus, 231. *See also* Schuyler, Hartley & Graham
Hasler, Jane, 103–4
Haven, James L., 232
haversacks. *See* equipment (infantry and cavalry)
Hawkins, Dexter, 152
hay. *See* forage
Hazard, Augustus G., 132
Hazard Powder Co., 118
Heg, Hans Christian, 15
Heidelbach, Kuhn & Rindskopf, 124
Heidelbach, Seasongood & Co., 17
Heineman, E. S. & Co., 18
Henry Simons & Co., 110, 119–20, 232, 263n. 38
Herman Boker & Co., 122, 128, 195, 231
Higgins. *See* E. S. Higgins & Co.
Hill, James J., 219
Hodges, Henry C., 69
Holabird, Samuel, 42, 69
Holenshade. *See* J. C. C. Holenshade; Jacob W. Holenshade
Holt, Joseph, 157–58, 166
Holt-Owen commission, 158–59, 199–200
Holton, James, 25
Hopkins, James M., 141–42
horses and mules: contracting and subcontracting for, 109–10, 127, 140; in Mexican War, 44–45; numbers procured, 1, 140; postwar numbers in army, 204; postwar sales of, 194; prices of, 127, 228; procured by Massachusetts, 18; procurement policy regarding, 101, 108, 140–47; suppliers of, 66, 140–41, 186–89; in Utah expeditions, 52–53. *See also* subcontracting
horseshoes, 53, 116–17
Hosmer, A. A., 169
Hotchkiss & Sons, 118–19, 232
Houston, W. C., 123, 234
Howard, George T., 50
Howard, Jacob, 183
Howes, Hyatt & Co., 233
Hubbard. *See* Gilbert Hubbard & Co.
Hughes, Joseph B., 123
Hurtt, Francis (Frank), 113, 163, 166

Illinois, 11, 13–14, 30–31
Illinois State Journal, 60
income tax, 133–34
Indemnification Act of 1861, 10, 29
Indiana, 10, 13–14, 21–22, 24, 31
inflation, 93–94, 130–32, 228–29
inspections and inspectors, 110–11, 128–29, 175

J. C. C. Holenshade, 120, 176, 232
J. & W. Lyall, 98
Jacob W. Holenshade, 120, 133, 176, 232
Jacobs, Sarah, 40, 81
James B. Harmer & Co., 233
James H. Freeland & Co., 122, 124–25, 215, 234–35
Jeffersonville (Ind.) depot, 196
Jenkes, Thomas A., 210–12
Jenkins, Walworth, 69–70, 90, 188
Jenkins Lane & Sons, 233
Jenks. *See* Alfred Jenks & Son
Jesup, Thomas S.: on antebellum spending, 47–48; death of, 62; as quartermaster general, 39–40, 66, 70; and Utah expeditions, 51–52, 54–55, 61
Jewett. *See* Albert Jewett & Co.
Jews. *See* anti-Semitism
John Boylan & James B. Boylan, 122, 235
Johnson, Andrew, 178, 201–2
Johnston, Joseph E., 62
Joint Select Committee on Retrenchment, 202, 210–11
Jones, D. Floyd, 20

Jones, George W., 122
Jones, J. W., 155
Jones Bros., 122, 234
Judge Advocate General (office of), 166, 184–85, 201. *See also* courts-martial; Holt, Joseph
Justice, Philip, 200

Kearney, Stephen W., 45
Kelley, William D., 94, 98, 156, 175
Kentucky, 31
Kimball & Robinson, 120, 233
Knap, Charles. *See* Fort Pitt Works foundry
knapsacks. *See* equipment (infantry and cavalry)
Knights of Labor, 219–20, 223
Kohner, Joseph, 18
Kohner, Marcus, 18, 19
Kunkel, Hall & Co., 96, 129, 131–32, 234

L. J. & I. Phillips, 17
L. & M. Stone, 131
Laflin, Smith & Boies, 118
Lane, A. T., 234
Lane. *See* Jenkins Lane & Sons
Latshaw, Henry J., 166
law, procurement, 108–9, 135
Le Duc, William G., 58
Ledyard, Henry B., 217
Lee, George W., 60
Lee, Joseph, 235
Lee, Robert E., 63
Levering, William, 16
Lewis, Boardman & Wharton, 122, 125–26, 131, 134, 234–35
Lewis & Hanford, 92
Lexington, Ky., 166
Lincoln, Abraham, 8, 10, 59, 153, 170, 173, 178, 183; assassination of, 193; Browning and, 89; calls for troops, 7, 31, 108; election of, 62; Grant's order on Jews revoked by, 180; and patronage, 34; as portrayed in popular print, 176–77; re-election of, 132; and seamstresses, 72, 93, 98, 100, 103; Spicer and Smoot aided by, 199

Lincoln, Mary, 66
Livingston, Bell & Co., 199
Logan, John, 196, 204–5, 212
Long, Anna, 93–94, 258n. 49
Looking Backward, 222–23
Lord, James, Jr., 121
Louisville, 76–77, 84, 86, 90, 186
Lovejoy, Owen, 54, 156
Lowe, Percival G., 204
Lyall. *See* J. & W. Lyall

Mack, Henry, 215
Mack, Stadler & Glaser, 124, 235
"make or buy" decision, 12, 73, 105
manpower mobilization, 7
Mansur, Isaiah, 21
Marshall, W. L., 171–72
Martin, John T.: as art collector, 216, 280–81n. 89; business operations of, 124; and certificates of indebtedness, 132; and politics, 179; postwar activities of, 215–16; and prices for uniforms, 228; as top Northern contractor, 122, 215–16, 235
Massachusetts, 13–15, 18
McClay, Wilfred M., 222
McClellan, George B., 43, 176
McClung, David W., 60, 131, 138
McComb, Henry S., 122, 125–26, 128, 134, 178–79, 215
McKim, William W., 69, 87, 105, 128, 175
McKinstry, Adelaide, 202
McKinstry, Justus: antebellum career of, 166–67; arrest of, 168; background of, 42; Blair's conflict with, 154; in cartoon, 155; clothing halls in Saint Louis created by, 86; criticized, 154–56, 158, 167–68, 182; courts-martial of, 167, 169–70, 179; dismissal of, 170; postwar career of, 170, 201–2; praise of, 167; *Vindication* of, 168
Mears, William, 16
Medical Department, 75
Meigs, Louisa Rodgers, 63
Meigs, Montgomery C., 71, 148, 188, 206; and advertising, 109, 136; antebellum career of, 63–64; background of, 42; Belger warned

INDEX

by, 171; and Boston depot, 83; on cash shortages, 102–5; characterized, 64; on claims, 197; and contracting out, 73, 100–101, 104–5; and contractors, 127, 130–32, 142–43; on employing soldiers' relatives, 90; end to state purchases ordered by, 26–28; on forage purchases, 138–39; and geographical distribution of war work, 82, 140, 142; guidelines for main depots issued by, 108; on horseshoes, 116; on inspectors, 175; on middlemen, 151; named quartermaster general, 63; postwar career of, 206–7; postwar retrenchment urged by, 203; quartermasters supported by, 161–63, 170; report on Illinois spending by, 30; Republican officials and, 65–66, 175; Saint Louis depot investigated by, 154; and seamstresses, 88–89, 97–98; on termination of contracting, 193; on uniforms, 23; wage increases approved by, 94

Meigs, Samuel E., 87–88
mercantile enterprise, 121
Metropolitan Railroad (Boston), 18
Mexican War, 8, 35, 40, 44–46, 167
Michigan, 13, 24
middleman problem, 1, 135–47, 149–60, 168–74, 219–20
militarization, 209, 218, 224–25
Miller, Morris, 55, 68–69, 94
Miller, Samuel, 200
Milwaukee, 77
Minnesota's 2nd Infantry regiment, 8
Missouri, 31
Missouri Republican, 156
Montgomery, Alexander, 28, 45, 59, 249n. 52
Moody, M. K., 164–65
Moonlight, Thomas, 212
Moore, Elizabeth, 90
Moore, Henry D., 130
Moore, James, 84–85
Moorhead, J. M., 110, 141
Moorhead, James K., 81
Morford, Henry, 179–81
Morgan, Edwin D., 8, 19–20, 23, 152
Morgan, George D., 152–53, 155–56, 164

Morgan, J. P., 269n. 14
Morris, Mary, 90
Morton, Oliver, 15, 17, 21, 27–28, 32
Moulton, Charles W., 58–59, 69, 175, 207
Mount, C. B., 234
mules. *See* horses and mules
Mundell, John, 121, 233
Murphy, Frank, 21
Murphy, Thomas, 234
Myers, Frederick, 26
Myers, Leonard, 94, 164
Myers, William, 55, 68–69, 114, 203, 206

Napoleonic wars, 151, 172–73, 182–83
Nashville, 76–77, 84, 178, 194, 196
Nation, 211
National Association of Wool Manufacturers, 195
National Labor Union, 220
Navy Department, 38, 57, 194
navy yards, 1, 12, 74, 76, 92
Naylor & Co., 122, 231
Neal, Charles, 24
nepotism, 152–53, 164–65
Nevins, Allan, 64
New Jersey, 13, 31, 249n. 61
New Mexico territory, 46–47
New Orleans, 77, 84
New York (city), 65, 77, 83, 152–53, 171, 180
New York (state), 9–10, 13, 19–21, 24, 151
New York Herald, 150
New York Sun, 95, 99, 181
New York Tribune, 64, 65, 113–14, 130, 149–51, 156, 167–68
Newark Daily Advertiser, 182
North American & U.S. Gazette, 181
North American Review, 211
North Carolina, 5
Norwich Arms Co., 193–94, 231
Nott, Charles, 199

Ohio, 9, 13–14, 26, 31, 151
Ohio & Mississippi Railroad, 59, 101, 165
Olcott, H. S., 188
oligopoly, 117–19

O'Neill, Charles, 164
Opdyke, George, 122, 176, 180–81, 235
open-market purchasing, 136–47
ordnance, 118–19, 232
Ordnance Department: and ammunition procurement, 76; armories and arsenals, 74; contracts audited, 158–59; expenditures of, 38; heavy ordnance purchases by, 118; inspectors, 127; new arsenals approved by, 31; postwar sales by, 195; reorganization of, 37; small arms procurement, 75
Oriental Powder Co., 118
Owen, Robert Dale, 15, 23, 158

Pacific Mail Steamship Co., 49
Page, Joseph F., 124, 235
Panic of 1857, 132
Parrott. See Robert P. Parrott & Co.
Parsons, Louis B., 59, 101, 161, 165, 207
patents, 117–18
patronage, 1, 10
Pay Department, 38
Peacock. See Adonijah Peacock & Son
Peay, Robert W., 141
Peddie & Morrison, 21
Peeble, W. P., 61
Pendleton, George H., 81
Pendleton Act, 209–10
Pennsylvania, 9, 12–15, 24–25
People's Party. See Populists
Perkins, Charles Eliot, 217–18
Perkins, Simon, Jr., 59
Perry, Alexander J., 102, 105
Philadelphia, 47–48, 65, 77, 83, 94. See also Schuylkill Arsenal
Philadelphia Evening Bulletin, 99–100
Phillips. See L. J. & I. Phillips
Phoenix India Rubber Co., 117, 122. See also Union India Rubber Co.
Phoenix Iron Co., 119
Pierce Bros. & Co., 18
Pittsburgh, 77, 81
Polhemus, Theodore, Jr., 122. See also Fox & Polhemus
Polk, James K., 40, 61

Pomeroy. See R. M. Pomeroy & Co.
Populists, 219–21, 223
pork, 1, 86
Post Office Department, 78, 211, 220, 223
Potter, Joseph, 206
Poultney, Thomas, 231
Pratt, Mary, 96
Proctor & Gamble, 215
procurement law, 108–9, 135
producerism, 1, 147, 191–92, 219. See also middleman problem
profitability of military contracting, 127–34
profiteering, 149–59, 222
projectiles. See ordnance
Providence India Rubber Co., 117
Providence Tool Co., 120, 231
public enterprise, 1, 106; in military supply system, 73–78; and postwar politics, 219, 223; and quartermasters, 35, 84–86; and seamstresses, 94–98; and states, 12–15
Putnam's Magazine, 213

Quartermaster General (office of), 67, 76, 197. See also Jesup, Thomas S.; Johnston, Joseph E.; Meigs, Montgomery C.
Quartermaster's Department, 1, 2, 9; antebellum activities and organization of, 35, 37–38, 41, 46–47; attractions of, 43; contracting practices of, 109–15; employment of civilians by, 1, 78, 194; expanded in 1861, 57–58; expenditures of, 57; officers of, 69; officers' cash shortages, 101–5; officers' complaints of neglect, 160–62; officers and party politics, 61; officers' postwar careers, 205–8; officers' use of vouchers, 111–13; open-market purchases by, 135–36, 188; postwar activities of, 197, 204–5; Treasury Department and, 101–3; and Utah expeditions, 51–54
Quincy, Ill., 76, 89

R. M. Pomeroy & Co., 233
Radical Republicans, 155–56
railroads, 135, 204–5, 217–18. See also Ohio &

INDEX

Mississippi Railroad; Union Pacific Railroad; United States Military Railroad
Ramsey, Alexander, 199
Randall, Alexander, 9–10
Read, Drexel & Co., 130
Rech, Jacob, 232
regulars, 57–58, 67–68
Remington. *See* E. Remington & Sons
Republican Party, 6, 22, 161; and civil service reform, 210; and patronage, 34–35, 159–60, 162, 175–79; and regular officers, 162–64; and volunteer officers, 59–60. *See also* Radical Republicans
Richardson, William A., 30
RM&W. *See* Russell, Majors & Waddell
Robert P. Parrott & Co., 93, 118–19, 122–23, 232
Robinson. *See* H. Robinson & Co.
Robinson, Henry L., 69
Rockefeller, John D., 116, 216–18
Rockhill & Wilson, 124, 235
Rodman, Thomas J., 118–19
Rose, Alvin, 122, 234–35
Rose, Hannah, 93
Rosecrans, William, 93
Ross, Lewis, 202
rubber blankets, 17–18, 117
Rubber Clothing Co., 17–18
Rucker, Daniel, 67–69, 195–96, 206
Running the Machine, 176, 177
Russell, Majors & Waddell, 51–56

Saint Louis, 47–48, 77, 83; clothing halls at, 72, 75–76, 86–89; complaints of neglect of, 81; depot investigated and audited, 153–55, 157–58, 167–68; martial law in, 93
Samuel Buckley & Co., 15
Samuel Sykes & Co., 18
Samuel Walker & Co., 18
San Antonio, 46, 48
San Francisco, 47–48, 77
Santa Fe, 47–48, 77
Sarmiento & McGrath, 22, 248n. 38
Saxonville Mills, 179, 234
Schenck, Robert C., 144

Schofield, Sevill, 234
Schuyler, George, 231
Schuyler, Hartley & Graham, 15, 195, 231
Schuylkill Arsenal: and Mexican War, 44; postwar construction at, 196; seamstresses and other civilians employed by, 75, 86, 90, 97–98, 163, 257n. 27
Schwab, Charles, 219
Scientific American, 151
Scott, Thomas A., 88–89, 113, 135
seamstresses, 92, 124; in Cincinnati, 100; in New York (city), 95–96, 99–100; organized campaign against contracting out, 91–100; in Philadelphia, 12, 90, 93–94, 100; and piece rates, 95–96, 99; in Saint Louis, 87–88, 102. *See also* clothing industry
Sears, Zenas, 233
Seidman, Rachel, 93
Select Committee on Government Contracts. *See* Van Wyck Committee
Seligman & Co., 129
Seminole wars, 37
Seven Days' campaign, 159
Seward, William H., 27–28
sewing machines, 92, 95. *See also* clothing industry
Seyfert, McManus & Co., 232
Sharps Rifle Manufacturing Co., 120, 122, 193–94, 231
Shaw, Edward T., 234
Sheridan, Philip, 203
Sherman, John, 59
Sherman, William T., 43, 132, 162, 203, 213
Shiloh, 159
shipbuilding industry, 76. *See also* vessels
shoddy, 25, 149–50, 176, 179–82
shoes. *See* boots and shoes
Sibley, Ebenezer S., 41
Simons. *See* Henry Simons & Co.
Simpkinson, John, 121, 133, 215, 233
Skelton, William, 61
Slade, Smith & Co., 122, 125–26, 235
small arms, 15, 75, 120, 231
Smith, Edward A., 140–41, 146, 186–87
Smith, J. Frailey, 125

INDEX

Smith, Thomas C., 20–21
Smoot, Samuel, 188–89, 199–200
South. *See* Confederacy
South Boston Iron Works. *See* Cyrus Alger & Co.
Southern Claims Commission, 197
Speed family, 188
Spicer, John, 141, 188–89, 199–200
Sprague, William, 26
Springfield, Ill., 13
Springfield (Mass.) Armory, 12, 47, 74, 194, 223–24
Springfield rifle-musket, 75
Stadler Bros., 17, 124
Stanton, Edwin M., 71, 113, 141–42, 148, 158, 160; becomes secretary of war, 66; Belger dismissed by, 173, 201; on contracting out, 73; and contractors, 129–30; correspondence with Meigs, 138, 161; Dickerson's removal urged by, 163; on imports, 83; Philadelphia depot pressured by, 178; and railroads, 135; and seamstresses, 90, 94, 97–99; on state purchasing, 31–32; and Steubenville, Ohio, 89; and Yates, 82
Stanton, M. Hall. *See* Anspach & Stanton
Starr Arms Co., 231
Stetler, John K., 184–85
Steubenville, Ohio, 76, 89
Stevens, Thaddeus, 156, 269n. 14
Stevens, William H., 127
Stewart. *See* A. T. Stewart & Co.
Stone. *See* L. & M. Stone
Stone, Chisholm & Jones, 117
Stoneman, George, 143
strikes, 92–93, 102–3
Strong, George Templeton, 64
subcontracting, 121, 126, 186, 261–62n. 25; attractions of, 127; in clothing industry, 92, 96, 124; in horse and mule industry, 127, 141, 157; in small arms industry, 75–76; in state supply systems, 17; in wagon industry, 119
Subsistence Department, 38, 75, 86, 184, 197
Sumner, Charles, 210
surplus, 194–96, 204

Swift, Gustavus, 116
Swords, Charlotte, 45
Swords, Thomas: antebellum career of, 41, 55–56; background of, 42; bribes rejected by, 165; and certificates of indebtedness, 114; contracting methods of, 109; in Mexican War, 45–46; on procurement practices, 90, 143, 188; retirement of, 206; as wartime Louisville and Cincinnati depot chief, 68–69
Sykes. *See* Samuel Sykes & Co.
Sylvis, William, 220
Symonds, Henry, 86

T. & E. Batcheller, 120
tailors, 83, 92. *See also* clothing industry
taxes. *See* income tax
teamsters, 52, 78, 102
Temple, Daniel, 233
Temple, William, 233
tents, 7, 9, 16, 44, 89, 125–28, 132
Texas, 47
textiles. *See* woolen goods
Thayer, Sylvanus, 41
Thomas, Charles: antebellum career of, 41, 51, 55; as acting quartermaster general, 67; Baltimore depot investigated by, 171; contracting methods of, 109; on horse purchases, 142; as Philadelphia depot chief, 94, 100, 109; Saint Louis clothing halls investigated by, 89
Thomas, John L., 222
Thomas, Lorenzo, 154
Thompson, John A., 199
Tod, David, 59
Tomes, Robert, 181–82
Tompkins, Daniel D., 41
Tourgée, Albion W., 212
Tracy, Augustus E., 163
Tracy. *See* E. Tracy & Co.
Treasury Department, 31, 111; advised by Cooke, 131; antebellum practices of, 111; and army supply officers, 60, 67–68, 113; autonomy of, 40; cash shortages of, 101; internal revenue bureau of, 133; settling

of postwar claims by, 196–97. *See also* certificates of indebtedness; government bonds
Tredegar Iron Works, 118
Tredick, Stokes & Co., 123, 126
Tredway, William W., 16, 17
Trowbridge, Wilcox & Co., 18
Turkey, 195
Turner, Horace, 166
Twain, Mark, 221
Tyler, Stone & Co., 130

uniforms, 7, 9; cost of, 91; early non-uniformity of, 23; and Illinois, 22; in Mexican War, 8, 44; and New York (state), 20; and Pennsylvania, 12–13, 25; Quartermaster's Department procurement of, 76, 82, 91; postwar sales of, 195; production at Cincinnati and Saint Louis clothing halls, 89; soldiers' allowances for, 91; suppliers of, 123–25, 235; trousers, 1, 228; for Utah expedition, 53. *See also* clothing industry; Schuylkill Arsenal
Union India Rubber Co., 117, 122, 133. *See also* Phoenix India Rubber Co.
Union Pacific Railroad, 204–5, 215, 217
United States Army: antebellum condition and size of, 7, 36, 42–43, 46; Civil War size of, 7, 57; expenditures of, 38; line and staff tensions in, 160; and party politics, 60–64; postwar police actions by, 220; postwar return to the West, 202–6; postwar size of, 193, 203; regulations of, 70
United States Christian Commission, 31
United States Commissioner of Agriculture, 195
United States Court of Claims, 197–201
United States Military Academy, 41, 61
United States Military Railroad, 76, 78, 194
United States Military Telegraph, 194
United States Sanitary Commission, 31, 64
United States Supreme Court, 200–1
Utah expeditions, 51–55

Vajen, John, 21–22
Van Lear, John, 212
Van Slyke, Napoleon Bonaparte, 145
Van Vliet, Stewart: antebellum career of, 55–56; and contracts for horses, 140; and mission to Utah, 52; as New York City depot officer, 68–69; postwar career of, 208; self-promotion by, 160–61
Van Wyck, Charles H., 152, 157
Van Wyck, T. J., 122
Van Wyck committee, 152–159, 167–68, 172, 183–84, 189, 199
Vance, Zebulon, 5
Vanity Fair, 25, 150
vessels, 194
Vinton, David H.: antebellum career of, 41, 46, 55; funds requested by, 127; as New York city depot chief, 68–69, 97; on open-market purchasing, 136; on paperwork, 139; retirement of, 206; and wages, 94, 98
volunteers, 7, 58–60
vouchers, quartermasters', 111–113, 129–32

wages, 14, 94–96, 227–28
wagon covers, 110
wagons, 44–45, 52–53, 119–20, 204–5, 229, 232
Walker, Francis A., 223
Walker. *See* Samuel Walker & Co.
War Department, 8–9; Brown-Moody relationship investigated by, 164–65; Cincinnati clothing procurement investigated by, 104–5; contracting law interpreted by, 136, 140; end of purchasing ordered by, 193; and horses, 143–44; and Indian affairs, 37; and patronage, 40; and seamstresses, 97–100. *See also* Cameron, Simon; Stanton, Edwin M.
War of 1812, 7, 8, 213
Ware & Taylor, 174, 233
warehousing, 192, 196
Washburn, Israel, 27, 28
Washburne, Elihu, 154, 200–201
Washington, D.C., 48, 65, 77, 84–86, 194. *See also* Quartermaster General (office of)
Watertown Arsenal, 74, 79

Watervliet Arsenal, 74
Watson, Samuel, 61
Weed, Thurlow, 180–81
Weigley, Russell, 64
Welch, Benjamin, Jr., 20–21
Welles, Gideon, 152–53, 164
Welsh, John, 16
West Point. *See* United States Military Academy
West Point Foundry. *See* Robert P. Parrott & Co.
Whig Party, 61
Whitaker. *See* William Whitaker & Sons
White, Leonard D., 40
White, William H., 185–86
Whitten, Hopkins & Co., 18
Wilde & Co., 92
William Whitaker & Sons, 112, 121, 128–29
Williams, John, 11

Wilmore, J. C., 187
Wilson, Childs & Co., 111, 119–20, 133, 175, 232
Wilson, Henry, 101, 142, 213
Wilson, J. H., 188–89
Wisconsin, 13, 15, 22–23, 25–26, 174
women, 14, 74, 76. *See also* seamstresses
Wood, Asa D., 140
Wood, John, 11, 19, 22
Woodbridge, Frederick E., 212
woolen goods, 83, 121–23, 125, 130, 195, 235. *See also* blankets; uniforms
Wormer, Daniel, 188–89, 199–200
Wright, George B., 11, 26

Yates, Richard, 22, 29–30, 32, 60, 82
Yeager, Martha, 93–94, 100
Young, Brigham, 51–53

Zunz, Olivier, 208

www.ingramcontent.com/pod-product-compliance
Lightning Source LLC
Chambersburg PA
CBHW051629230426
43669CB00013B/2230